T0228950

International Conference
SHALLOW CRACK FRACTURE
MECHANICS, TOUGHNESS TESTS
AND APPLICATIONS

Cambridge, UK, 23-24 September 1992

Michael G Dawes, PhD, CEng, FWeldI
Principal Fracture Consultant, Engineering and Materials Group, TWI

Conference Technical Director

ABINGTON PUBLISHING
Woodhead Publishing Ltd in association with The Welding Institute
Cambridge, England

Published by Abington Publishing
Woodhead Publishing Limited, Abington Hall, Abington
Cambridge CB1 6AH, England
www.woodheadpublishing.com

First published 1993

© 1993 Woodhead Publishing Limited

Conditions of sale
All rights reserved. No part of this publication may be reproduced or transmitted in any form or by any means, electronic or mechanical, including photocopying, recording or any information storage and retrieval system, without permission in writing from the publisher.

British Library Cataloguing in Publication Data
A catalogue record for this book is available from the British Library

ISBN 978-1-85573-122-6

FOREWORD

Fracture mechanics theory and practice has developed to the stage where there are now many widely recognised national standards for fracture mechanics toughness tests. These standards have been greatly influenced by the early development of the ASTM E399 K_{Ic} test method, which involves test specimens having crack depth (a) to specimen width (W) ratios of about 0.5. However, within the last decade there has been an increasing awareness that the use of such deeply notched specimens may result in extreme under- or over-estimates of the true fracture toughness associated with shallow surface cracks in welded structures. Since this knowledge has serious implications for structural integrity and safety, it has stimulated work towards a better understanding of the factors that control fracture toughness, and also the new procedures that are necessary to extend the current standards for applications to shallow cracks in specific situations. This work was exemplified by the recent completion of a large TWI/Edison Welding Institute research project, which led to the idea for this conference.

Thus, the objective of this conference was to provide a timely and first international state-of-the-art review of shallow crack problems and tests, and their application to metal structures.

The conference comprised four sessions:

Session A: Fundamentals
Session B: Fracture toughness
Session C: Applied conditions
Session D: Fracture assessments

The first contribution to Session A, Paper 13, provides a general introduction to the conference and poses a series of questions that need to be answered. The answers are provided in the context of a brief post-conference overview. The subsequent papers in Session A cover the fundamentals of fracture characterising parameters, such as K, CTOD and J; the quantification of geometric constraint using Q and T stress approaches; the application of these and other analytical and numerical approaches (including the well known Anderson and Dodds, and Beremin local approach) to cleavage and ductile crack extension; and a simple engineering approach for characterising shallow crack fracture toughness under explosive loading.

Session B includes papers on the development of shallow crack testing procedures; shallow crack test results for base metals and welded joints; considerations of weld yield strength overmatching, and fatigue precracking compared to EDM notching.

Session C contains papers on 2D and 3D finite element analysis of applied or driving force values of K, CTOD and J for base metals, welded laboratory test specimens and welded T-butt welds. Papers 2, 14 and 28 give proposed CTOD and J estimations formulae for deep and shallow crack SENB specimens. Paper 29 shows the influence of weld strength mismatch on crack tip constraint in SENB specimens. This session also includes a contribution (Paper 35) on the emission and reflection of surface (Rayleigh) waves following pop-in at the crack tip.

Session D covers fracture assessments for both ductile and brittle (cleavage) crack extension. The papers consider components as diverse as large through-thickness, surface and edge cracked laboratory specimens, cracked broad flanged beams and pipes in four-point bending, and a pipe elbow loaded in cantilever in-plane bending.

At this state-of-the-art of shallow crack mechanics tests and analyses, it is significant that none of the papers in Session D include fracture toughness data associated with the shallow crack SENB specimen designs that are featured so prominently in Sessions A, B and C. This may be seen as an indication of the relatively recent and rapid evolution of shallow crack test methods, which leads one to expect that further significant advances in this subject area will occur prior to the 2nd International Conference on Shallow Crack Fracture Mechanics Toughness Tests and Applications, which is planned tentatively for a venue in Europe in 1995.

The success of this conference was made possible by the authors, a body of experts that devoted their time and wisdom to reviewing the papers, the session chairmen, the Conference Organiser, Tony Gray, and his staff at TWI. All of these people are gratefully acknowledged.

M G Dawes
Principal Fracture Consultant
Engineering and Materials Group
TWI

Shallow Crack Fracture Mechanics, Toughness Tests and Applications

CONTENTS

3D elastic-plastic finite element analysis for CTOD and J in SENB, SENAB and SENT specimen geometries (Paper 14)
R H LEGGATT and J R GORDON

The limits of applicability of J and CTOD estimation procedures for shallow-cracked SENB specimens (Paper 28)
Y-Y WANG and J R GORDON

The influence of weld strength mismatch on crack tip constraint in single edge notch bend specimens (Paper 29)
M T KIRK and R H DODDS Jr

Emission and reflection of surface waves from shallow surface notches (Paper 35)
C THAULOW, M HAUGE and J SPECHT

Shallow surface cracks in welded T butt plates — a 3D linear elastic finite element study (Paper 22)
B FU, J V HASWELL and P BETTESS

SESSION D: Fracture assessments

Prediction of overmatching effects on fracture of cracked stainless steel welds (Paper 4)
C ERIPRET and P HORNET

An experimental investigation of the transferability of surface crack growth characteristics (Paper 5)
J FALESKOG, F NILSSON and H ÖBERG

The influence of microstructural variation on crack growth (Paper 9)
V GENSHEIMER DEGIORGI, P MATIC, G C KIRBY III and D P HARVEY II

The deformation and failure at −30 °C of cracked broad flange beams of ordinary structural steel in bending (Paper 10)
K ERIKSSON, F NILSSON and H ÖBERG

Experimental and analytical investigations into the mechanical behaviour of pipe with short through-wall cracks (Paper 32)
Ph GILLES, D MOULIN and D GOUTERON

Application of K_J rule to different specimens and to a cracked elbow (Paper 18)
C COUTEROT, Y WADLER and P TAUPIN

SESSION A: Fundamentals

The requirement for shallow crack toughness tests (Paper 13)
M G DAWES and J R GORDON

A fracture mechanics approach based on a toughness locus (Paper 31)
C F SHIH and N P O'DOWD

A two parameter approach to the fracture behaviour of short cracks (Paper 25)
J W HANCOCK

Characterisation of constraint effect on cleavage fracture using T-stress (Paper 33)
Y-Y WANG and D M PARKS

A framework for predicting constraint effect in shallow notched specimens (Paper 30)
T L ANDERSON and R H DODDS Jr

Constraint based analysis of shallow cracks in mild steel (Paper 7)
J D G SUMPTER and A T FORBES

Interpretation of shallow crack fracture mechanics tests with a local approach of fracture (Paper 27)
M Di FANT and F MUDRY

The effect of constraint on the ductile-brittle transition (Paper 21)
I J MACLENNAN and J W HANCOCK

The effect of constraint on fracture of carbon and low alloy steel (Paper 3)
BRIAN COTTERELL, SHANG-XIAN WU and YIU-WING MAI

A general discussion of short crack fracture (Paper 16)
J R MATTHEWS, J F PORTER, C V HYATT and K J KARISALLEN

A crack stability approach to inelastic dynamic shallow-crack fracture (Paper 11)
L N GIFFORD, J W DALLY and A J WIGGS

TWI

PAPER 13

The requirement for shallow crack fracture toughness tests

M G Dawes, PhD, CEng, FWeldI (TWI, UK) and J R Gordon, BSc, PhD, CEng (EWI, USA)

SUMMARY

This paper provides a brief review of the conference papers. It is presented in the context of five important questions that were asked at the start of the conference, and a post–conference consideration of the most appropriate answers. The five questions were:

1.   What are shallow crack fracture toughness tests?

2.   Why do we need shallow crack fracture tests?

3.   Can we predict shallow crack fracture test results?

4.   How do we conduct shallow crack fracture tests?

5.   How do we apply the results?

The answers to the first two questions provide an introduction to the conference. The answers to the subsequent questions refer to the papers in Session A – Fundamentals, Session B – Fracture Toughness, Session C – Applied Conditions, and Session D – Fracture Assessments.

It is concluded that the recent advances in the understanding and quantification of geometric constraint, and its influence on ductile and brittle crack extension, will lead to rapid progress in the development and standardisation of shallow crack fracture mechanics test methods and assessment procedures. These will provide a basis for more accurate and economic determinations of the fracture resistance and reliability of engineering structures containing shallow cracks.

WHAT ARE SHALLOW CRACK FRACTURE MECHANICS TOUGHNESS TESTS?

Figure 1 shows the single edge notched bend (SENB) and compact specimen designs that are used most widely with the present British Standards Institution (BSI) and American Society for Testing and Materials (ASTM) standard fracture mechanics toughness test methods (1–6). In order to maximise geometric constraint, and encourage the onset of crack extension, the test specimens have relatively deep fatigue precracked notches (crack depth, a, to specimen width, W, ratio a/W ≥ 0.45) and are tested in predominantly bending loading.

Thus, the most obvious definition of a shallow crack toughness test is one involving specimens with a/W values < 0.45. Other definitions of shallow crack tests have been associated with a/W ratios below which there is a significant loss of constraint. This can be indicated by such effects as a significant increase in fracture toughness, yielding to the notched face of the test specimen, and the Q stresses (7) or T stresses (8) becoming negative. Alternatively, any specimen having absolute crack lengths less than about 2mm would surely be associated with a shallow crack test.

WHY DO WE NEED SHALLOW CRACK TESTS?

Brittle fractures in steel structures are often associated with surface cracks of only a millimetre or so in depth located in heterogeneous regions having significantly lower toughness than the surrounding base metal. These situations are generally associated with welded joints, flame cut edges, coated, clad, locally heat treated or environmentally degraded materials. In order to measure the fracture toughness in such situations it is very important for the notch in the test specimen to be in the same orientation and microstructure as the crack in the material concerned.

Unfortunately, it is generally not possible to measure the fracture toughness in these situations using the standard BSI and ASTM deep notch (a/W ≥ 0.45) test methods (1–6). This may be explained with reference to Fig 2a. This depicts a 2mm deep surface crack in a relatively brittle region of a 20mm thick section, and the relatively small cross-section test specimen that is implied by the requirement for a/W ≥ 0.45. Clearly, even for a very brittle steel, say K_{Ic} = 30 MPa m$^{0.5}$, such small specimen dimensions will generally not satisfy the standard test requirement for a valid K_{Ic} result (1,3) of a, and B ≥ 2.5$(K_{Ic}/\sigma_{YS})^2$, which for this example would require a steel having the very high yield strength of σ_{YS} ≥ 1060 MPa.

The small specimen shown in Fig 2a would also be unacceptable in the most recently published standards for elastic plastic fracture mechanics (EPFM) crack tip opening displacement (CTOD) and J test methods (1,6), which require the specimens to have both a/W ≥ 0.45, and the dimension B equal to the full section thickness (t, Fig 2a) of interest.

The above observations have led many people to conclude that the full thickness shallow crack (a/W < 0.45) specimen in Fig 2b will provide more appropriate measurements of the fracture toughness for a surface crack. It has also been widely concluded that the use of three point single edge notched bend (SENB) specimens in this application will match or overmatch the plastic constraint associated with the surface crack, and result in safe determinations of fracture toughness.

Numerous EPFM studies have been made of how the toughness in three point SENB specimens varies as a function of a/W. Particular attention has been given to resistance to cleavage fracture within the ductile to brittle transition temperature range in ferritic steels, as shown schematically in Fig 3 and 4. These figures represent the behaviour of reasonably homogeneous materials, and indicate that the most deeply notched specimens are associated with the lowest values of cleavage fracture toughness. Unfortunately, in heterogeneous materials having a variation in mechanical properties with position across the section thickness, shallow notch specimens can give lower values of fracture toughness than deep notch specimens, as shown schematically in Fig 5. This situation typifies brittle fractures from weld toe cracks, where cleavage fracture initiation in the weld HAZ leads to a rapid increase in strain rate at the running crack tip, and a concomitant and sometimes catastrophic reduction in the fracture toughness of the surrounding base metal.

Thus, the indiscriminate application of the present standard deep notch tests (1–6) may result in gross over or under–estimates of the true fracture toughness associated with shallow cracks in heterogeneous materials. The application of the deep notch tests may also result in gross under–estimates of the fracture toughness associated with shallow cracks in homogeneous materials. Clearly, before we can make safe and realistic assessments of specific shallow cracks in structures, we may need to carry out shallow crack fracture toughness tests on specimens of the type indicated in Fig 2b, although, not always in the full thickness of interest, which may be massive.

CAN WE PREDICT SHALLOW CRACK FRACTURE TOUGHNESS?

In the last few years there have been significant advances in the understanding and quantification of triaxial constraint and its influence on such single parameter measures of fracture toughness as K_{Ic}, δ_c and J_c. These advances are of particular interest for explaining and predicting the geometry dependence (see Fig 4) of the EPFM toughness parameters δ_c and J_c, where here the subscript 'c' indicates cleavage fracture following less than 0.2mm of stable crack extension (1,6). The latest research on this type of fracture, and fracture following more than 0.2mm of ductile crack extension, with or without cleavage, was covered mainly in Session A of this conference (7–16).

Shih and O'Dowd describe the elastic plastic Q stress approach (7), which when combined with J provides a two parameter (J–Q) description of crack tip deformation and constraint. Similarly, Hancock describes the earlier and simpler elastic T stress approach (8), which provides a J–T description of deformation and constraint. Both two parameter approaches have been linked with critical cleavage fracture stress models (7,9,13), or simply indexed with individual experimental toughness results (8,11), and have shown an encouraging ability to predict variations in geometry with J_c. However, as Sumpter found (11), the Q stress approach can give accurate predictions of J_c for a wider range of specimen geometries than the T stress approach.

Anderson and Dodds (10) summarise a scaling model for predicting the effect of crack tip constraint on cleavage fracture toughness. The model is based on detailed elastic plastic finite element analyses (FEAs) and assumes that fracture is controlled by a critically stressed volume ahead of the crack tip. The model has been applied to three–point SENB specimens having a wide range of a/W ratios. It has been very successful in adjusting shallow notch δ_c and J_c toughness values to the lower values associated with deep notch specimens (10,11), and has resulted in the well known Anderson and Dodds (A & D) size requirements for geometry independent toughness values in standard deep notch (a/W ≥ 0.45) three–point SENB specimens (see [10] and [11] in Ref 10). However, it is important to note that values of δ_c and

J_c meeting the A&D size requirements generally correspond to lower shelf values, as shown in Fig 6, and such values will be considered too small for many structural applications.

At the present stages of development, the Q stress (7) and A & D scaling approaches (10) are only recommended for application to such values of fracture toughness as δ_c and J_c (1,6), where $\leq$ 0.2mm of stable ductile crack extension is exhibited. However, this situation may change, since Hancock's paper (8) indicates that the T stress approach can also be used to predict how the toughness for the onset and propagation of stable crack extension varies with specimen geometry.

Di Fant and Mudry (12) have used the local approach to fracture to predict the critical displacements for cleavage fracture, and the ductile/brittle transitions for SENB and single edge notched tension (SENT) specimens having a/W ratios of 0.05 and 0.5. They made excellent predictions of the experimental results for the SENB specimens, but overestimated the critical displacements for the SENT specimens. Since the local approach is very sensitive to yield strength, they suggested that the overestimates for the SENT specimens may have been caused by variability in the yield strength of the test material.

Cotterell, Wu and Mai (14) describe studies of the influence of constraint on the fracture of carbon and low alloy steels. They tested SENB specimens having a/W values ranging from 0.1 to 0.5 and concluded that constraint has a significant influence on the critical values of CTOD and J for the onset of both stable crack extension and cleavage fracture.

Matthews, Porter, Hyatt and KarisAllen (15) studied the onset and propagation of stable crack extension in SENB specimens having a/W ratios ranging from 0.07 to 0.95. Although values of J for the onset of crack extension increased with decreasing a/W, there were no observable differences in stretch zone size between deep and shallow notched specimens. They suggest that the point at which the plastic zone breaks to the notched surface of the specimen be used to define the region between true shallow crack (strain dominated) fracture and deep crack (J dominated) fracture.

All the papers referred to above were concerned with static or slowly applied loading, whereas the paper by Gifford, Dally and Wiggs (16) is concerned with explosive loading. These authors describe a simple engineering approach for establishing the shallow crack depth limit at which the dynamic fracture mode changes from stable to unstable at or just beyond the limit state in bending. The approach is applied to structural elements in the material and thickness of interest, and thereby considers automatically the correct conditions of crack tip constraint.

It may be concluded that, at least for slowly applied load SENB tests on reasonably homogeneous materials and cleavage fracture (Fig 4), it is now possible to either predict EPFM toughness values (12) as a function of a/W (12), or adjust known values of toughness for different values of a/W (7,8,10). Overall, the Q stress approach (7) seems to provide the best prospect for adjusting known fracture mechanics toughness values for a particular specimen geometry to the appropriate values for other cracked geometries. However, for **heterogeneous** materials, and the variations in toughness depicted in Fig 5, there seems to be no justifiable alternative at present to using full thickness SENB specimens having the cross-sections indicated in Fig 2b.

HOW DO WE CONDUCT SHALLOW CRACK TESTS?

The four papers in conference Session B represent recent practical experience and applications of shallow crack fracture mechanics tests on ferritic steels (17–20). Each paper includes at least one set of three–point SENB tests on square section specimens (see Fig 2b), and only in one paper (20) is the dimension W (= 100mm) significantly less than the full thickness of original section, which in this case is believed to have been >200mm. Also, except for the use of shallow cracks, all the specimen preparation, instrumentation and testing procedures met or exceeded the requirements in the BSI and ASTM standard CTOD test methods (1,6).

Dawes, Slater, Gordon and McGaughy (17) describe the development shallow crack test specimen fatigue precracking and instrumentation for base metal SENB, SENT and single edge notched arc bend (SENAB) specimens having a/W ratios within the range 0.05 to 0.5, and absolute crack depths as small as 1.25mm in 25mm thick sections.

The paper by Thaulow and Paauw (18) describes how they determined the weld HAZ fracture toughness of single pass bead–in–groove welds using 50 x 50mm section SENB specimens having a/W values of 0.06.

Rak, Koçak, Golesorkh and Heerens (19) explain how they investigated the fracture toughness of repair welds in 28 x 28mm section SENB specimens having shallow (a/W = 0.16) and deep (a/W = 0.5) notches. They found, as might be expected, that the standard relationships for determining CTOD (1,6) had to be modified to include a different plastic rotation factor, r_p, for the shallow notch specimens.

Theiss, Rolfe and Shum (20) describe equivalent CTOD and J fracture toughness tests on A 533B steel SENB specimens having shallow (a/W = 0.1) and deep (a/W = 0.5) notches, W = 100mm and B = 50, 100 and 150mm. These authors also investigated r_p and concluded that the standard value (6) could be used for both notch depths. They also observed similar values of fracture toughness from specimens having B = 50, 100 and 150mm, but a tendency for less data scatter with the larger values of B.

Further important information about how to conduct shallow crack fracture mechanics tests is provided by six papers in conference Session C (21–26).

Four papers (21–24) describe either 2D or 3D elastic plastic finite element analyses of the relationships between CTOD and J and the global parameters load, load–line displacement and crack mouth opening displacement (CMOD) in deep and shallow crack SENB specimens. The analyses show that small changes in the values of r_p in the standard CTOD estimation equations (1,6) can disguise significant changes in the plastic components of CTOD, especially in shallow crack specimens in low work hardening materials (22,23). This has led to the development of new CTOD estimation formulae, which are a function of a/W and the stress strain behaviour of the material. It is now agreed by Kirk and Dodds (22), Leggatt and Gordon (23), and Wang and Gordon (24), that the most accurate estimates of CTOD may be determined from the area under a load versus CMOD record. Similarly, these authors also agree (22–24) that the most accurate estimates of J may be determined from the area under the load versus CMOD record. This would dispense with the need to measure load–line displacement, which would greatly simplify the instrumentation in tests on SENB specimens.

The second paper by Kirk and Dodds (25) shows how weld yield strength mismatch in SENB specimens can influence the parameter J_{SSY}, which is used in the Anderson and Dodds (10) scaling model for predicting the effect of crack tip constraint on cleavage fracture toughness, as discussed earlier.

Another factor to take into consideration when conducting a shallow crack test is the interpretation of pop-in behaviour (1,6). This is investigated in the paper by Thaulow, Hauge and Specht (26), who use dynamic FEAs to predict the emission and reflection of Rayleigh waves from the crack tip. This is ongoing work, and at this stage no firm conclusions can be drawn concerning the propensity for the reflected stress waves to arrest cleavage crack extension.

It may be concluded that the papers in this conference provide considerable guidance on how to conduct shallow crack fracture mechanics tests.

HOW DO WE APPLY THE RESULTS?

To apply the results of the shallow crack tests in fracture assessments it is necessary to equate the fracture toughness (adjusted for the constraint as necessary) to the applied or driving force for the cracked body concerned. Thus, the appropriate fracture toughness could be equated with the driving force for the surface cracks in the welded T butt plates described in the paper by Fu, Haswell and Bettess (27).

Because of the relatively recent and rapid development of shallow crack fracture toughness test methods and constraint adjustment procedures, it was generally too early for these advances to have been taken into consideration in the papers in Session D (28-33).

CONCLUSIONS

The recent advances in the understanding and quantification of geometric constraint, and its influence on ductile and brittle crack extension, is leading to rapid progress in the development and standardisation of shallow crack fracture mechanics test methods and assessment procedures. These will provide a basis for more accurate and economic determinations of the fracture resistance and reliability of engineering structures containing shallow cracks.

REFERENCES

1   BS 7448:Part 1:1991: 'Fracture Mechanics Toughness Tests, Part 1. Method for Determination of K_{Ic}, Critical CTOD and Critical J values of Metallic Materials'. (This standard replaced BS 5447:1977:K_{Ic} tests, and BS 5762:1979: CTOD tests).

2   BS 6729:1987: 'British Standard Method for Determination of the Dynamic Fracture Toughness of Metallic Materials'.

3 ASTM E399–90: 'Standard Test Method for Plane–Strain Fracture Toughness of Metallic Materials'.

4 ASTM E813–89: 'Standard Test for J_{Ic}, a Measure of Fracture Toughness'.

5 ASTM E1152–87: 'Standard Test Method for Determining J–R Curves'.

6 ASTM E1290–89: 'Standard Test Method for Crack– tip Opening Displacement (CTOD) Fracture Toughness Measurement'.

7 Shih C F and O'Dowd N P: 'A Fracture Mechanics Approach based on a Toughness Locus'. Paper 31, Proceedings of the 1st International Conference on Shallow Crack Fracture Mechanics Tests and Applications, TWI, Cambridge, 23–24 September 1992.

8 Hancock J W: 'A Two Parameter Approach to the Fracture Behaviour of Short Cracks'. Paper 25, Ibidem.

9 Wang Y–Y and Parks D M: 'Characterisation of Constraint Effect on Cleavage Fracture using the T–stress'. Paper 33, Ibidem.

10 Anderson T L and Dodds R H: 'A Framework for Predicting Constraint Effects in Shallow Notched Specimens'. Paper 30, Ibidem.

11 Sumpter J D G and Forbes A T: 'Constraint Based Analysis of Shallow Cracks in Mild Steel'. Paper 7, Ibidem.

12 Di Fant M and Mudry F: 'Interpretation of Shallow Crack Fracture Mechanics Tests with a Local Approach to Fracture'. Paper 27, Ibidem.

13 MacLennan I J and Hancock J W: 'The Effect of Constraint on the Ductile–brittle Transition'. Paper 21, Ibidem.

14 Cotterell B, Wu S–X and Mai Y–W: 'The Effect of Constraint on Fracture of Carbon and Low Alloy Steel'. Paper 3, Ibidem.

15 Matthews J R, Porter J F, Hyatt C V and KarisAllen K J: 'A General Discussion of Short Crack Fracture'. Paper 16, Ibidem.

16 Gifford L N, Dally J W and Wiggs A J: 'A Crack Stability Approach to Inelastic Dynamic Shallow–crack Fracture. Paper 11, Ibidem.

17 Dawes M G, Slater G, Gordon J R and McGaughy T H: 'Shallow Crack Test Methods for the Determination of K_{Ic}, CTOD and J fracture Toughness'. Paper 15, Ibidem.

18 Thaulow C and Paauw A J: 'Shallow Surface Notch – Experience from Bead in Groove Testing'. Paper 19, Ibidem.

19 Rak I, Koçak M, Golesorkh M and Heerens J: 'CTOD Toughness Evaluation of Hyperbaric Repair Welds with Shallow and Deep Notch Specimens'. Paper 20, Ibidem.

20 Theiss T J, Rolfe S T and Shum D K M: 'Shallow-crack Toughness Results for Reactor Pressure Vessel Steel'. Paper 37, Ibidem.

21 Sprock A and Dahl W: '3D Elastic-plastic Finite Element Analysis for the Determination of the J-integral in Single Edge Notched Bend Specimens'. Paper 36, Ibidem.

22 Kirk M T and Dodds R H: 'J and CTOD Estimation Equations for Shallow Cracks in Single Edge Notch Bend Specimens'. Paper 2, Ibidem.

23 Leggatt R H and Gordon J R: '3D Elastic-plastic Finite Element Analysis for CTOD and J in SENB, SENAB and SENT Specimen Geometries'. Paper 14, Ibidem.

24 Wang Y-Y and Gordon J R: 'The Limits of Applicability of J and CTOD Estimation Procedures for Shallow Cracked SENB Specimens'. Paper 28, Ibidem.

25 Kirk M T and Dodds R H: 'The Influence of Weld Strength Mismatch on Crack Tip Constraint in Single Edge Notch Bend Specimens'. Paper 29, Ibidem.

26 Thaulow C, Hauge M and Specht J: 'Emission and Reflections of Surface Waves from Shallow Surface Notches'. Paper 35, Ibidem.

27 Fu B, Haswell J V and Bettess P: 'Shallow Surface Cracks in Welded T Butt Plates – a 3D Linear Elastic Finite Element Study'. Paper 22, Ibidem.

28 Eripret C and Hornet P: ' Prediction of Overmatching Effects on Fracture of Cracked Stainless Steel Welds'. Paper 4, Ibidem.

29 Faleskog J, Nilsson F and Oberg H: 'An Experimental Investigation of the Transferability of Surface Crack Growth Characteristics'. Paper 5, Ibidem.

30 DeGiorgi V G, Matic P, Kirby G C and Harvey D P: 'The Influence of Microstructural Variation on Crack Growth'. Paper 9, Ibidem.

31 Eriksson K, Nilsson F and Oberg H: ' The Deformation and Failure at −30° C of Cracked Broad Flange Beams of Ordinary Structural Steel in Bending'. Paper 10, Ibidem.

32 Gilles P H, Moulin D and Gouteron D: 'Experimental and Analytical Investigations of a Pipe with a Short Through-wall Crack'. Paper 32, Ibidem.

33 Couterot C, Wadier Y and Taupin P: 'Application of K_J Rule to Different Specimens and to a Cracked Elbow'. Paper 18, Ibidem.

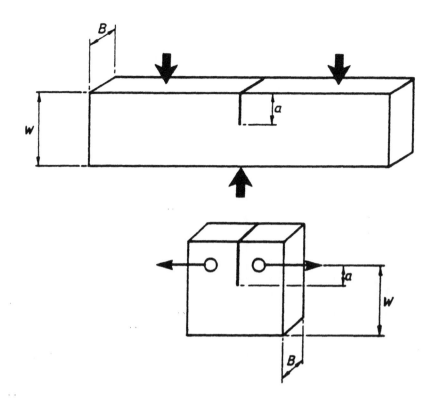

Fig.1 Standard designs of test specimen.

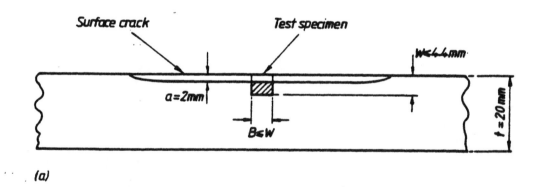

(a)

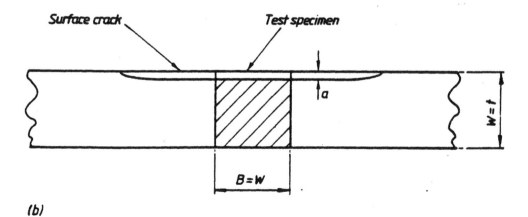

(b)

Fig.2 Extraction of SENB specimens associated with surface cracks in discrete microstructures:

a) Deep notch (a/W= 0.5) specimens;   b) Shallow notch specimens.

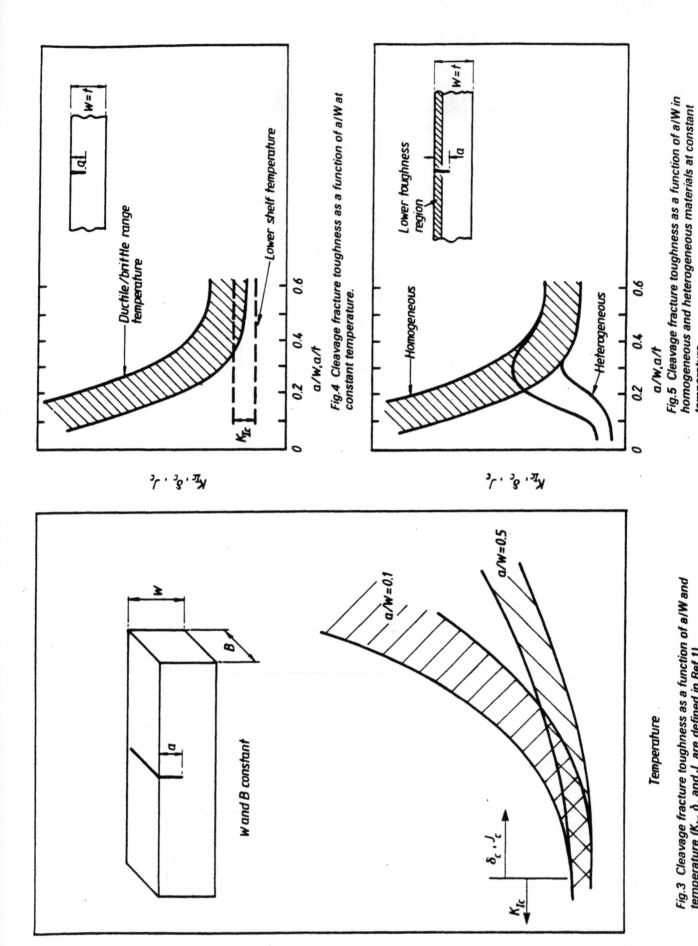

Fig.4 Cleavage fracture toughness as a function of a/W at constant temperature.

Fig.5 Cleavage fracture toughness as a function of a/W in homogeneous and heterogeneous materials at constant temperature.

Fig.3 Cleavage fracture toughness as a function of a/W and temperature (K_{Ic}, δ_c and J_c are defined in Ref.1).

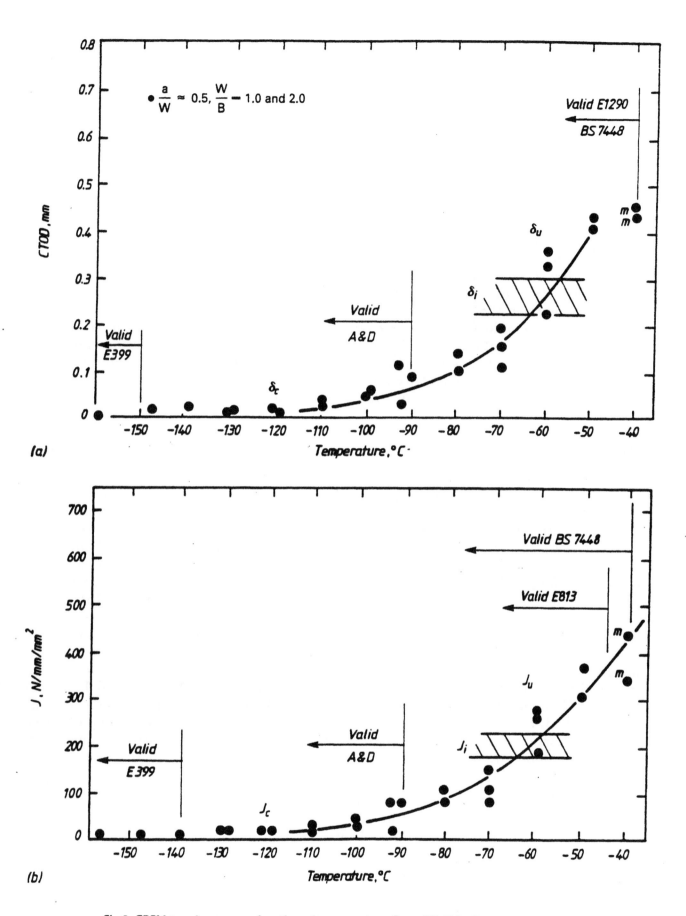

Fig.6 EPFM toughness as a function of temperature. For a BS 4360 Grade 50C steel in 25mm thickness.

A fracture mechanics approach based on a toughness locus

C F Shih, PhD (Brown University) and N P O'Dowd, PhD (California Institute of Technology, USA)

SUMMARY

A family of self-similar fields provides the two parameters required to characterize the full range of high- and low-triaxiality crack tip fracture states. The two parameters J and Q have distinct roles: J sets the size scale over which large stresses and strains develop, while Q scales the near-tip stress distribution relative to a high triaxiality reference stress state. Careful investigations show that Q accurately describes the evolution of near-tip stress triaxiality over a wide range of loading and crack geometries. The $J - Q$ theory provides a framework which allows the toughness locus to be measured and utilized in engineering applications. Methods for evaluating Q in fully yielded crack geometries and a scheme to interpolate for Q over the entire range of yielding are presented. An indicator of the robustness of the $J - Q$ fields is introduced; Q as a field parameter and as a pointwise measure of stress level are discussed.

INTRODUCTION

A two-parameter fracture theory can be motivated by considering the progression of plastic states as loading on a cracked body is increased. At low loads, the near-tip stresses and deformations evolve according to a self-similar field. This field, characterized by a high level of stress triaxiality, also describes the evolution of the near-tip stresses and deformations in certain crack geometries as plastic flow progresses from well-contained yielding to large scale yielding. While this high triaxiality field is only one out of many possible states that can exist under fully yielded conditions, it is the only field that has received careful study until recently. When the high triaxiality field (Hutchinson (17), Rice and Rosengren (18), Rice and Johnson (19), McMeeking (20)) scaled by the $J-$ integral (Rice (16)) prevails over distances comparable to several crack tip openings, J alone sets the near-tip stress level and the size scale of the zone of high stresses and large deformations. Considerable efforts have been directed to establishing, for different crack geometries, the remote deformation levels which ensure that the near-tip behavior is uniquely measured by J (McMeeking and Parks (21), Shih and German (22)). The end result is a framework, based on J and the high triaxiality crack tip field, for correlating crack growth over a range of plane strain yielding conditions (see review articles by Hutchinson (23), Parks (24)) and for relating critical values of the macroscopic parameter J_{Ic} to fracture mechanisms operative on the microscale (see review article by Ritchie and Thompson (25)).

Arguments that a single parameter might not suffice to characterize the dissimilar near-tip states of fully yielded crack geometries have been raised by McClintock (26). He noted that non-hardening plane strain crack tip fields of fully yielded bodies are not unique but exhibit levels of stress triaxiality that depend upon crack geometry. Though high stress triaxiality is maintained in geometries involving predominantly bending on the uncracked ligament, the level of crack tip stress triaxiality in geometries dominated by tension loads generally decreases as yielding progresses into the fully plastic state (21, 22).

There is some merit to the latter point of view and it is the purpose of this paper to show that this viewpoint can be properly reconciled by a two-parameter $J - Q$ fracture theory

(O'Dowd and Shih (1,2), Shih *et al.* (3), Xia *et al.* (27)) and a similar approach (Li and Wang (4), Sharma and Aravas (5)). Another possible approach is the $J - T$ theory (Betegón and Hancock (6), Al-Ani and Hancock (7), Parks (8), Hancock *et al.* (9), Wang (10)). Under the circumstances where it is applicable, the $J - T$ theory can be shown to be equivalent to the $J - Q$ theory. The toughness scaling approach of Dodds *et al.* (12) can also be shown to be consistent with the $J - Q$ theory (see Kirk *et al.* (11), Dodds and Shih (30)).

J-Q THEORY

Consider a cracked body of characteristic dimension L. The scale of crack tip deformation is measured by J/σ_0 where σ_0 is the material's tensile yield stress. At a sufficiently low load, $L \gg J/\sigma_0$. Then it can be shown from dimensional grounds that all near-tip fields are members of a single family of crack tip fields. Each member field is characterized by its level of deformation as measured by J/σ_0 and by its level of crack tip stress triaxiality as measured by Q which also identifies that field as a particular member of the family. For example, the self-similar solution of Rice and Johnson (19) and McMeeking (20) (and the HRR field (17,18)) can be regarded as the $Q = 0$ member field.

Q-Family of Fields

It proves convenient to construct the Q−family of fields using a modified boundary layer formulation in which the remote tractions are given by the first two terms of the small-displacement-gradient linear elastic solution, Williams (13),

$$\sigma_{ij} = \frac{K_I}{\sqrt{2\pi r}} \tilde{f}_{ij}(\theta) + T \delta_{1i} \delta_{1j}. \tag{1}$$

Here r and θ are polar coordinates centered at the crack tip with $\theta = 0$ corresponding to a line ahead of the crack as shown in Fig. 1.

Fields of different crack tip stress triaxialities can be induced by applying different combinations of K and T. From dimensional considerations, these fields can be organized into a family of crack tip fields parameterized by T/σ_0:

$$\sigma_{ij} = \sigma_0 \overline{f}_{ij} \left(\frac{r}{J/\sigma_0}, \theta; T/\sigma_0 \right). \tag{2}$$

That is, the load parameter T/σ_0 provides a convenient means to investigate and parameterize specimen geometry effects on near-tip stress triaxiality under conditions of well-contained yielding. Indeed, such studies have been carried out by Betegón and Hancock (6), Bilby *et al.* (14) and Harlin and Willis (15). Nevertheless, the result in [2] cannot have general applicability since the elastic solution [1], upon which the T−stress is defined, is an asymptotic condition which is increasingly violated as plastic flow progresses beyond well-contained yielding.

Recognizing the above limitation O'Dowd and Shih (1,2), hereafter referred to as OS, identified members of the family of fields by the parameter Q which arises naturally in the plasticity analysis. OS write:

$$\sigma_{ij} = \sigma_0 f_{ij} \left(\frac{r}{J/\sigma_0}, \theta; Q \right), \quad \epsilon_{ij} = \epsilon_0 g_{ij} \left(\frac{r}{J/\sigma_0}, \theta; Q \right), \quad u_i = \frac{J}{\sigma_0} h_i \left(\frac{r}{J/\sigma_0}, \theta; Q \right). \tag{3}$$

The additional dependence of f_{ij}, g_{ij} and h_i on dimensionless combinations of material parameters is understood. The form in [3] constitutes a one-parameter family of self-similar solutions, or in short a Q-family of solutions.

Difference Field and Near-Tip Stress Triaxiality

Using the modified boundary layer formulation and considering a piece-wise power law material, OS generated the full range of small scale yielding solutions, designated by $(\sigma_{ij})_{SSY}$. Stress distributions associated with several values of T/σ_0 are shown in Fig. 2. OS considered the difference field defined by

$$(\sigma_{ij})_{\text{diff}} = (\sigma_{ij})_{SSY} - (\sigma_{ij})_{HRR}, \qquad [4]$$

where $(\sigma_{ij})_{HRR}$ is the HRR field. They systematically investigated the difference fields within the forward sector, $|\theta| < \pi/2$, of the annulus $J/\sigma_0 < r < 5J/\sigma_0$, since this zone encompasses the microstructurally significant length scales for both brittle and ductile fracture (25). Remarkably, the difference fields in the forward sector displayed minimal dependence on r. Noting this behavior, OS expressed the difference field within the forward sector in the form

$$(\sigma_{ij})_{\text{diff}} = Q\sigma_0\hat{\sigma}_{ij}(\theta), \qquad [5]$$

where the angular functions $\hat{\sigma}_{ij}$ are normalized by requiring $\hat{\sigma}_{\theta\theta}(\theta = 0)$ to equal unity. Moreover, the angular functions within the forward sector exhibit these features: $\hat{\sigma}_{rr} \approx \hat{\sigma}_{\theta\theta} \approx$ constant and $|\hat{\sigma}_{r\theta}| \ll |\hat{\sigma}_{\theta\theta}|$, see Figs. 3, 4 and 5 in (1).

Thus difference fields within the sector, $|\theta| < \pi/2$ and $J/\sigma_0 < r < 5J/\sigma_0$, correspond effectively to spatially uniform hydrostatic stress states. Therefore Q defined by

$$Q \equiv \frac{\sigma_{\theta\theta} - (\sigma_{\theta\theta})_{HRR}}{\sigma_0} \qquad \text{at} \quad \theta = 0, \quad r = 2J/\sigma_0 \qquad [6]$$

is a natural, obvious measure of near-tip stress triaxiality, or crack tip constraint, relative to a reference high triaxiality stress state. The distance chosen for the definition of Q lies just outside the finite strain blunting zone.

OS also considered the difference field relative to a reference stress field given by the standard small scale yielding solution, $(\sigma_{ij})_{SSY;T=0}$, which is driven by K alone. That is,

$$(\sigma_{ij})_{\text{diff}} = (\sigma_{ij})_{SSY} - (\sigma_{ij})_{SSY;T=0}. \qquad [7]$$

In this case the difference field in the forward sector matches a spatially uniform hydrostatic stress state even more closely. Thus an alternative definition of Q is

$$Q \equiv \frac{\sigma_{\theta\theta} - (\sigma_{\theta\theta})_{SSY;T=0}}{\sigma_0} \qquad \text{at} \quad \theta = 0, \quad r = 2J/\sigma_0. \qquad [8]$$

The $Q-$values reported in this paper are based on the definition in [8]. Representative stress distributions of the Q-family of fields can be found in (1-3).

Choice of Reference Field

The value of Q is affected slightly by the choice of reference field. Thus a small increment (or decrement) must be applied to the $Q-$values if the reference field is changed from $(\sigma_{\theta\theta})_{HRR}$ to $(\sigma_{\theta\theta})_{SSY;T=0}$, or vice versa.

In practice it really does not matter whether we use $(\sigma_{\theta\theta})_{HRR}$ [6], or $(\sigma_{\theta\theta})_{SSY;T=0}$ [8], for the definition of Q so long as it is applied consistently. In other words, the evaluation and tabulation of $Q-$ solutions for test specimens, the determination of the toughness locus from test data, and subsequent application of such data to predict fracture in structural components should be based on the same reference field. Numerical studies of several crack geometries show that [8] is preferable, when as part of the study, it is desired to establish the robustness of the $J - Q$ theory (28, 29, 30). We should also point out that differences between the finite on the first argument in [3] and works within deformation plasticity theory and an elastic power-law hardening material, then one obtains a series in $r/(J/\sigma_0)$:

$$\sigma_{ij} = \sigma_0 \left(\frac{J}{\alpha \epsilon_0 \sigma_0 I_n r} \right)^{1/(n+1)} \tilde{\sigma}_{ij}(\theta; n) + \underbrace{\text{second order term + higher order terms}}_{\text{Difference Field}} \qquad [13]$$

where the first term is the HRR field and the second- and higher-order terms constitute the difference field.

Xia *et al.* (27) have obtained a four term expansion for the series in [13] for an elastic power-law hardening material with $n = 10$. They have successfully matched the four term series to the radial and angular variations of the difference field given in Fig. 3 and Fig. 5 in O'Dowd and Shih (1). Furthermore, they found that the collective behavior of the second-, third- and fourth-order terms in the forward sector, $|\theta| < \pi/2$, is effectively equivalent to a spatially uniform hydrostatic stress state, so that [13] can be approximated by the simpler forms in [11] or [12].

SMALL SCALE YIELDING: Q-T RELATION

Within the modified boundary layer formulation [1], Q can be shown to depend on T alone (1,2). For a piecewise power-law hardening material,

$$Q = F(T/\sigma_0; n). \qquad [14a]$$

Curves of Q vs. T/σ_0 for $n = 5$, 10, 20 and ∞ materials ($E/\sigma_0 = 500$ and Poisson's ratio $\nu = 0.3$) are provided in Fig. 3 (28). These Q values, based upon the definition in [8], were determined by small strain analyses; essentially identical results were obtained from finite strain analyses. It can be seen that Q is a monotonically increasing function of T/σ_0. Furthermore, the crack tip stress triaxiality can be significantly lower than the standard small scale yielding state (the $Q = 0$ state) but cannot be elevated much above it. For $|T/\sigma_0| < 0.5$, the $Q - T$ relationship is not sensitive to n. Moreover, the $Q - T$ relationship is effectively independent of E/σ_0 and ν. In Table I, we have tabulated Q and Q' keeping only two places beyond the decimal point. Note that Q' is essentially zero everywhere except when $|T/\sigma_0| \approx 1$.

TABLE I. Values of Q and Q' for several values of T/σ_0

T/σ_0	−1.0	−0.75	−0.5	−0.25	0.0	0.5	1.0
$n = 3$, Q	−0.78	−0.57	−0.37	−0.16	0.0	0.27	0.41
Q'	−0.02	−0.01	0.0	0.0	0.0	−0.01	−0.04
$n = 5$, Q	−0.98	−0.74	−0.46	−0.20	0.0	0.27	0.36
Q'	−0.03	0.0	0.0	0.0	0.0	0.0	−0.02
$n = 10$, Q	−1.23	−0.92	−0.54	−0.23	0.0	0.21	0.24
Q'	−0.03	0.0	0.0	0.0	0.0	0.0	0.0
$n = 20$, Q	−1.48	−1.06	−0.55	−0.19	0.0	0.17	0.20
Q'	−0.02	0.02	0.0	0.0	0.0	0.0	0.0
$n = \infty$, Q	−1.84	−1.17	−0.60	−0.21	0.0	0.12	0.12
Q'	0.03	0.03	0.0	0.0	0.0	0.0	0.0

Using a least squares fit, the curves in Fig. 3 can be closely approximated by

$$Q = a_1 \left(\frac{T}{\sigma_0} \right) + a_2 \left(\frac{T}{\sigma_0} \right)^2 + a_3 \left(\frac{T}{\sigma_0} \right)^3 \qquad [14b]$$

strain and small strain $(\sigma_{\theta\theta})_{\text{SSY};T=0}$ fields are negligible at radial distances greater than about $2J/\sigma_0$, see Fig. 2 in (1).

There is possibly an advantage to using $(\sigma_{\theta\theta})_{\text{SSY};T=0}$ as the reference field. We note that $(\sigma_{\theta\theta})_{\text{SSY};T=0}$ can be evaluated for an actual stress-strain relation. Therefore, the analyses in the fracture application can be carried out for the actual stress-strain relation of the material. By contrast, the reference field $(\sigma_{\theta\theta})_{\text{HRR}}$ is defined for an elastic power-law hardening material and the calculations in the application also should employ an elastic power-law hardening relation.

Variation of Q with Distance

Because Q scales the difference field relative to a reference stress state, it provides a sensitive measure of the evolution of near-tip stress triaxiality. It also can be used to detect changes in the stress triaxiliaty that do not conform to the structure identified through analyses using the modified boundary layer model. For this purpose, we consider $Q(\bar{r})$ defined by

$$Q(\bar{r}) = \frac{\sigma_{\theta\theta}(\bar{r}) - [\sigma_{\theta\theta}(\bar{r})]_{\text{SSY};T=0}}{\sigma_0}, \qquad [9]$$

where $\bar{r} \equiv r/(J/\sigma_0)$, all evaluated ahead of the crack ($\theta = 0$). Note that $(\sigma_{\theta\theta})_{\text{SSY};T=0}$ is chosen as the reference field.

The mean gradient of Q over distances $1 < \bar{r} < 5$,

$$Q' = \frac{Q(\bar{r} = 5) - Q(\bar{r} = 1)}{4}, \qquad [10]$$

can be used to monitor changes in the spatial distribution of the hoop stress that do not conform to a spatially uniform difference field. For example, over the range of MBL loadings considered in this paper, the largest magnitude of Q' is less than 0.04 (see Table I); this means that $(\sigma_{\theta\theta})_{\text{diff}}$ is effectively unchanging, varying by less than $0.16\sigma_0$ over the interval $1 < \bar{r} < 5$. When $|Q'|$ is much larger than 0.1, Q should be regarded as a *pointwise* measure of the stress level at the distance $2J/\sigma_0$ ahead of the crack tip. The evolution of Q' in finite width crack bodies as the load level increases can be used to delineate the range of applicability of the $J - Q$ approach.

Simplified Forms for Engineering Applications

Two simplified representations for the Q-family of fields within the forward sector have been proposed by OS. The first is

$$\sigma_{ij} = (\sigma_{ij})_{\text{HRR}} + Q\sigma_0\delta_{ij}, \qquad [11]$$

where δ_{ij} is the Kronecker delta. This form is consistent with [6]. The second form is

$$\sigma_{ij} = (\sigma_{ij})_{\text{SSY};T=0} + Q\sigma_0\delta_{ij}, \qquad [12]$$

which is consistent with [8]. The physical interpretation of [11] and [12] is this: *negative (positive) Q values mean that the hydrostatic stress is reduced (increased) by $Q\sigma_0$ from the $J-dominant$ state, or $Q = 0$ stress state.* This interpretation is precise when $|Q'| \ll 1$.

Our numerical studies show that [12] provides a more accurate representation of the full range of near-tip fields so that a fracture methodology based on [12] and [8] has a greater range of validity.

Difference Field and Higher-Order Terms of the Asymptotic Series

The higher-order asymptotic analysis of Li and Wang (4) and Sharma and Aravas (5) has been extended by Xia, Wang and Shih (27). Indeed, if one assumes a product dependence

For $n = 3$, $a_1 = 0.6438$, $a_2 = -0.1864$, $a_3 = -0.0448$; $n = 5$, $a_1 = 0.7639$, $a_2 = -0.3219$, $a_3 = -0.0906$; $n = 10$, $a_1 = 0.7594$, $a_2 = -0.5221$, $a_3 = 0.0$; $n = 20$, $a_1 = 0.7438$, $a_2 = -0.6673$, $a_3 = 0.1078$; $n = \infty$, $a_1 = 0.6567$, $a_2 = -0.8820$, $a_3 = 0.3275$. Betegón and Hancock (6) and Wang (10) have provided relations correlating T with the near-tip hoop stress, which can be rearranged into the form given in [14b]. It must be emphasized that the one-to-one correspondence between Q and T/σ_0 applies to MBL loadings and does not apply to fully yielded states in general.

Equivalence of J-T and J-Q Approaches under Small Scale Yielding

We have discussed two approaches to specifying families of Mode I plane strain elastic-plastic crack tip fields. The first utilizes the elastic $T-$stress associated with the second term of Williams' expansion [1] to correlate the elastic-plastic crack tip fields of varying stress triaxiality. The second uses the Q parameter, defined as in [6] or [8], to quantify the near-tip stress level relative to a high triaxiality reference stress state. Indeed, a description of near-tip states by J and Q is strictly equivalent to that phrased in terms of K and T when small scale yielding conditions applies, since K and J are related by

$$J = \frac{1 - \nu^2}{E} K_I^2,$$ [15]

and Q and T are related through [14]. *It also follows that a two-parameter description of near-tip states based on J and T is equivalent to that based on J and Q. However, this equivalence does not hold under fully yielded conditions.*

We should point out that the $J - Q$ fields can exist over the entire range of plastic yielding and this does not depend on the existence of the elastic field [1]. By contrast, T is undefined under fully yielded conditions.

EVOLUTION OF CONSTRAINT IN FINITE-WIDTH GEOMETRIES

$Q-$solutions over the full range of loading and hardening exponents have been obtained for several crack geometries (1-3, 28-30). Readers are referred to these publications for the results. Representative solutions are reported below.

Tension Geometry — Center-Cracked Panel

Figure 4a shows the evolution of Q for a shallow crack, crack length to width ratio $a/W = 0.1$, evaluated at $r/(J/\sigma_0) = 1, 2, 3, 4$ and 5 for $n = 10$. It can be seen the stress triaxiality decreases gradually and approaches a steady-state value at fully yielded conditions. Note that Q is effectively independent of r. It varies slowly and linearly with r under fully yielded conditions. Indeed, $|Q'| < 0.03$ over the broad range of loading considered, that is, Q is a robust field parameter for describing the evolution of stress triaxiality. Figure 4b shows the behavior of Q for a deep crack, $a/W = 0.8$. The trends seen in Fig. 4 are representative of the stress triaxiality behavior for the other a/W ratios and strain hardening exponents.

The open circles in Fig. 4a and 4b are the stress triaxiality predicted by the $T-$stress. The prediction is accurate at low loads. However, at fully yielded conditions, the stress triaxiality is incorrectly predicted. In the case of $a/W = 0.1$, T underestimates the stress triaxiality by about $0.5\sigma_0$. For a deep crack, T overestimates the stress triaxiality by a similar amount. More results for the center-cracked panel can be found in (28).

Bend Geometry — Three-Point-Bend Bar

Q solutions for the three-point-bend bar over the full range of loading and a/W ratios are given in (28, 29). Figure 5a shows the evolution of stress triaxiality for $a/W = 0.4$ and $n = 10$. High triaxiality is maintained for deformations less than about $J/(a\sigma_0) = 0.01$. Loss of triaxiality becomes rapid when the global bending stress field, impinges on the near-tip region $r \approx 2J/\sigma_0$. This occurs at about $J/(a\sigma_0) = 0.02$ corresponding to the ASTM limit for a valid J_{IC} test. The trends exhibited in Fig. 5a are also seen for other a/W ratios (28, 29). We also have observed that Q vs. $J/(a/\sigma_0)$ is not sensitive to n. The lack of sensitivity on n can be exploited in developing a $J - Q$ engineering fracture methodology.

The evolution of $Q(\bar{r})$ over distance $1 \leq r/(J/\sigma_0) \leq 5$ is shown in Fig. 5b for several deformation levels as measured by $J/a\sigma_0$. Observe that Q is effectively independent of r, i.e. $Q' \approx 0$, for $J/a\sigma_0 < 0.03$. At higher deformation levels Q varies linearly with distance; $Q' = 0.1$ at $J/a\sigma_0 = 0.04$. Beyond this load level, Q' increases rapidly with load — under such conditions, Q only provides a pointwise measure of the stress level.

Stress triaxiality predictions by the T-stress are shown by the open circles in Fig. 5a. It can be seen that T predicts the stress triaxiality correctly under contained yielding but overestimates the stress triaxiality when large scale yielding prevails. The numerical investigations in (28, 29) have shown that T overestimates the near-tip stress triaxiality for some geometries and underestimates it in other cases so that there is not a consistent trend. *Stated in another way, a T-stress fracture methodology could be conservative for some geometries and a/W ratios and non-conservative in others suggesting that the application of such an approach may be difficult.*

Crack-Front Stress Triaxiality in Surface-Cracked Plates

The plate is loaded by remote tension σ^∞ and the results presented here are taken from (30). A sectional view of the plane containing a part-through crack is shown in the insert in Fig. 6; the geometric parameters of the flaw are indicated. $\phi = 0°$ and $90°$ correspond to lines along the free surface and directly ahead of the crack front. Let r be the radial distance ahead of (normal to) the crack front and measured in the plane containing the crack. Figure 6a shows the relaxation of stress triaxiality with increasing deformation, as measured by $J_{\text{local}}/a\sigma_0$, on different points near the crack front. J_{local} is the pointwise value of J evaluated at different locations on the crack front. Q is evaluated at distance, $r/(J_{\text{local}}/\sigma_0) = 2$, ahead of the crack front. Even at relatively low loads, it can be seen the stress triaxiality near the free surface ($\phi = 2.4°$) has been reduced to a level about equal to the yield stress.

Figure 6b shows the behavior of near-tip stress triaxiality along $\phi = 90°$, i.e. directly ahead of the surface crack. Over the entire range of loadings investigated, the variation of Q with distance is minimal. The behavior of stress triaxiality along a line near the free surface, corresponding to $\phi = 17°$, is shown in Fig. 6c. At high loads, $\sigma^\infty/\sigma_0 > 1.15$, Q exhibits a slight dependence on distance. This can be anticipated since the free surface is expected to induce strong gradients in the fields of neighboring region when large scale yielding prevails.

APPROXIMATE METHODS FOR EVALUATING Q

Q Estimates Under Contained Yielding

Leevers and Radon (31), Sham (32) have obtained T solutions for a number of crack geometries. The solutions are expressible in the form

$$T = \frac{K}{\sqrt{\pi a}}\Sigma(\text{geometry}), \quad \text{or} \quad T = \sigma^\infty h_T(\text{geometry}), \qquad [16]$$

where Σ and h_T are dimensionless geometric factors and σ^∞ is a representative stress magnitude. Now combine the second result in [16] with [14a] to get

$$Q_{SSY} = F_\sigma(\sigma^\infty/\sigma_0; \text{geometry}, n). \qquad [17]$$

F_σ depends on the normalized load, geometry, n, and dimensionless combinations of material parameters though the dependence on the latter is expected to be weak. Pointwise values of T along 3-D crack fronts of a single-edge crack plate and a plate containing a surface flaw have been obtained by Nakamura and Parks (33).

Q Estimates Under Fully Yielded Conditions

Under fully plastic conditions, Q can be determined from fully plastic crack solutions used in simplified engineering fracture analysis, Kumar *et al.* (34). The 'fully plastic' Q for a crack geometry of characteristic dimension L has the form (3),

$$Q_{FP} = H_Q \left(\frac{J}{\sigma_0 L}; \text{geometry}, n \right). \qquad [18]$$

The dimensionless function H_Q depends on the load $J/\sigma_0 L$, dimensionless groups of geometric parameters and n. In certain geometries, the stress triaxiality changes negligibly under fully yielded conditions; for these cases, a further simplification of the form in [18] is possible.

An efficient numerical method for generating fully plastic crack solutions is described by Shih and Needleman (35). This method can be used to evaluate Q_{FP} which can then be catalogued in a handbook much like that of the $J-$Handbook (34).

A Scheme to Interpolate Between Q_SSY and Q_FP

Since the dependence of Q_{SSY} on the generalized load $\mathcal{L}$ is known from [17], its derivative $dQ_{SSY}/d\mathcal{L}$ can be determined. Moreover, the relationship between J and $\mathcal{L}$ is known for small scale yielding: $J \propto K_I^2 \propto \mathcal{L}^2$. Therefore $dQ_{SSY}/d(J/\sigma_0 L)$ is available as well.

Under fully plastic conditions, Q is given by [18] and its derivative $dQ_{FP}/d(J/\sigma_0 L)$ also can be evaluated from [18]. In addition, one can determine $dQ_{FP}/d\mathcal{L}$ since $J/(\sigma_0 L) \propto (\mathcal{L}/\mathcal{L}_0)^{(n+1)}$.

By the above procedures, Q and its derivative at limiting load states of small scale and fully plastic yielding — Q_{SSY}, $dQ_{SSY}/d\mathcal{L}$ and Q_{FP}, $dQ_{FP}/d\mathcal{L}$ — are known. Q's for intermediate load states can now be obtained by interpolation. Alternatively, Q can be regarded as a function of J and interpolated using Q_{SSY}, $dQ_{SSY}/d(J/\sigma_0 L)$ and Q_{FP}, $dQ_{FP}/d(J/\sigma_0 L)$. This procedure is illustrated in Fig. 7. Both procedures are under investigation (28).

FRACTURE TOUGHNESS LOCUS

Kirk *et al.* (11) have measured cleavage toughness data for A515 steels at room temperature. They tested edge-cracked bend bars with thicknesses $B = 10, 25.4$ and 50.8 mm and various crack length to width ratios. The measured toughness data is plotted against Q in Fig. 8. The trend of the data supports the concept of a $J - Q$ toughness locus.

Triaxiality effects on fracture toughness can be predicted by using the $J - Q$ field in conjunction with a fracture criterion based on the attainment of a critical stress, $\sigma_{22} = \sigma_c$, at a characteristic distance, $r = r_c$, Ritchie *et al.* (36). Within the $J - Q$ annulus, the normal stress ahead of the crack is given by [11] or more accurately by [12]. Assume that r_c is within the $J - Q$ annulus. Apply the RKR fracture criterion to [11] to get

$$\frac{\sigma_c}{\sigma_0} = \left(\frac{J_C}{\alpha \epsilon_0 \sigma_0 I_n r_c} \right)^{1/(n+1)} \tilde{\sigma}_{22}(\theta = 0) + Q. \qquad [19]$$

Therefore one can solve for J_C as a function of Q for selected values of σ_c and r_c. Now designate the toughness value at $Q = 0$ as J_O, and use [19] to arrive at

$$J_C = J_O \left(1 - \frac{Q \sigma_0}{\sigma_c} \right)^{n+1}. \qquad [20]$$

Figure 8 shows the variation of J_C with Q (dash line) for $\sigma_c = 3.5\sigma_0$, $J_O = 40$ kPa·m and $n = 5$. The relation in [20] captures the trend of the experimental data, that is, cleavage toughness depends sensitively on constraint.

ACKNOWLEDGEMENTS

This investigation is supported by grants from the Office of Naval Research and the Nuclear Regulatory Commission.

REFERENCES

(1) O'Dowd, N. P. and Shih, C. F., *Journal of the Mechanics and Physics of Solids*, Vol. 39, 1991, pp. 989-1015.

(2) O'Dowd, N. P. and Shih, C. F., *Journal of the Mechanics and Physics of Solids*, Vol. 40, 1992, pp. 939-963.

(3) Shih, C. F., O'Dowd, N. P. and Kirk, M. T., "A Framework for Quantifying Crack Tip Constraint," 1991. To be published in ASTM-STP.

(4) Li, Y. C. and Wang, T. C., *Scientia Sinica (Series A)*, Vol. 29, 1986, pp. 941-955.

(5) Sharma, S. M. and Aravas, N., *Journal of the Mechanics and Physics of Solids*, Vol. 39, 1991, pp. 1043-1072.

(6) Betegón, C. and Hancock, J. W., *Journal of Applied Mechanics*, Vol. 58, 1991, pp. 104-110.

(7) Al-Ani, A. M. and Hancock, J. W., *Journal of the Mechanics and Physics of Solids*, Vol. 39, 1991, pp. 23-43.

(8) Parks, D. M., "Engineering Methodologies for Assessing Crack Front Constraint," *Proceedings of 1991 SEM Spring Conference on Experimental Mechanics*, 1991, pp. 1-8.

(9) Hancock, J. W., Reuter, W. G. and Parks, D. M., "Constraint and Toughness Parameterized by T, " 1991. To be published in ASTM-STP.

(10) Wang, Y.-Y., "On the Two-Parameter Characterization of Elastic-Plastic Crack-Front Fields in Surface-Cracked Plates," 1991. To be published in ASTM-STP.

(11) Kirk, M. T., Koppenhoefer, K. C., and Shih, C. F., "Effect of Constraint on Specimen Dimensions Needed to Obtain Structurally Relevant Toughness Measures," 1991. To be published in ASTM-STP.

(12) Dodds, R. H., Jr., Anderson, T. L. and Kirk, M. T., *International Journal of Fracture*, Vol. 48, 1991, pp. 1-22.

(13) Williams, M. L., *Journal of Applied Mechanics*, Vol. 24, 1957, pp. 109-114.

(14) Bilby, B. A., Cardew, G. E., Goldthorpe, M. R. and Howard, I. C., *Size Effects in Fracture*, 1986, Institution of Mechanical Engineers, London, pp. 37-46.

(15) Harlin, G. and Willis, J.R., *Proceedings of the Royal Society*, Vol. A 415, 1988, pp. 197-226.

(16) Rice, J. R., *Journal of Applied Mechanics*, Vol. 35, 1968, pp. 379-386.

(17) Hutchinson, J. W., *Journal of the Mechanics and Physics of Solids*, Vol. 16, 1968, pp. 13-31.

(18) Rice, J. R. and Rosengren, G. F., *Journal of the Mechanics and Physics of Solids*, Vol. 16, 1968, pp. 1-12.

(19) Rice, J. R. and Johnson, M. A. in *Inelastic Behavior of Solids*, (M.F. Kanninen et al., Eds.), McGraw-Hill, New York, 1970, pp. 641-671.

(20) McMeeking, R. M. *Journal of the Mechanics and Physics of Solids*, Vol. 25, 1977, pp. 357-381.

(21) McMeeking, R. M. and Parks, D. M., in *Elastic-Plastic Fracture Mechanics ASTM STP 668*, American Society for Testing and Materials, Philadelphia, 1979, pp. 175-194.

(22) Shih, C. F. and German, M. D., *International Journal of Fracture Mechanics*, Vol. 17, 1981, pp. 27-43.

(23) Hutchinson, J. W., *Journal of Applied Mechanics*, Vol. 50, 1983, pp. 1042-1051.

(24) Parks, D. M., in *Defect Assessment in Components – Fundamentals and Applications*, ESIS/EGF9 Mechanical Engineering Publications, London, 1991, pp. 205-231.

(25) Ritchie, R. O. and Thompson, A. W., *Metallurgical Transactions A*, Vol. 16A, 1985, pp. 233-248.

(26) McClintock, F. A., in *Fracture: An Advanced Treatise, Vol. III*, (H. Liebowitz, Ed.), Academic Press, New York, 1971, pp. 47-225.

(27) Xia, L., Wang, T. C. and Shih, C. F., "Higher-Order Analysis of Crack Tip Fields in Elastic Power-Law Hardening Materials," 1992. To appear in *Journal of Mechanics and Physics of Solids*.

(28) O'Dowd, N. P. and Shih, C. F., "Two-Parameter Fracture Mechanics: Theory and Applications." 1992. Submitted for publication.

(29) Kirk, M. T., Dodds, R. H., Jr. and Anderson, T. L., "Approximate Techniques for Predicting Size Effects on Cleavage Fracture Toughness (J_c)." 1992. Submitted for publication.

(30) Dodds, R. H., Jr and Shih, C. F., "J-Q and Micromechanics Treatment of Constraint in Ductile Fracture Mechanics," for presentation at IAEA/CSNI Specialists' Meeting, Oak Ridge, TN, October 1992.

(31) Leevers, P. S. and Radon, J. C., *International Journal of Fracture*, Vol. 19, 1982, pp. 311-325.

(32) Sham, T. L., *International Journal of Fracture*, 1991, Vol. 48, 1991, pp. 81-102.

(33) Nakamura, T. and Parks, D. M., "Determination of Elastic $T-$Stress along 3-D Crack Fronts Using an Interaction Integral," 1991. To be published.

(34) Kumar, V., German, M. D. and Shih, C. F., "An Engineering Approach for Elastic-Plastic Fracture Analysis," EPRI Topical Report NP-1931, Electric Power Research Institute, Palo Alto, Calif., July, 1981.

(35) Shih, C. F. and Needleman, A., *Journal of Applied Mechanics*, Vol. 51, 1984, pp. 48-56, pp. 57-64.

(36) Ritchie, R.O., Knott, J.F. and Rice, J.R., *Journal of Mechanics and Physics of Solids*, Vol. 21, 1973, pp. 395-410.

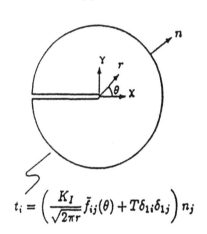

Figure 1. Conventions at a crack tip.

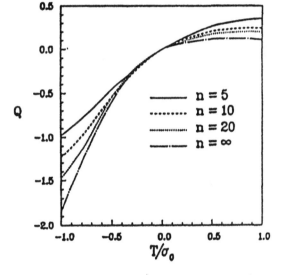

Figure 3. Variation of Q with T/σ_0.

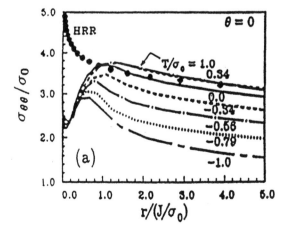

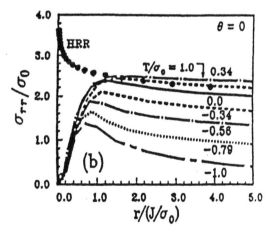

Figure 2. Distribution of hoop stress and radial stress directly ahead of crack tip for several values of T/σ_0. $n = 10$ material $(E/\sigma_0 = 500, \nu = 0.3)$.

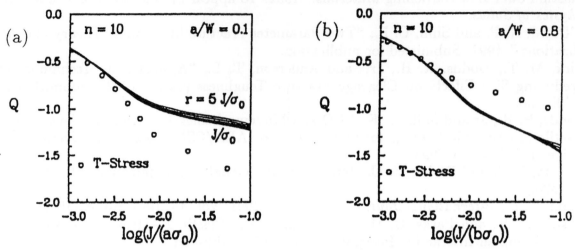

Figure 4. Center-cracked panel. Evolution of Q with increasing J for $n = 10$; Q is evaluated at $r/(J/\sigma_0) = 1, 2, 3, 4$ and 5. (a) shallow crack, (b) deep crack.

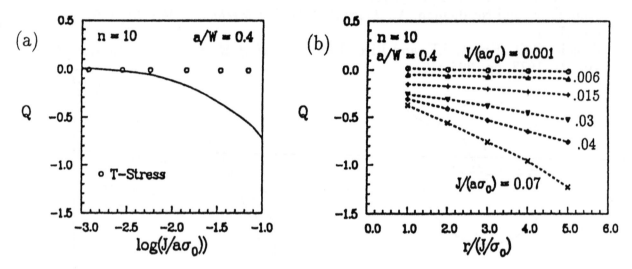

Figure 5. Three-point bend bar. Evolution of Q with increasing J for $n = 10$. (a) Q is evaluated at $r/(J/\sigma_0) = 2$. (b) Q is evaluated at several positions ahead of crack tip for several deformation levels.

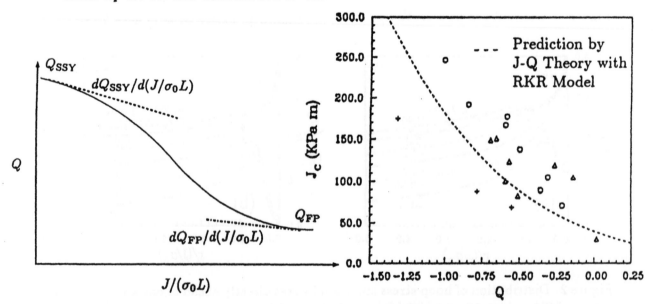

Figure 7. Procedure for interpolating between Q_{SSY} and Q_{FP}.

Figure 8. Toughness data plotted against Q for ASTM A515 Grade 70 steels [11].

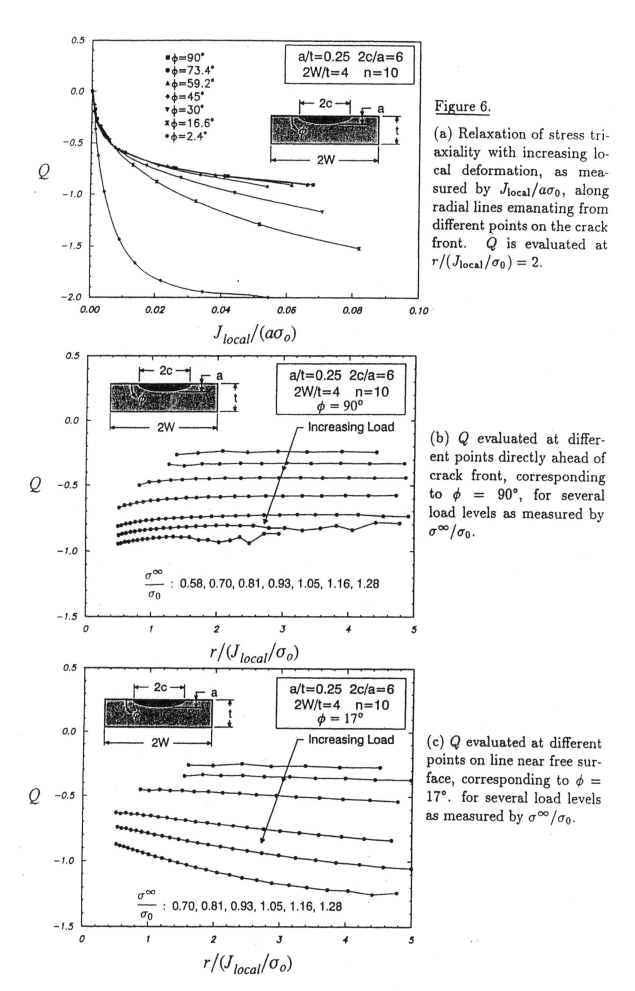

Figure 6.

(a) Relaxation of stress triaxiality with increasing local deformation, as measured by $J_{local}/a\sigma_0$, along radial lines emanating from different points on the crack front. Q is evaluated at $r/(J_{local}/\sigma_0) = 2$.

(b) Q evaluated at different points directly ahead of crack front, corresponding to $\phi = 90°$, for several load levels as measured by σ^∞/σ_0.

(c) Q evaluated at different points on line near free surface, corresponding to $\phi = 17°$. for several load levels as measured by σ^∞/σ_0.

PAPER 25

A two parameter approach to the fracture behaviour of short cracks

J W Hancock, BSc, PhD (Department of Mechanical Engineering, University of Glasgow, UK)

SUMMARY

The loss of constraint, and subsequent enhanced toughness of short crack geometries evolves from small scale yielding and the non-singular component of crack tip stress field denoted T. Transitions from deeply cracked to shallow crack behaviour are identified with the change in sign of the T stress from positive to negative. This provides a convenient definition of mechanically short cracks, which develop unconstrained flow fields. Modified boundary layer formulation of small scale yielding fields, based on K and T, have been correlated with fully plastic solutions. Finally, failure criteria have been expressed in the form of J-T loci for a wide range of cracked geometries including short and deeply cracked bars in bending, centre cracked panels, CTS specimens and surface cracked plates

INTRODUCTION

Fracture Mechanics attempts to ensure the structural integrity of engineering plant which may contain cracks or flaws or defects. The usual approach is based on the measurement of a single parameter which quantifies a lower bound material toughness. Material data is typically obtained from deeply cracked bend specimens which develop a high level of crack tip constraint. This provides a lower bound estimate of toughness, and ensures a conservative approach when applied to structural defects. Although this approach is safe, it is often unnecessarily conservative, and may lead to the imposition of prohibitive repair and inspection policies when applied to short cracks which develop unconstrained flow fields. This approach to defect assessment also denies a fundamental scientific objective of fracture mechanics, that data should be able to be transferred from one geometry to another. In this context it has long been recognised that the fracture toughness of fully plastic specimens is geometry dependent. In particular shallow fully plastic cracks are associated with markedly higher values of toughness than deeply notched bend specimens.

SHALLOW CRACKS

Cracks may be short in the sense that the crack length or the extent of the plasticity is comparable to micro-structural dimensions such as the grain size. In this context the micro-mechanics of short fatigue cracks have been extensively discussed. However the present discussion is motivated towards mechanically short cracks, in which the plastic zone is no longer small compared to the crack length, but is still large in relation to microstructural size scales.

Deeply cracked geometries exhibit deformation fields in which the plasticity is confined to the uncracked ligament. As a result the size of the uncracked ligament is the only significant structural dimension, and the fully plastic flow field is independent of the (a/W) ratio. Such behaviour is exemplified by a deeply cracked bend bar in which plasticity can be represented by a slip line field proposed by Green (1), denoted by the solid lines in Figure 1a. Deformation consists of a plastic hinge which is independent of the exact (a/W) ratio as long as plasticity is confined to the uncracked ligament. In contrast shallow cracked bend bars exhibit plasticity to the cracked face as shown by the extension to the slip line field proposed by Ewing(2), and indicated by broken lines in Figure 1a. These two solutions may be compared with the development of plasticity in shallow and deeply cracked bend bars shown in Figure 1b and 1c after the numerical solutions of Al-Ani and Hancock (3). For a weakly hardening material plasticity was confined to the uncracked ligament for a/w ≥ 0.3 thus defining deep crack behaviour for this loading. For edge cracked bars subject to a uniformly distributed applied force, a similar behaviour was found in which deep crack behaviour applied in the range a/w ≥ 0.55.

A similar situation applies in double edge cracked bars, as illustrated in Figure 2. For a/w ≥ 0.9, plasticity is confined to the uncracked ligament, when deformation has been described by the Prandtl field (4) shown in Figure (2b). In contrast shallow crack behaviour pertains when plasticity extends out to the edge of the cracked body as shown in the slip line field of Ewing and Hill (5) shown in Figure 2a, and confirmed by the etched plastic zones (6) for shallow and deeply double edge cracked bars shown in Figures 2c,d

McClintock (7) originally noted the lack of uniqueness of such fully plastic flow fields, and the implication of geometry dependent toughness. Hancock and co-workers (3, 8, 9) have argued that the lack of uniqueness is not associated with the sudden development of the fully plastic flow field but rather evolves gradually from small scale yielding and in particular the non-singular T stress associated with the elastic field which is geometry dependent.

T STRESS

The elastic crack tip field can be expressed as an asymptotic series in cylindrical co-ordinates (r,θ) about the crack tip following Williams (10). Interest is restricted to the first two terms of this series, which are non-zero at the crack tip

$$\begin{bmatrix} \sigma_{11} & \sigma_{12} \\ \sigma_{21} & \sigma_{22} \end{bmatrix} = \frac{K}{\sqrt{2\pi r}} \begin{bmatrix} f_{11}(\theta) & f_{12}(\theta) \\ f_{21}(\theta) & f_{22}(\theta) \end{bmatrix} + \begin{bmatrix} T & 0 \\ 0 & 0 \end{bmatrix} \qquad [1]$$

The first term in the expansion is singular at the crack tip and embodies the stress intensity factor K. In the notation of Rice (11), the second term in the expansion is denoted T, and consists of a uniaxial tensile or compressive stress parallel to the crack flanks. The T stress has been tabulated for through cracked geometries (12, 13, 14), and for surface cracked panels (15). T is either expressed simply as a stress concentration factor $\left(\frac{T}{\sigma}\right)$, so that T is obtained by simply multiplying an applied stress by a tabulated constant, or is given in the form of a biaxiality parameter, B,

$$B = \frac{T\sqrt{\pi a}}{K} \qquad [2]$$

Figures 1 and 2e show the biaxiality parameter, B, plotted as a function of (a/W) for edge cracked bend bars in bending (12), and double edge cracked bars in tension (16). The salient point is that deep crack behaviour as defined by the fact that plasticity is restricted to the uncracked ligament is broadly characterised by positive or zero values of the T stress. Short crack behaviour which leads to a loss of crack tip constraint characterised by negative T stresses in which plasticity extends to the cracked face. This now provides a unifying feature of all short crack behaviour. Geometries which evolve from small scale yielding to an unconstrained flow field are thus all identified by negative T stresses.

ELASTIC-PLASTIC CRACK TIP FIELDS

In order to examine the stress field within the plastic zone crack tip fields in small scale yielding computational models of crack tip plasticity have been developed (17, 8) in which the displacements corresponding to the first two terms of the Williams expansion (10) are applied as boundary conditions to the outer elastic field. These are referred to as modified boundary layer formulations. Within the plastic zone such analyses produce a family of fields which are characterised by J and T. Fields with positive T stresses approach the HRR field, and are basically insensitive to T. In contrast, fields with negative T stresses show a loss of crack tip constraint. Fields with negative T stresses lose crack tip constraint within the plastic zone.

Crack tip plasticity has been widely discussed, for non-hardening materials in terms of plane strain slip line fields. Rice and Tracey (20) have shown that for a non-hardening material under plane strain conditions, the crack tip stress state must satisfy either

$$\frac{\partial \sigma_{r\theta}}{\partial \theta} = 0 \quad \text{or} \quad \frac{\partial (\sigma_{rr} + \sigma_{\theta\theta})}{\partial \theta} = 0 \qquad \qquad \text{[3]}$$

In sectors in which the first condition applies the equilibrium equations combined with the yield criterion leads to stress states which are represented by centred fans, such as that shown in Figure 3. In regions where the second condition applies the crack tip stress states correspond to a homogeneous deformation field in which the cartesian stresses are independent of the angular co-ordinate θ, and the slip lines are straight. Within the angular span of the plastic zone at the crack tip the slip line field can only consist of assemblies of centred fans and constant stress regions. The total angular span can thus only comprise elastic wedges, centred fans and constant stress regions in which the slip lines are straight.

If plasticity is *assumed* to surround the crack tip there is only one possible form of solution which is the Prandtl field shown in Figure 3a, which is the limiting case of the HRR field for perfect plasticity. At the crack tip $(r = 0)$ the assumption that plasticity surrounds the tip thus lead to a unique solution. In this situation there can be no higher order terms in existence *at the tip*, and the HRR field is the complete solution at $r = 0$. It is worth noting that the field features a tensile stress on the crack flanks within the plastic zone, and it is perhaps not surprising that this should evolve from fields with a tensile T stress in the surrounding elastic field. Once plasticity surrounds the tip, any further increase in the T stress does not change the stresses in the plastic field. This explains the observation (7) that J dominance is characterised by positive T stresses, but that any further increase in the T stress does not change the local field once plasticity is established on the crack flanks.

In contrast, compressive T stresses significantly change the crack tip deformation. The effect of compressive T stresses is to decrease the angular span of the centred fan, giving an incomplete Prandtl field in which plasticity does not envelop the tip, and an elastic wedge emerges on the crack flanks, as shown in Figures 3.b,c,d which are taken from Du and Hancock (9). As the fan is the region of strain concentration in the small strain solution, this corresponds to the forward rotation of the lobes of the plastic zone in small scale yielding. The diamond ahead of the tip maintains a uniform stress state in which the hydrostatic stress is less than the fully constrained Prandtl value, differing from it in the forward sectors by a simple hydrostatic stress.

Sham (21) has analysed elastic-perfectly plastic crack tip deformation with the possibility that plasticity does not entirely surround the crack tip. The fields are of the form shown in Figure 3 and comprise a homogeneous region ahead of the crack which is contiguous with a centred fan and an elastic wedge on the crack flanks. The analysis is based on isotropic incompressible elasticity, satisfies equilibrium, compatibility and the yield criterion, and continuity of stresses across the elastic plastic boundary. The stress field within the elastic wedge is not that of the fully elastic solution. It is however worth pointing out that as the crack tip is approached asymptotically, the elastic stresses have a strength of order r^0, like the fully elastic stress field.

It is important to note that assemblies of this form are not restricted to small scale yielding. Such assemblies must also apply in full plasticity. In fact the only way in which an unconstrained flow field can rise in either small scale yielding or in full plasticity for a perfectly plastic material is that plasticity does not surround the tip and the crack flanks are elastic. Thus the flow field of short cracks and other unconstrained flow fields such as centre cracked panels in full plasticity is that the loss of crack tip constraint is associated with an elastic wedge on the crack flanks. The elastic stress distribution in the wedge determines its angular span and also fixes the stress state which occurs directly ahead of the crack. In fully plastic flow fields there is thus a physical significance of the vestige of the original elastic stress field. In non-hardening plasticity the nature of the elastic stress field fixes the nature and constraint of the stress field ahead of the crack tip. Thus for unconstrained flow fields in which plasticity spreads across the ligament, the crack tip constraint is fixed by the elastic stress distribution within the wedge of unyielded material on the crack flank.. The loss of constraint and the enhanced toughness of short cracks arises from the non-singular elastic stresses on the crack flank in both small scale yielding and full plasticity.

The ability of modified boundary layer formulations to describe contained yielding fields is not surprising, however remarkably Betegón and Hancock (8) and Al-Ani and Hancock (3) were able to correlate modified boundary layer formulations with full field solutions of a wide range of geometries into full plasticity. In order to correlate fully plastic solutions with small scale yielding solutions using modified boundary conditions, T was calculated in the same manner from the applied load or equivalently from the elastic component of J, in both cases.

The ability of an elastic parameter to correlate fully plastic flow fields of such a diverse range of geometries can be explained qualitatively. At infinitesimally small loads, plasticity is only a minor local perturbation of the leading term of the elastic field, allowing crack tip deformation to be represented by single parameter characterisation in a boundary layer formulation with the K field displacements imposed on the boundaries. As the load increases within contained yielding, the outer elastic field can be characterised by K and T, both of which are rigorously defined. Within the plastic zone the crack tip field now evolves in a way that is correctly represented by a modified boundary layer formulation with both K and T as boundary conditions. The initial evolution of the crack tip field is thus rigorously determined by T. Geometries with negative T stresses start to lose crack tip constraint, while those which have positive T stresses maintain the character of the small scale yielding field. For simplicity it is appropriate to restrict discussion to small geometry change perfect plasticity when the appropriately non-dimensionalised crack tip field reaches a steady state, independent of deformation. At limit load, the value of T calculated from the applied load or equivalently the elastic component of J, also remains constant. When T, as calculated from the limit load, is used to make contact with the modified boundary layer formulations, the predicted stress field also reaches a steady state.

The justification for using T to correlate modified boundary layer formulations with full elastic-plastic full fields solutions is that within small scale yielding it gives rigorously correct solutions which start to evolve in the correct manner towards full plasticity. In full plasticity the method produces fields which reach steady state and moreover have been shown to match full field solutions for a very wide range of geometries. The approach has been criticised by O'Dowd and Shih (22, 23, 24). It is clear that T is an elastic concept, and that in full plasticity Hancock and co-workers have used it as a device for relating fully plastic and small scale yielding solutions. In both small scale yielding and full plasticity, loss of constraint is due to the higher order terms in the non-linear solutions as by discussed by Li and Wang (18) and Sharma and Aravas (19) and used by Shih and co-workers. However such solutions are all based on the assumption that plasticity surrounds the crack tip. However fields such as those shown in Figure 3, and discussed by Du and Hancock((9) and Sham (21), are significantly different in that the loss of constraint is associated with an elastic wedge on the crack flanks. In this case the second order term at the crack tip in the non-hardening case is simply a hydrostatic term with an r^0 dependence.

It is clear that T is an appropriate, simple and robust device to determine the nature and development of such terms. In this sense calculations based on T may be likened to calculating the elastic component of J from K in a fully plastic field, in this sense T may be thought of as affecting the elastic contribution to the second term in the non-linear asymptotic series. The basis of J-T fracture mechanics is that the major contribution to the second term in the non-linear expansions can be associated with T. The method provides a simple and robust method of analysing the loss of constraint associated with shallow cracks and other unconstrained flow fields. More detailed comparisons of the approaches has been given in recent reviews by Parks (25, 26).

EXPERIMENTS

Brittle failure

The effect of (a/W) ratio on the toughness of a brittle experimental steel tested in three point bending by Betegón (27, 28) is shown in Figure 4. The results are expressed in the form of a J-T locus in Figure 5. For the deeply cracked geometries (a/W ≥ 0.3) the toughness is independent of geometry as failure occurs at the same level of constraint in basically similar flow fields under flow fields which evolve from small scale yielding fields with a positive T stress. However, for a/W ≤ 0.3 plasticity extends initially to the cracked face and constraint is lost resulting in flow fields which evolve from compressive T stress fields resulting in an increased toughness for shallow cracks.

Ductile failure

Hancock, Reuter and Parks (29) tested a wide range of through and part through cracked geometries of an A710 steel which fails in a ductile manner. In this, and other tough materials the start of crack tip extension is capable of a number of rather arbitrary definitions. Here failure is defined at small amounts of crack extension as shown in Figure 6. The results show a significant effect of constraint on toughness after small amounts of crack growth. Thus the J and CTOD values for centre cracked panels are approximately 4 times greater than that of highly constrained deeply cracked bend bars and CTS specimens at crack extensions of $200\mu m$,.

The toughness of all the geometries is correctly ordered by the T stress. Geometries with positive T stresses show a basically geometry independent toughness as the crack tip are capable of single parameter characterisation by J or CTOD. Similarly, geometries with negative T stresses show a geometry dependent toughness in which the crack tip constraint and associated toughness are correctly parameterised by T. The reason for the strong effect of the amount of crack extension on the constraint sensitivity of toughness is due to the effect of constraint on the slope of the resistance curves shown in Figure 7. Similarly the T stress correctly orders the resistance to crack tip tearing. Geometries with negative T stresses show an enhanced resistance to tearing which is correctly parameterised by T. As a simple specific illustration it is appropriate to compare the behaviour of surface cracked panel (a/2c = 0.5) tested in tension and the edge cracked bar (a/W = 0.1) tested in three point bending. Both configurations have similar T stresses and exhibit closely similar behaviour in terms of initiation and toughness. It would be difficult to have made such a connection without the use of a constraint based methodology

ACKNOWLEDGEMENTS

Acknowledgements are due to the support and encouragement of Dr J Sumpter through MoD grant D/ERI/9/4/2048/59/ARE(D). Thanks are due to Hibbitt Karlsson and Sorensen Inc for access to ABAQUS under academic license.

REFERENCES

(1)   Green A.P. and Hundy B.B. 1956 J. Mech Phys Slds Vol 4,128
(2)   Ewing D.(1968),J. Mech Phys Slds, Vol 16, 205
(3)   Al-Ani, A. A. and Hancock, J. W., (1991), J. Mech Phys Slds Vol. 39 pp. 23-34.
(4)   Prandtl,L.1920 Nachr.Ges.Wiss.,Gottingen,74
(5)   Ewing D.J.F and Hill R. 1967 J.Mech Phys. Solids 15,115
(6)   Hancock J.W. and Cowling M.J.
(7)   McClintock F.A., (1971), "Plasticity Aspects of Fracture", "Fracture" Vol 3, pp. 47-225, ed Liebowitz H,. Academic Press, London.
(8)   Betegón, C. and Hancock, (1991), J. W.,J.Appl.Mech. Vol 58, pp.104-110.
(9)   Du Z-Z and Hancock J.W. (1990), Vol.39, pp.555-567
(10) Williams, M. L.,(1957), Vol. 24, pp.111-114.
(11) Rice, J. R., (1974) , J. Mech Phys Slds, Vol. 22, pp.17-26.
(12).Sham, T.- L. (1991), Int. J. Fracture., 48, 81
(13) Leevers, P. S. and Radon, J. C.,(1983), Int J Fracture, Vol. 19, pp. 311-325.
(14) Kfouri A.P. (1986) Int .J.Fracture, Vol 30,pp.301-315
(15) Parks D.M. and Wang Y.Y.,(1991) submitted to Int. J. Fracture.
(16) Nekkal A 1991 MSc Thesis University of Glasgow
(17) Bilby, B. A., Cardew, G. E., Goldthorpe, M. R. and Howard, I. C., (1986), in "Size effects in Fracture", I. Mech. E., London, pp. 37-46
(18) Li, Y. and Wang, Z.,(1986), Scientia Sinica (Series A), Vol. 29, pp. 942-955.
(19) Sharma S.M. and Aravas N.J. (1990) J. Mech Phys Slds. in press.
(20) Rice J.R. and Tracey D.M 1973 in "Numerical and Computational Methods in Structural Mechanics"
(21) Sham T-L 1990 private communication
(22).O'Dowd N.P. and Shih C.F.1991 J. Mech Phys SldsVol 39,989
(23).O'Dowd N.P. and Shih C.F.1991 J. Mech Phys Slds in press
(24).Shih C.F, O'Dowd N.P.and Kirk M.T. (1991), ASTM Symposium on Constraint Effects in Fracture, Indianapolis.
(25) Parks D.M.(1991)"Engineering Methodolgies for Assessing Crack Front Constraint" in proc Spring Meeting of the Soc Experimental mechanics Milwaukee USA
(26) Parks D.M (1991) in "Topics in Deformation and Fracture" ed Argon A. Springer Verlag
(27) Betegón C.(1990), PhD Thesis, Department of Construction, University of Oviedo, Spain
(28) Betegón C. and Hancock J.W.,(1990) pp 999-1002, ECF8 Turin (ed D. Firrao), EMAS, Warley , U.K.
(29) Hancock, J. W., Reuter, W. A. and Parks, D. M.,(1991), ASTM Symposium on Constraint Effects in Fracture, Indianapolis

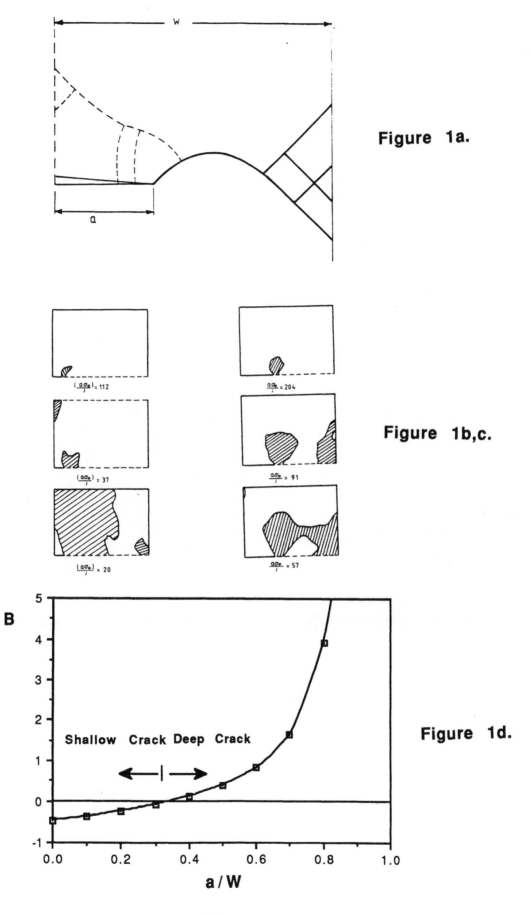

Figure 1a.

Figure 1b,c.

Figure 1d.

Figure 1

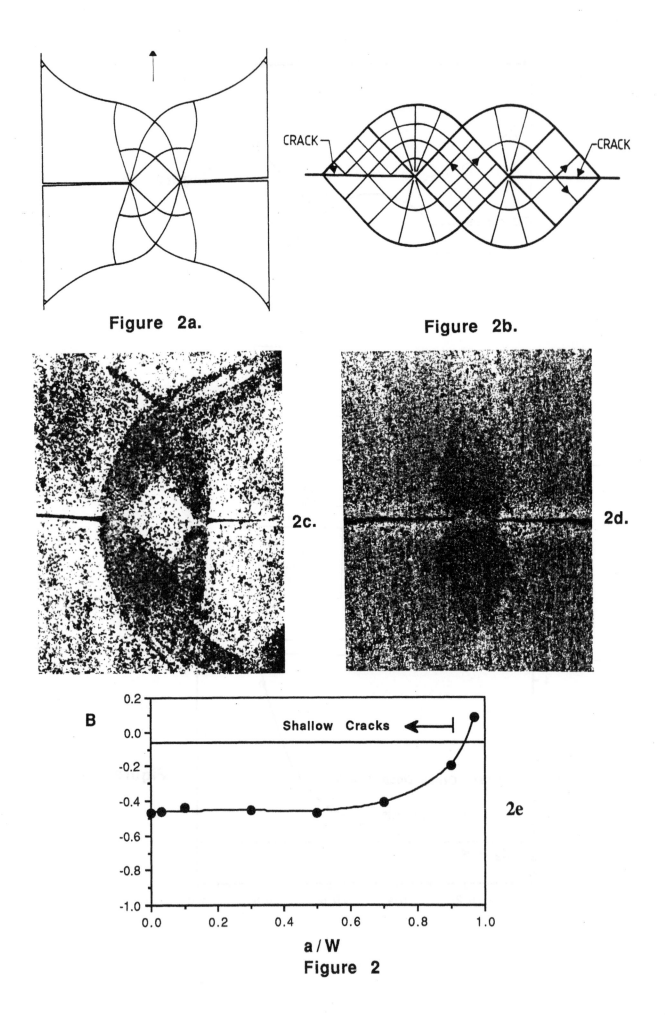

Figure 2a.

Figure 2b.

2c.

2d.

2e

Figure 2

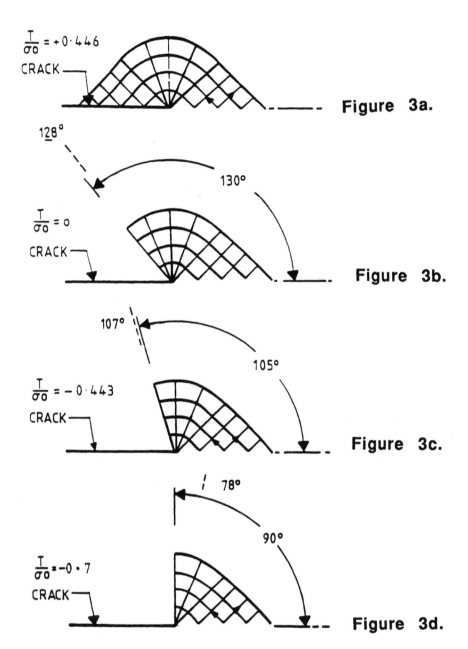

$\frac{T}{\sigma_0} = +0.446$

CRACK

Figure 3a.

128°

130°

$\frac{T}{\sigma_0} = 0$

CRACK

Figure 3b.

107°

105°

$\frac{T}{\sigma_0} = -0.443$

CRACK

Figure 3c.

78°

90°

$\frac{T}{\sigma_0} = -0.7$

CRACK

Figure 3d.

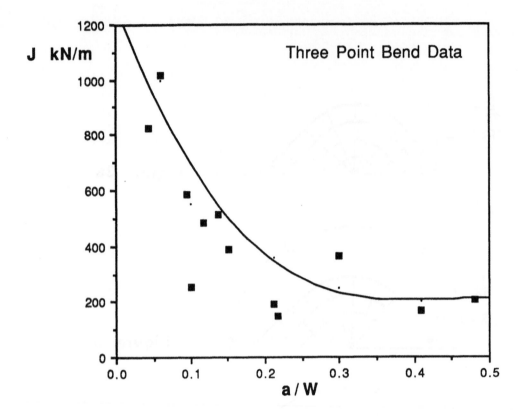

Figure 4.

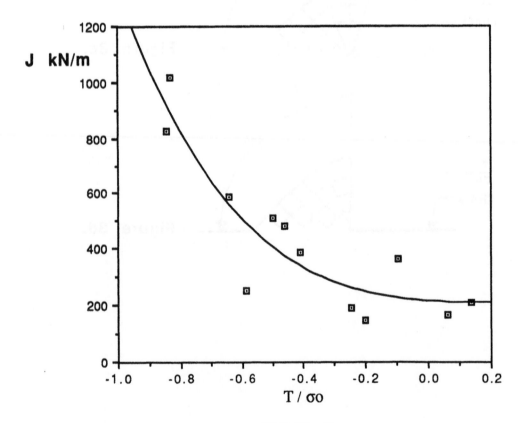

Figure 5.

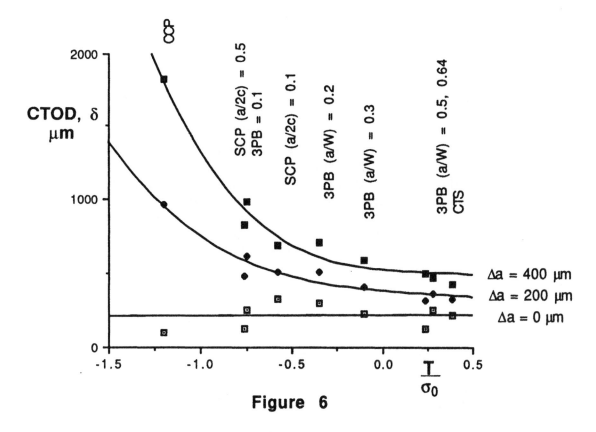

Figure 6

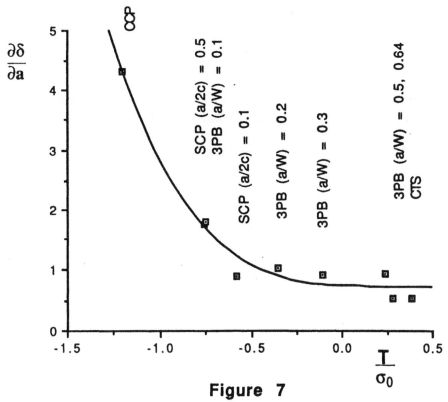

Figure 7

SHALLOW CRACK FRACTURE MECHANICS,
TOUGHNESS TESTS AND APPLICATIONS

TWI

Cambridge, UK
23-24 September 1992

PAPER 33

Characterisation of constraint effect on cleavage fracture using T-stress

Y-Y Wang, BS, MS, PhD (Edison Welding Institute, USA) and
D M Parks, BS, PhD, Prof (Massachusetts Institute of Technology, USA)

SUMMARY

The second term in Williams' (1) expansion of Mode-*I* crack-tip stress fields in an elastic body, or *T*-stress, has shown great promise as a unified constraint parameter in characterising plane strain elastic-plastic crack-tip fields at large scale yielding. In this paper, plane strain elastic-plastic crack-tip fields of a blunted crack at a constant value of K_I and various values of *T*-stress were analysed. The stress fields were incorporated into a local cleavage fracture model to determine the *T*-stress effect on macroscopic cleavage fracture toughness. The local cleavage fracture model is based on weakest link statistics using a microscale Weibull model. Significant variation of cleavage fracture toughness with respect to *T* was found at a constant value of K_I. The model was able to qualitatively correlate experimentally observed constraint effects on cleavage fracture toughness.

INTRODUCTION

Metallurgical and Statistical Aspects of Cleavage Fracture

Cleavage fracture in polycrystalline metals can be either transgranular or intergranular (2). In most ferritic steels, the dominant cleavage failure mode is transgranular cleavage cracking along well defined, low index crystallographic planes (2,3,4,5). The cleavage fracture process in these steels starts with slip-induced microcracks initiated in carbide particles located primarily on grain boundaries (6). The microcracks then propagate across the carbide/ferrite interface, and subsequently across the ferrite/ferrite grain boundary (6,7,3). Typically, the critical step is the propagation of microcracks from carbides into the surrounding ferrite grains.

It is generally accepted that the nucleation of microcracks in carbide particles requires some degree of plastic deformation in the surrounding ferrite matrix (6,7,8,3). If the stress produced by the plastic deformation is sufficiently high (or the strength of the carbide particle is relatively low), a micro-crack would be nucleated inside the carbide particle. The effective particle strength is given in terms of the particle diameter d_p and the critical strain energy release rate for a dynamically propagating crack G_{cf} (8),

$$S - \left[\frac{\pi E G_{cf}}{2(1 - v^2)d_p} \right]^{1/2} \qquad [1]$$

Here S can also be considered as a (stress-dimensioned) measure of resistance to crack propagation.

Cleavage fracture near a relatively sharp crack, where the crack tip opening displacement, or CTOD, is much smaller than the microstructural dimensions, is controlled by the magnitude of tensile stress and the availability of carbide particles. Ritchie, Knott, and Rice (9) postulated that the local tensile stress must exceed the fracture stress over a microscopically-significant (characteristic) distance ahead of an initially sharp crack tip (denoted RKR model hereafter). The existence of the characteristic distance in the RKR model reflects the competition between the rapidly decaying local tensile stress and the larger number of eligible particles participating in the cracking, as distance from the crack tip increases. In addition to the difference in particle size (thus difference in particle strength), various particles may participate in the cleavage process in different capacities, due to the variation of tensile stress magnitude, orientation of particles, and other factors (4). Such a large range of variations may be treated statistically.

Cleavage fracture has been analysed from the weakest link statistics (10,11,12,5,13,15). Since plasticity is the prerequisite of crack initiation, only carbide particles within the plastic zone were considered to be able to initiate cracking. The resultant microcracks were treated as completely independent cracks (non-interacting). The final failure was postulated to occur when the microcracks propagate unstably (and dynamically) into the surrounding ferrite matrix. The strength of the carbide particles determines the propagation of the microcracks in the carbides into the surrounding matrix. The three-parameter Weibull distribution (15,16) may be used for the particle strength distribution $g(S)$ (the number of particles per unit volume having the strength S),

$$\int_{\sigma_u}^{\sigma} g(S)dS = \left(\frac{\sigma - \sigma_u}{S_0}\right)^m fN, \qquad [2]$$

where σ is a tensile stress (often taken as the largest principal stress), m is the Weibull shape factor, σ_u is the lowest threshold stress at which cleavage can be initiated, S_0 is a scaling parameter with a dimension of stress, N is the number of particles per unit volume, and f is the fraction of those particles able to participate in cleavage process. According to the weakest link statistics, the total failure probability ϕ in a 2-D planar geometry is obtained as (5,17),

$$\ln(1 - \phi) = - fNb \left(\frac{\sigma_0}{S_0}\right)^m \int^A \left(\frac{\sigma - \sigma_u}{\sigma_0}\right)^m dA, \qquad [3]$$

where b is the characteristic sampling distance along a 2-D unit-length crack front, δA is the "infinitesimally" small area in the plane perpendicular to the crack front (11), A is the area within which the material satisfies both yield condition and $\sigma \geq \sigma_u$, and σ_0 is the yield stress. For detailed derivation, see (18). For a given material, f, N, b, σ_0, S_0, m, and σ_u can be considered constant.

Constraint Effect on Near-Crack-Tip Fields

Under small scale yielding (SSY) conditions, the near-crack-tip stress fields are uniquely related to the stress intensity factor K_I, or the J-integral. However, such one-to-one relationship does not, in general, exist at moderate to large scale yielding, e.g., (19,20). Experimental evidence has shown that such non-uniqueness may result in different fracture toughness values (21,22). The extent of deviation of the near-crack-tip fields from SSY solutions is related to the crack tip "constraint", which, broadly speaking, is a measure of the magnitude of near-crack-tip hydrostatic stress, and is specimen geometry and loading conditions dependent. The elastic T-stress has been found to be a unified constraint parameter capable of correlating such deviation for a variety of specimens and cracked structures for loads ranging up to large scale yielding (23,24,25,26). For a comprehensive review on the subject, see (27). Figure 1, taken from (28), shows the contours of maximum principal stress near the crack tip of a plane strain domain at a constant value of K_I (or J) and various values $\tau = T/\sigma_0$. Referring to Eq. [3], it is expected that the T-stress would affect the cleavage fracture probability. In this paper, the T-stress effect on the near-crack-tip fields of a blunted crack are analysed. The near-crack-tip fields at various values of T are incorporated into the weakest link model to examine the T-stress effect on the cleavage fracture toughness.

NEAR-CRACK-TIP FIELDS AT VARIOUS VALUES of T

Finite Element Model and Formulation

The near-crack-tip region was simulated by a circular domain of outer radius R shown in Fig. 2. An initial blunted crack tip with a radius of $l_o/2$, was introduced to simulate the finite strain zone. The radius of the semi-circular arc at the tip is much smaller than the outer radius of the domain ($l_o/R = 2 \times 10^{-6}$). Using the symmetry conditions on the plane of $Y = 0$, a finite element mesh was generated, as shown in Fig. 3. The mesh had 40 rings of elements radially, and 28 fans of elements circumferentially, except that the first ring had 27 fans. The entire model had 1119 plane strain, reduced integration elements (29).

To accurately simulate the stress fields in the near-crack-tip region, large geometry change formulation was used, and isotropic hardening J_2 flow theory plasticity was adopted. The material is assumed to obey power law hardening stress strain relation. The uniaxial relation can be written as,

$$\frac{\epsilon}{\epsilon_O} = \frac{\sigma}{\sigma_O} \qquad \text{for } \sigma \leq \sigma_O \, ;$$

$$\frac{\epsilon}{\epsilon_0} = \left(\frac{\sigma}{\sigma_O}\right)^n \qquad \text{for } \sigma > \sigma_O \, . \tag{4}$$

Here σ_O is the yield stress, $\epsilon_O = \sigma_O/E$ is the reference yield strain (E is Young's modulus), and n is the strain hardening exponent. This stress-strain relation was multilinearised for input in the ABAQUS program.

In-plane displacement boundary conditions,

$$u_i = \frac{K_I}{E} \sqrt{\frac{R}{2\pi}} \, f_i(\theta, v) + \frac{T}{E} R g_i(\theta, v) \, , \tag{5}$$

were applied on the outer boundary of the domain shown in Fig. 3. Here v is Poisson's ratio, $f_i(\theta, v)$ are the angular variations of the cartesian displacement components of the plane strain elastic singular field, and $g_i(\theta, v)$ are the angular variations of the displacements arising from the (plane strain) T-term. The exact expressions for $f_i(\theta, v)$ and $g_i(\theta, v)$ are listed in Table 1. The displacements contributed by K_I and T were ramped up proportionally to reach a fixed value of K_I and various values of T. The maximum radius of the plastic zone in all cases was less than $R/10$.

Table 1. The functional forms of $f_i(\theta, v)$ and $g_i(\theta, v)$

Fields	x-Component	y-Component
f_i	$(1 + v)(3 - 4v - \cos \theta) \cos (\theta/2)$	$(1 + v)(3 - 4v - \cos \theta) \sin (\theta/2)$
g_i	$(1 - v)(1 + v) \cos \theta$	$-v(1 + v) \sin \theta$

Figure 4 shows the variation of normalised crack-opening stress (σ_{yy} at $\theta = 0$) vs. normalised distance at various values of τ. The stresses were taken from extrapolated stresses at the nodes on the line $\theta = 0$, and radial distance r was based upon the deformed geometry with the zero point at the deformed crack tip. The J was calculated from the applied K_I using the identity $J = K_I^2(1 - v^2)/E$. The thick solid line at $\tau = 0$ is the stress profile at SSY. At $r < \sim J/\sigma_0$, the crack-opening stress increases with increasing radial distance from crack tip. The ascending crack-opening stress inside this region is the manifestation of the build-up of hydrostatic stress with increasing distance from the free surface. Substantial stress reduction is seen at negative τ; moderate stress elevation is observed at positive τ. The stress deviation from SSY stress accelerates with decreasing τ, while the stress seems to approach a plateau at high positive τ.

Under conditions of single parameter characterization (or SSY), there is a unique relation between the CTOD and J (30). However, such relation appears to be constraint dependent (31,18). Using Rice's 45 degree intersection definition of CTOD (32), the CTOD values at each τ were obtained from the crack opening profiles (18). At $\tau = 0$, CTOD = $0.59J/\sigma_0$, which is about 18% greater than the value of 0.51 given in (33). The CTOD values, normalised by the value at $\tau = 0$, are shown in Fig. 5. The normalised CTOD is minimal at $\tau = 0$, and increases with the increase of $|\tau|$. For comparison, the CTOD value of HRR singularity field corresponding to the current material is also shown. The constraint effect on CTOD is greater for negative values of τ than for the positive values of τ. Overall, the constraint effect is relatively small.

T-STRESS EFFECT ON THE CLEAVAGE TOUGHNESS

The relationship between the constraint parameter T and the macroscopic cleavage fracture toughness may be obtained from Eq. [3] based on dimensional analysis, similar to the relation in (5) based on the first-term eigen-expansion of the elastic crack-tip fields. The ratio of the cleavage fracture toughness at any τ, J_{IC}^{τ}, to the cleavage fracture toughness at $\tau = 0$, $J_{IC}^{\tau=0}$, can be expressed as

$$\frac{J_{IC}^{\tau}}{J_{IC}^{\tau=0}} = \sqrt{\frac{\int^A \left(\frac{\sigma - \sigma_u}{\sigma_0}\right)^m dA\big|_{\tau=0}}{\int^A \left(\frac{\sigma - \sigma_u}{\sigma_0}\right)^m dA\big|_{\tau}}} . \qquad [6]$$

The detailed derivation may be found in (18). Using the near-crack-tip stress fields obtained with a constant J and various values of τ, numerical integration in the form of Eq. [6] was carried out with a range of parameters σ_u/σ_0 and m. The numerical integration was performed using the maximum principal stresses at the integration points and the area dA associated with the stresses. The integration domain A was the area inside the plastic zone and satisfying $\sigma \geq \sigma_u$. Details of the numerical integration can be found in (18).

Figure 6 shows the cleavage fracture toughness normalised by the cleavage fracture toughness at $\tau = 0$, plotted against the value of τ for a fixed Weibull shape factor m and $n = 10$. The predicted cleavage fracture toughness increases significantly with the decrease in the value of τ. For $\sigma_u/\sigma_0 = 1.5$, the fracture toughness at $\tau = -0.9$ is about 8 times of that at $\tau = 0$.

COMPARISON WITH EXPERIMENTS

Systematic experimental studies on the constraint effect on cleavage fracture toughness have generally been lacking, even though such effect has been documented for a few individual cases (22). To demonstrate the usefulness of the model, the model-predicted cleavage toughness variation was compared with some limited experimental data available.

Betegón (34) performed three-point-bend (TPB) tests on specimens with various crack depths. All the specimens failed by cleavage fracture at room temperature. The varying crack depth provided large variation of crack tip constraint (thus a large range of τ value). The value of J-integral at final failure and the value of τ corresponding to the final failure load were obtained. The cleavage fracture toughness showed a significant increase at large negative values of τ. We tentatively normalised the toughness value by the toughness value at $\tau = 0$ ($200 \ kJ/m^2$), as shown in Fig. 7. Also shown in Fig. 7 is a fitted curve based upon the current model with $n = 7$, $\sigma_u/\sigma_0 = 1.75$, and $m = 2.0$. The model-fitted correlation between the cleavage fracture toughness and τ closely follows the experimentally obtained trend. It should be noted, however, that the material constants are merely fitted values. They cannot be determined from the limited experimental data.

To prove T as a unifying constraint parameter, correlation such as that shown in Fig. 7 should be valid for different types of specimens. Unfortunately, experimental data is not sufficient at present time to verify such J_{Ic} - τ locus. Recently, fracture tests were performed on mild steel using TPB and centre-cracked-tension (CCT) specimens (35). All the specimens failed by cleavage fracture at room temperature. The shallow-cracked 3PB specimen and the deeply-cracked CCT specimen were designed to have the same negative value of τ at limit load. If the J_{Ic}-τ cleavage locus is unique, specimens with the same τ should fail at the same J_{Ic}. The experiment showed a clear dependence of J_{Ic} on τ. We normalised the cleavage fracture toughness by the toughness value at $\tau = 0$ ($J_{Ic}^{\tau=0} = 0.042 \ MJ/m^2$) and tentatively plotted the model-predicted trend for $n = 10$, $\sigma_u/\sigma_0 = 2.0$, and $m = 2$, as shown in Fig. 8. The model-predicted trend appears to agree with the experimentally observed variation.

DISCUSSION

We have presented a micro-mechanistically based model which incorporates the variation of near-crack-tip stress fields under various crack-tip constraints, as parameterised by τ. The model qualitatively fits the experimentally observed cleavage fracture toughness variation with respect to the crack-tip constraint. However, difficulties remain in the determination of the material/microstructural parameters, such as σ_u/σ_0, m etc. These parameters may be extracted from the distribution of carbide particles based upon the relation between particle strength and carbide particle size, Eq. [1] (5,14). However, such method has not been tested on a sufficient number of materials to examine its consistency. Further application of the present model is also relying on the critical examination of the applicability of the J_{Ic}-τ leavage fracture locus on specimens with various geometries and loading configurations.

ACKNOWLEDGEMENT

This work was supported by the Office of Basic Energy Sciences, Department of Energy, under grant No. DE-FG02-85ER13331. Computation was performed on an Alliant FX-8 computer obtained under D.A.R.P.A. Grant No. N00014-86-K-0768, on the Cray-2 at MIT, and on Sun workstations obtained under ONR Grant No. 0014-89-J-3040. The ABAQUS finite element program was made available under academic license from Hibbitt, Karlsson, and Sorensen, Inc., Pawtucket, RI. The authors would like to acknowledge useful discussions with Professors A. Argon, F. McClintock, J. Hancock and Dr. C. Betegón. The support and encouragement of Doctors J. R. Gordon and P. Dong at Edison Welding Institute in the final preparation of this paper is appreciated.

REFERENCES

(1)   Williams, M. L., 1957, *Journal of Applied Mechanics*, Vol. 24, pp. 111-114.

(2)   Hahn, G. T., Averbach, B. L., Owen, W. S., and Cohen, M., 1959, in *Fracture*, Averbach, et al., Eds., M.I.T. Press, Cambridge, MA, pp. 91-116.

(3) Knott, J. F., 1973, *Fundamentals of Fracture Mechanics*, Butterworths, London.

(4) Hahn, G. T., 1984, *Metallurgical Transactions A*, Vol. 15A, pp. 947-959.

(5) Lin, T., Evans, A. G., and Ritchie, R. O., 1986, *Journal of the Mechanics and Physics of Solids*, Vol. 34, pp. 477-497.

(6) McMahon, Jr., C. J., and Cohen, M., 1965, *Acta Metallurgica*, Vol. 13, pp. 591-604.

(7) Smith, E., 1966, in *Physical Basis of Yield and Fracture*, Institute of Physics and Physical Society, London, pp. 36-45.

(8) Curry, D. A., and Knott, J. F., 1978, *Metal Science*, Vol. 12, pp. 511-514.

(9) Ritchie, R. O., Knott, J. F., and Rice, J. R., 1973, *Journal of the Mechanics and Physics of Solids*, Vol. 21, pp. 395-410.

(10) Curry, D. A., and Knott, J. F., 1979, *Metal Science*, Vol. 13, pp. 341-345.

(11) Evans, A. G., 1983, *Metallurgical Transactions A*, Vol. 14A, pp. 1349-1355.

(12) Beremin, F. M., 1983, *Metallurgical Transactions A*, Vol. 14A, pp. 2277-2287.

(13) Lin, T., Evans, A. G., and Ritchie, R. O., 1986, *Acta Metallurgica*, Vol. 34, pp. 2205-2216.

(14) Lin, T., Evans, A. G., and Ritchie, R. O., 1987, *Metellurgical Transactions A*, Vol. 18A, pp. 641-651.

(15) Weibull, W., 1938, *Ingenioersvetenskapakad, Handl.*, Vol. 151, pp. 45.

(16) Weibull, W., 1939, *Ingenioersvetenskapakad, Handl.*, Vol. 153, pp. 55.

(17) Matthews, J. R., Shack, W., and McClintock, F. A., 1976, *Journal of the American Ceramic Society*, Vol. 59, pp. 304-308.

(18) Wang, Y.-Y., 1991, "A Two-Parameter Characterization of Elastic-Plastic Crack Tip Fields and Applications to Cleavage Fracture," Ph.D. Thesis, Department of Mechanical Engineering, Massachusetts Institute of Technology, May, 1991.

(19) McMeeking, R. M., and Parks, D. M., 1979, in *Elastic-Plastic Fracture*, ASTM STP 668, American Society for Testing and Materials, Philadelphia, pp. 175-194.

(20) Shih, C. F., and German, M. D., 1981, *International Journal of Fracture*, Vol. 17, No. 1, pp. 27-43.

(21) Begley, J. A., and Landes, J. D., 1976, *International Journal of Fracture*, Vol. 12, pp. 764-766.

(22) Hancock, J. W., and Cowling, M. J., 1986, *Metal Science*, August--September 1980, pp. 293-304.

(23) Al-Ani, A. M., and Hancock, J. W., 1991, *Journal of the Mechanics and Physics of Solids*, Vol. 39, No. 1, pp. 23-43.

(24) Betegón, C., and Hancock, J. W., 1991, *Journal of Applied Mechanics*, Vol. 58, pp. 104-110.

(25)   Parks, D. M., 1991a, in *Defect Assessment in Components -- Fundamentals and Applications*, ESIS/EGF9, J. G. Blauel and K.-H. Schwalbe, Eds., Mechanical Engineering Publications, London, pp. 206-231.

(26)   Wang, Y.-Y., "On the Two-Parameter Characterization of Elastic-Plastic Crack-Front Fields in Surface-Cracked Plates," presented at the Proceedings ASTM Symposium on Constraint Effects in Fracture, Indianapolis, May 8-9, 1991, to appear in ASTM STP 1171.

(27)   Parks, D. M., 1992, "Advances in Characterization of Elastic-Plastic Crack-Tip Fields," to appear in *Topics in Fracture and Fatigue*, McClintock Festschrift.

(28)   Wang, Y.-Y., and Parks, D. M., 1990, "Evaluation of the Elastic *T*-stress in Surface-Cracked Plates Using Line-Spring Method," MIT report, August, 1990, to appear in *International Journal of Fracture*.

(29)   Hibbitt, Karlsson and Sorensen, Inc., 1988, *ABAQUS User's Manual*, Version 4.7-1, Hibbitt, Karlsson and Sorensen, Inc., Pawtucket, RI.

(30)   Shih, C. F., 1981, *Journal of the Mechanics and Physics of Solids*, Vol. 29, pp. 305-326.

(31)   O'Dowd, N. P., and Shih, S. F., 1991, "Family of Crack-Tip Fields Characterized by a Triaxiality Parameter: Part II --- Fracture Applications," Brown University Report, August, 1991, to appear, *Journal of the Mechanics and Physics of Solids*.

(32)   Tracey, D. M., 1976, *Journal of Engineering Materials and Technology*, Vol. 98, pp. 146-151.

(33)   Shih, C. F., 1983, "Tables of Hutchinson-Rice-Rosengren Singular Field Quantities," Division of Engineering, Brown University, Providence, RI, June 1983.

(34)   Betegón, C., 1990, Ph.D. Thesis, University of Oviedo, Spain.

(35)   Sumpter, J., 1991, "An Experimental Investigation of the *T*-stress Approach," presented at the ASTM Symposium on Constraint Effects in Fracture, Indianapolis, May, 1991, to appear in the ASTM STP 1171.

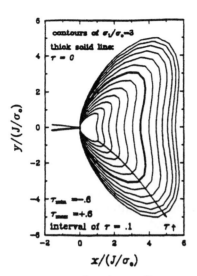

Fig. 1 Contours of $\sigma_1/\sigma_0 = 3$ near a crack tip.

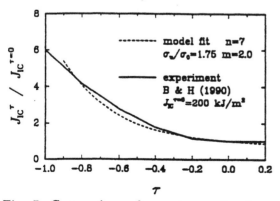

Fig. 3 FE mesh near the blunted crack tip.

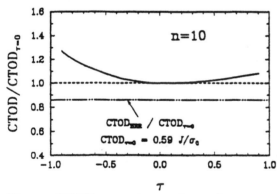

Fig. 5 CTOD at various values of τ,
normalized by the CTOD at $\tau = 0$.

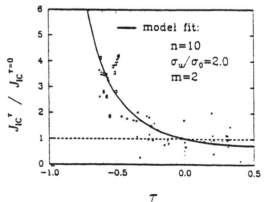

Fig. 7 Comparison of experimentally-observed
fracture toughness variation and the fitted model.

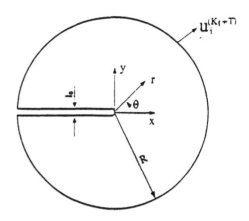

Fig. 2 Circular domain with a blunted crack tip.

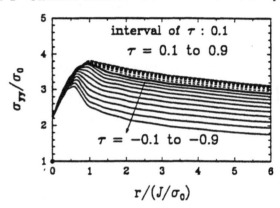

Fig. 4 Stress profiles at various values of τ.

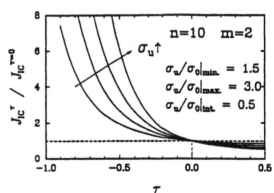

Fig. 6 Variation of cleavage fracture toughness
vs. τ at various values of σ_u.

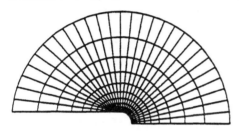

Fig. 8 Cleavage fracture toughness variation vs.
τ: experiment and fitted model prediction.

TWI

PAPER 30

A framework for predicting constraint effects in shallow notched specimens

T L Anderson, BSc, MSc, PhD, PE (Texas A&M University) and R H Dodds Jr, BSc, MSc, PhD (University of Illinois)

SUMMARY

This article outlines a model for predicting the effect of crack tip constraint on cleavage fracture toughness. The model assumes that cleavage is governed by a critical stress volume ahead of the crack tip. Very detailed elastic-plastic finite element analysis is required to infer the crack tip stress fields in the geometry of interest. This model allows cleavage fracture toughness data to be corrected for constraint loss, which is particularly pronounced in shallow notched specimens. Predictions from the constraint model are in excellent agreement with experimental data for a wide range of steels.

INTRODUCTION

Classical fracture mechanics theory assumes that a single parameter, such as the stress intensity factor, J contour integral, or crack tip opening displacement (CTOD) uniquely defines the conditions at the tip of a crack; a critical value of K, J or CTOD at fracture (i.e, the fracture toughness) is assumed to be a material constant. When this single-parameter assumption is valid, laboratory fracture toughness tests can be used to predict the fracture behaviour of large structures, much like tensile data are necessary to calculate yield loads in structures.

Numerous experimental studies, however, have demonstrated the inability of single-parameter fracture mechanics to predict structural behaviour. The classical fracture mechanics approach does not account for the loss in crack tip triaxiality that occurs under large-scale yielding conditions in geometries containing shallow cracks and in geometries loaded predominantly in tension. This loss of constraint has the most pronounced effect on cleavage fracture of ferritic steels in the low- to mid-transition region. The interaction between plastic deformation with a shallow crack causes the stresses in the vicinity of the crack tip to relax dramatically. Consequently, a higher applied J or CTOD is required to raise the stresses to a level sufficient to cause fracture. This produces a significant (factor of 3 to 6) increase in the fracture toughness for shallow cracked geometries, a beneficial phenomenon not currently accounted for in fracture testing standards or structural design/performance specifications.

A number of researchers have recently developed approaches to account for constraint effects (1-10). Most of the leading methodologies can be classified as either continuum or micromechanics approaches. The continuum approaches (1-5) are an extension of classical single-parameter fracture mechanics, in that they introduce a second parameter to characterize crack tip triaxiality. Micromechanics approaches (6-10) attempt to predict the effect of crack tip constraint on fracture toughness by performing a stress analysis on the geometry of interest and applying a local failure criterion. Both the continuum and micromechanics approaches play an important role. The two-parameter continuum approaches provide a framework for quantifying constraint, while the micromechanics approaches relate constraint to fracture behaviour.

This paper describes a micromechanics approach for predicting the effect of constraint loss on the cleavage fracture toughness of specimens and structures with shallow cracks. The details of this model and the accompanying numerical analysis are given elsewhere (6-8) and are only discussed briefly here. The present article compares predictions from this model with experimental data for several steels.

THE SCALING MODEL FOR CLEAVAGE

Cleavage fracture criterion

In order to quantify size effects on fracture toughness, one must assume a local failure criterion. In the case of cleavage fracture, a number of micromechanical models have recently been proposed (11-14), most based on weakest-link statistics. The weakest-link models assume that cleavage failure is controlled by the largest or most favourably oriented fracture–triggering particle. The actual trigger event involves a local Griffith instability of a microcrack which forms from a microstructural feature such as a carbide or inclusion; the Griffith energy balance is satisfied when a critical stress is reached in the vicinity of the microcrack. The size and location of the critical microstructural feature dictate the fracture toughness; thus cleavage toughness is subject to considerable scatter (14).

The Griffith instability criterion implies fracture at a critical normal stress near the tip of the crack; the statistical sampling nature of cleavage initiation (i.e., the probability of finding a critical microstructural feature near the crack tip) suggests that the volume of the process zone is also important. Thus the probability of cleavage fracture in a cracked specimen can be expressed in the following general form:

$$F = F[V(\sigma_1)] \qquad [1]$$

where F is the failure probability, σ_1 is the maximum principal stress at a point, and $V(\sigma_1)$ is the cumulative volume sampled where the principal stress is $\geq \sigma_1$. Equation [1] is sufficiently general to apply to any fracture process controlled by maximum principal stress, not just weakest link failure. For a specimen subjected to plane strain conditions, $V = BA$, where B is the specimen thickness and A is cumulative area on the x-y plane.

The J_{ssy} parameter

For small scale yielding, dimensional analysis shows that the principal stress ahead of the crack tip can be written as

$$\frac{\sigma_1}{\sigma_0} = f\left(\frac{J}{\sigma_0\, r}, \; \theta\right) \qquad [2]$$

It can be shown that the HRR singularity is a special case of Eq. [2]. When J dominance is lost, there is a relaxation in triaxiality; the principal stress at a fixed r and θ is less than the small scale yielding value.

Equation [2] can be inverted to solve for the radius corresponding to a given stress and angle:

$$r\,(\sigma_1/\sigma_0, \theta) = \frac{J}{\sigma_0}\, g\,(\sigma_1/\sigma_0, \theta) \qquad [3]$$

Solving for the area inside a specific principal stress contour gives

$$A\,(\sigma_1/\sigma_0) = \frac{J^2}{\sigma_0^2}\, h\,(\sigma_1/\sigma_0) \qquad [4]$$

where

$$h\,(\sigma_1/\sigma_0) = \frac{1}{2} \int_{-\pi}^{\pi} g^2\,(\sigma_1/\sigma_0, \theta)\, d\theta \qquad [5]$$

Thus for a given stress, the area scales with J^2 in the case of small scale yielding. Under large scale yielding conditions, the test specimen or structure experiences a loss in constraint, and the area inside a given principal stress contour (at a given J value) is less than predicted from small scale yielding:

$$A(\sigma_1/\sigma_0) = \phi \frac{J^2}{\sigma_0^2} h(\sigma_1/\sigma_0)$$

[6]

where ϕ is a constraint factor that is ≤ 1. Let us define an *effective* J in large scale yielding that relates the area inside the principal stress contour to the small scale yielding case:

$$A(\sigma_1/\sigma_0) = \frac{(J_{ssy})^2}{\sigma_0^2} h(\sigma_1/\sigma_0)$$

[7]

where J_{ssy} is the effective small scale yielding J; i.e., the value of J that would result in the area $A(\sigma_1/\sigma_0)$ if the structure were large relative to the plastic zone. Therefore, the ratio of the applied J to the effective J is given by

$$\frac{J}{J_{ssy}} = \sqrt{\frac{1}{\phi}}$$

[8a]

The small scale yielding J value (J_{ssy}) can be viewed as *the effective driving force for cleavage*, while J is the *apparent* driving force.

The J/J_{ssy} ratio quantifies the size dependence of cleavage fracture toughness. Consider, for example, a finite size test specimen that fails at $J_c = 200$ kPa m. If the J/J_{ssy} ratio were 2.0 in this case, a very large specimen made from the same material would fail at $J_c = 100$ kPa m. An equivalent toughness ratio in terms of CTOD can also be defined as

$$\frac{\delta}{\delta_{ssy}} = \frac{J}{J_{ssy}} \frac{m_{ssy}}{m}$$

[8b]

where $m = J/\sigma_0 \delta$ is a dimensionless J-CTOD ratio and m_{ssy} is the value of this ratio in small scale yeilding

Numerical analysis

Implementation of the scaling model for cleavage fracture requires numerical analyses that provide detailed descriptions of the crack tip stress fields. In the present study, plane strain elastic-plastic finite element analysis was performed on a number of configurations. The principal stress distribution in small scale yielding was compared to the corresponding stress fields in specimens with finite dimensions, in order to determine the J/J_{ssy} and δ/δ_{ssy} ratios for the finite specimens.

The material stress-strain behaviour was modelled with a Ramberg–Osgood power law expression:

$$\frac{\varepsilon}{\varepsilon_0} = \frac{\sigma}{\sigma_0} + \alpha \left(\frac{\sigma}{\sigma_0} \right)^n$$

[9]

where ε is strain, σ is stress, σ_0 is a reference stress, $\varepsilon_0 = \sigma_0/E$, and α and n are dimensionless constants. For the present study, $\alpha = 1.0$, $\varepsilon_0 = 0.002$, and $\sigma_0 = 414$ MPa (60 ksi); in this case σ_0 corresponds to the 0.2% offset yield strength, σ_{YS}. The strain hardening exponent, n, was assigned values of 5, 10 and 50, which correspond to high, medium and low work hardening, respectively. Small-scale yielding conditions were modelled by means of a modified boundary layer analysis, in which the boundary of a circular domain containing a crack is subject to the Mode I elastic singular fields. This configuration models a crack in an infinite body; i.e., a body in which the plastic zone is small compared to the crack size and other length dimensions.

Finite element meshes of single edge notched bend (SENB) specimens were generated with a/W = 0.05, 0.15, 0.25, and 0.50. Each of these meshes contained approximately 350 elements and 1200 nodes, with most of the elements and nodes concentrated near the crack tip.

All analyses were performed on stationary cracks. Consequently, the model in its current form is only applicable to cleavage fracture without significant prior stable crack growth.

RESULTS

Effect of specimen dimensions on J_c and δ_c

Figures 1 and 2 illustrate the effect of crack length, a/W and hardening exponent on the J/J_{ssy} ratio. Since a critical value of J_{ssy} represents a size-independent cleavage toughness, the J/J_{ssy} ratio quantifies the geometry dependence of J_c, the measured fracture toughness. For the deeply notched specimens (a/W = 0.5), J_c approaches the small scale yielding value when the ratio $a\sigma_0/J$ is greater than ~200, but the shallow notched specimens do not produce small scale yielding behaviour unless the specimen is very large relative to J/σ_0. The relative crack tip constraint increases as strain hardening rate increases, i.e., as n decreases.

The effect of specimen size on critical CTOD is shown in Fig. 3. The curves for the three hardening exponents converge and approach $\delta/\delta_{ssy} = 1.0$ when the a/δ ratio is greater than ~ 300.

Based in part on the curves in Figs. 1 to 3, the authors have previously recommended the following size requirements for cleavage fracture in deeply notched (a/W ≥ 0.5) bend and compact specimens (6):

$$B, b, a \geq \frac{200 \, J_c}{\sigma_Y} \tag{10}$$

or

$$B, b, a \geq 300 \, \delta_c \tag{11}$$

where B is the specimen thickness, b is the remaining ligament, and σ_Y is the flow stress. If a given test satisfies either of the above criteria, the resulting fracture toughness value can be viewed as size independent. That is, specimens that satisfy Eqs. [10] and [11] are sufficiently large so as not to be influenced by crack tip constraint effects. Note that Eq. [10] is eight times as severe as the J_{IC} size requirements in ASTM E 813-87. The reason for the stricter limits is that cleavage fracture is more sensitive to specimen size than is ductile fracture.

It must be emphasized that the above size criteria only apply to deeply notched bend specimens. As Fig. 1 implies, *it is virtually impossible to obtain a size-independent result in a shallow-notched specimen.* Size requirements for such specimens would be so restrictive as to render them useless for most practical applications.

When performing tests on shallow notched specimens or on deep notched specimens that do not meet the above size requirements, it is possible to *correct* the resulting fracture toughness data for constraint loss. A given J_c or δ_c value can be corrected down to the equivalent small scale yielding value by dividing by the J/J_{ssy} or δ/δ_{ssy} ratio, respectively. The authors have developed parametric equations that express these ratios as a function of specimen dimensions, flow properties, and strain hardening exponent [8].

Comparison with experiment

Figure 4 shows CTOD data obtained by Sorem (15) for A 36 steel at two temperatures in the transition region. The solid diamonds represent the experimental data, while the crosses indicate predicted small scale yielding (δ_{ssy}) values. Every specimen but one (the highest CTOD value for a/W = 0.15 at -43°C) failed by cleavage without significant prior stable crack growth. At both temperatures, the shallow notched specimens have a higher apparent toughness than the deep notched specimens, but the corrected values agree very well. Relatively small corrections are needed for specimens with a/W = 0.50, but the small scale yielding correction has a major effect when a/W = 0.15. The small scale yielding CTOD values appear to be less scattered than the uncorrected data.

Sumpter (5) performed fracture toughness tests on mild steel SENB specimens with a range of a/W values. The material was sufficiently brittle to fail by cleavage at room temperature. Figure 5(a) is a plot of J_c values versus a/W. These data exhibit considerable scatter, which is typical of cleavage fracture in ferritic steel. A least squares fit* through the data indicates an upward trend in toughness with increasing a/W, as expected. Figure 5(b) is a plot of corrected toughness values (J_{ssy}), which does not exhibit any significant trend with a/W; the least squares fit is nearly horizontal and the correlation coefficient is nearly zero, which implies that the critical J_{ssy} does not depend on a/W.

Kirk, et al. (9) varied both specimen size and a/W in fracture toughness tests on a plate of A 515 Grade 70 steel. The uncorrected and corrected data for this material are shown in Figs. 6(a) and 6(b), respectively. The uncorrected data exhibit the expected increase in toughness with decreasing a/W. The least squares fits of the 25 mm and 50 mm thick data indicate that the smaller specimens are more sensitive to the a/W ratio, and the toughness at a given a/W tends to be higher in the smaller specimens. (Insufficient data were available for the 10 mm specimens to make any inferences on trends.) All size and geometry effects are removed, however, in the corrected toughness values (Fig. 6(b)).

DISCUSSION

Figures 4 to 6 indicate excellent agreement between predictions from the scaling model and experimental data. Other data sets show equally good agreement, but space limitations preclude showing all data analysed so far.

The model in its current form is limited to cleavage fracture in plane strain without significant prior ductile crack growth. The authors are currently extending the model to account for three dimensional effects and cleavage preceded by stable tearing. We are applying the model to structural configurations such as semi-elliptical surface cracks. In addition, recent work has shown that the scaling model can be applied to cracks in weldments (16).

SUMMARY

- A model has been developed which corrects cleavage fracture toughness data for constraint loss. Predictions from this model are in excellent agreement with experimental data.

- Current efforts are aimed at extending this model to three-dimensional geometries and to the upper transition region.

REFERENCES

(1)   Betegon, C. and Hancock, J.W., "Two Parameter Characterisation of Elastic-Plastic Crack Tip Fields." to appear in *Journal of the Mechanics and Physics of Solids*, 1991.

---

* By using a least squares fit, the authors do not mean to imply that the toughness should vary linearly with a/W. The linear curve fit is merely a crude statistical tool for identifying a trend in the data.

(2) Wang, Y.-Y. and Parks, D.M., "Evaluation of the Elastic T-Stress in Surface-Cracked Plates Using the Line-Spring Method." Submitted for publication, 1991.

(3) O'Dowd, N.P. and Shih, C.F., "Family of Crack-Tip Fields Characterized by a Triaxiality Parameter: Part I–Structure of Fields." to appear in *Journal of the Mechanics and Physics of Solids*, 1991.

(4) O'Dowd, N.P. and Shih, C.F., "Family of Crack-Tip Fields Characterized by a Triaxiality Parameter: Part II–Fracture Applications." submitted to *Journal of the Mechanics and Physics of Solids*, 1991.

(5) Sumpter, J.D.G., "An Experimental Investigation of the T Stress Approach." Presented at the *ASTM Symposium on Constraint Effects in Fracture*, Indianapolis, May 8-9, 1991.

(6) Anderson, T.L. and Dodds, R.H., Jr., "Specimen Size Requirements for Fracture Toughness Testing in the Ductile-Brittle Transition Region." *Journal of Testing and Evaluation*, Vol. 19, 1991, pp. 123-134.

(7) Dodds, R.H. Jr., Anderson T.L. and Kirk, M.T., "A Framework to Correlate a/W Effects on Elastic-Plastic Fracture Toughness (J_c)." *International Journal of Fracture*, Vol. 88, 1991, pp. 1-22.

(8) Anderson, T.L. and Dodds, R.H. Jr., "Simple Constraint Corrections for Subsize Fracture Toughness Specimens." Presented at the ASTM International Symposium on Small Specimen Test Techniques and Their Application to Nuclear Reactor Vessel Thermal Annealing and Plant Life Extension, New Orleans, January 29-30. 1992.

(9) Kirk, M.T. , Koppenhoffer, K.C., Shih, C.F., "Effect of Constraint on Specimen Dimensions Needed to Obtain Structurally Relevant Toughness Measurements." Presented at the *ASTM Symposium on Constraint Effects in Fracture*, Indianapolis, May 8-9, 1991.

(10) Wallin, K, "Statistical Aspects of Constraint with Emphasis to Testing and Analysis of Laboratory Specimens in the Transition Region." Presented at the ASTM Symposium on Constraint Effects in Fracture, Indianapolis, IN, May 8-9, 1991.

(11) Lin, T., Evans, A.G. and Ritchie, R.O., "Statistical Model of Brittle Fracture by Transgranular Cleavage." *Journal of the Mechanics and Physics of Solids*, Vol 34, 1986, pp. 477-496.

(12) Wallin, K., Saario, T., and Törrönen, K., "Statistical Model for Carbide Induced Brittle Fracture in Steel." *Metal Science*, Vol. 18, 1984, p. 13.

(13) Beremin, F.M., "A Local Criterion for Cleavage Fracture of a Nuclear Pressure Vessel Steel." *Metallurgical Transactions*, Vol. 14A, 1983, p. 2277.

(14) Anderson, T.L. and Stienstra, D., "A Model to Predict the Sources and Magnitude of Scatter in Toughness Data in the Transition Region." *Journal of Testing and Evaluation*, Vol. 17, 1989, pp. 46-53.

(15) Sorem, W.A., "The Effect of Specimen Size and Crack Depth on the Elastic-Plastic Fracture Toughness of a Low-Strength High-Strain Hardening Steel." Ph.D. Dissertation, University of Kansas, Lawrence, KS, May 1989.

(16) Kirk, M.T. and Dodds, R.H. Jr., "The Influence of Weld Strength Mismatch on Crack-Tip Constraint in Single Edge Notch Bend Specimens." submitted to the *International Journal of Fracture*, 1992.

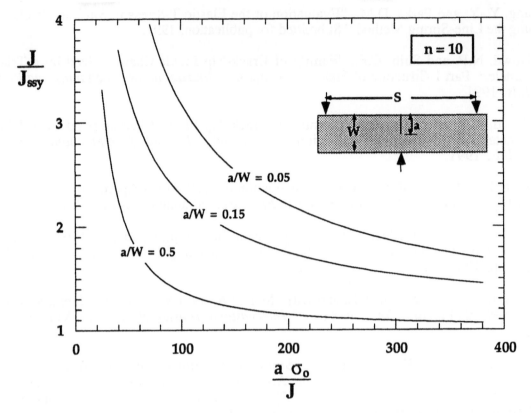

Figure 1   Effect of specimen size and a/W on the J/J_{ssy} ratio for n = 10.

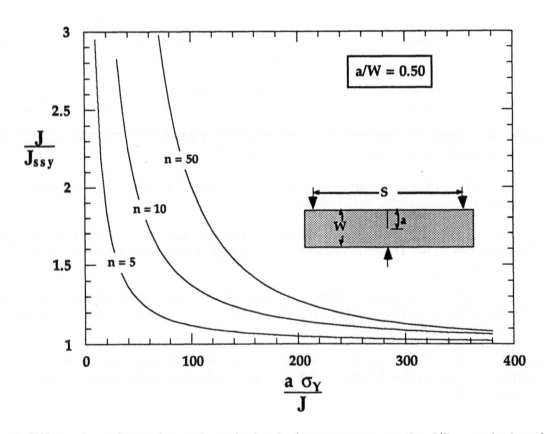

Figure 2 Effect of specimen size and strain hardening exponent on the J/J_{ssy} ratio for a/W = 0.5

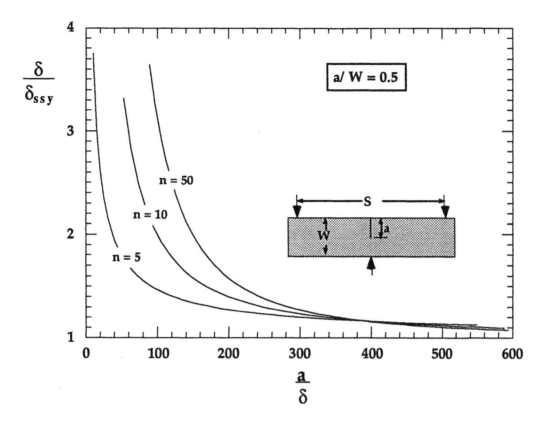

Figure 3 Effect of specimen size and strain hardening exponent on the δ/δ_{ssy} ratio for a/W = 0.5

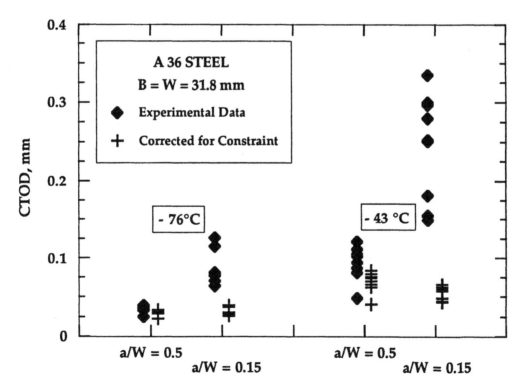

Figure 4 Comparison of experimental CTOD values for A 36 steel with CTOD corrected for constraint loss. Data were obtained from Sorem [15].

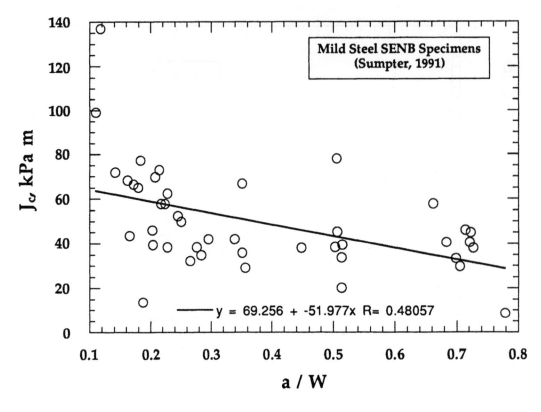

(a)   Experimental J_c values.

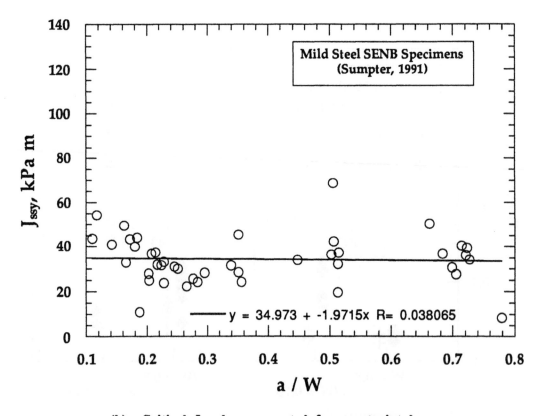

(b)   Critical J values corrected for constraint loss.

Figure 5 Fracture toughness data for a mild steel at room temperature. Data were obtained from Sumpter (5). The lines are least squares fits through the data.

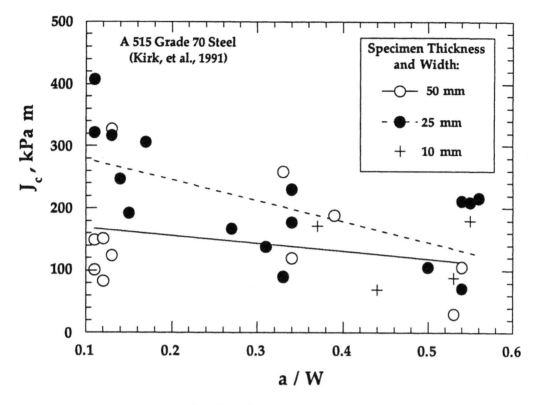

(a) Experimental J_c values.

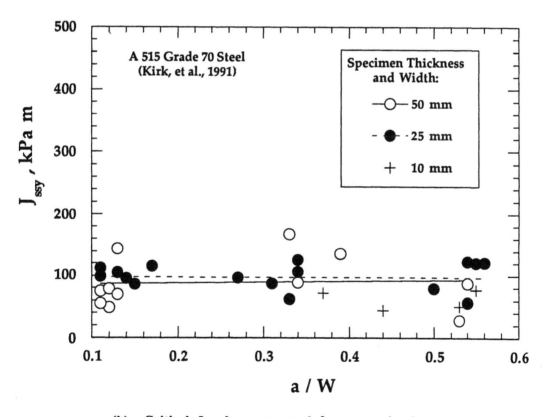

(b) Critical J values corrected for constraint loss.

Figure 6 Fracture toughness data for an A 515 Grade 70 steel plate at room temperature. Data were obtained from Kirk, et al. (9). The lines are least squares fits through the data.

TWI

PAPER 7

Constraint based analysis of shallow cracks in mild steel

J D G Sumpter, PhD and A T Forbes (Defence Research Agency, UK)

SUMMARY

Recent years have seen important advances in constraint based fracture prediction methods. These methods remove the conservatism of current procedures based on lower bound (high constraint) fracture toughness. This report presents data for mild steel at -50°C where failure is by J_c (cleavage with no prior stable crack growth). J_c shows clear geometry dependence, with higher values in shallow cracked three point bend (3PB) specimens, and in centre cracked tension (CCT) specimens, than in conventional, deeply cracked 3PB specimens. It is shown that these trends are predictable by elastic T-stress analysis, and may be quantified very precisely by more exact elastic-plastic constraint theories. It is concluded that these advanced methods will soon be useable in structural fracture analyses.

INTRODUCTION

It is often observed that shallow cracks in steel give higher J_c than deep cracks*, and that tension loaded specimens give higher J_c than bend specimens at an equivalent crack depth. This geometry dependence of J_c is clearly important when making critical defect size predictions for a structure, but until recently no practical framework has existed to use the enhanced toughness values in fracture-safe design. Efforts to explain geometry dependence of J_c now centre on the quantification of crack tip constraint (elastic-plastic tensile stress elevation directly ahead of the crack tip). One of the most promising parameters for indexing crack tip constraint is the elastic T-stress. No single set of authors can be credited with 'discovering' the relevance of T to elastic-plastic crack tip constraint, but influential papers have been those by Larsson and Carlson (1), Rice (2), Leevers and Radon (3), Bilby et al (4), Harlin and Willis (5), Al-Ani and Hancock (6), and Betegón and Hancock (7). The major advantage of T over other candidate constraint indexing procedures is that it may be evaluated by a purely elastic analysis.

The parameter T is the constant elastic stress which would remain close to a crack tip, parallel to the crack flanks, if the near tip infinite stress term were removed. Although T is a product of elastic stress analysis it has been found to be surprisingly accurate for indexing crack tip constraint well into the plastic loading regime. This statement is based on the observation that, when two dissimilar structures or test specimens have the same elastic T-stress, elastic-plastic finite element analyses show that they also develop very similar tensile stress elevation in the plastic zone ahead of the crack tip. Negative T, associated with shallow cracks and tension loading, promotes low crack tip constraint. Zero or positive T, associated with small scale yielding and deep cracks in bending, promotes high crack tip constraint. Since cleavage fracture is governed by the attainment of a critical crack tip tensile stress this is a fundamentally important observation. It can be postulated that combining T with J (J is a measure of the

---

* Discussion of toughness in this paper will be confined to the onset of cleavage instability with no prior stable crack growth. The estimated value of the J integral at the onset of cleavage fracture will be designated as J_c.

spatial extent of the elastic-plastic crack tip stress concentration) provides the necessary descriptive ability to predict cleavage fracture in any arbitrary geometry.

The truth of this postulate can be tested by finite element analysis or by experiment. The present paper concentrates on experiment. Data are presented for mild steel using two very different specimen geometries, which have the same negative elastic T-stress at plastic limit load; shallow cracked three point bend (3PB) and deeply notched centre cracked tension (CCT). If it can be shown that these two very different test geometries fail at a common J_c value, which is higher than that associated with the deeply cracked 3PB specimen (positive T-stress), then the usefulness of the J plus T approach will have been confirmed. Details of the calculation of T-stress for 3PB and CCT specimens were given by Sumpter (8). This paper extends the data in Sumpter (8) with further shallow notched 3PB tests. These data clearly establish a trend for J_c to increase with decreasing a/W (crack depth to specimen width ratio) in three point bending. The data allow a better comparison between J_c in 3PB and CCT specimens than was possible in Sumpter (8). The opportunity is also taken to analyse the data using alternative constraint indexing approaches suggested by Anderson and Dodds (9) and by Shih et al (10).

EXPERIMENTAL PROCEDURE

Material Properties

The test material was a 25 mm thick, low grade mild steel with chemical composition

0.19C   0.59Mn   0.04Si   0.01P   0.032S

Room temperature 0.2% proof strength was 235 MPa. At -50°C, the temperature relevant to the test programme, the material showed a pronounced upper yield stress of 315 MPa. This feature of the stress-strain curve was also reflected in the fracture toughness tests, particularly the CCT specimens, which were all performed at -50°C. At this temperature all specimens failed by cleavage fracture with no prior stable crack growth. Figure 1 shows the fracture surface of a shallow cracked 3PB specimen (a/W = .03) which failed at J_c = 0.23MN/m. Figure 2 shows the fracture surface of a CCT specimen which failed at J_c = 0.15MN/m.

Fracture Specimen Design

In this phase of the programme the broken halves of the CCT specimens tested by Sumpter (8) were used to manufacture a total of 35 further 3PB specimens. All specimens were originally manufactured and fatigue cracked with a depth W = 40 mm and a machine notch depth of 10 mm. Specimen thickness, B, was 25 mm in all cases. The notches in the 3PB specimens were located in material which had seen only elastic stresses in the CCT tests. Specimens were fatigue pre-cracked at a stress intensity range of around 25 MPa√m to produce 2.5 mm fatigue crack extension from the tip of the notch. Three point bend specimens with a range of crack depths down to as small as 0.8 mm were then manufactured by machining away different amounts of the notch to give a range of a/W between 0.03 and 0.4. This method of manufacture overcomes the considerable problems involved in growing fatigue cracks from a small notch in a low strength material, and ensures that fatigue pre-cracking conditions do not influence the trend of a/W on J_c.

Calculation of J

Values of J for all 3PB tests were calculated by splitting J into elastic and plastic components, J_e

and J_p

$$J = J_e + J_p \tag{1}$$

$$J = \frac{K^2}{E}(1-v^2) + \frac{\eta_p U_p}{B(W-a)} \tag{2}$$

where   K   is the elastic stress intensity factor
                 E   is Young's modulus
                 v   is Poisson's ratio
                 U_p is the area under the load versus load point displacement curve
                 B, W, a   are specimen thickness, width, and crack depth

$$\begin{aligned} \eta_p &= 2.0 &&& a/W \geq 0.282 \\ &= 0.32 + 12\,(a/W) - 49.5\,(a/W)^2 + 99.8\,(a/W)^3 && a/W < 0.282 \end{aligned} \tag{3}$$

The plastic component of load point displacement was taken as the plastic component of machine crosshead displacement. A number of checks were made to confirm the accuracy of this approach including measuring extraneous plastic displacements on a solid block of steel set in the 3PB test rig. The errors involved in equating machine and specimen plastic displacements were found to be negligible at the relatively small loads and large ratios of plastic to elastic displacement involved in this test series. For instance, at a/W = 0.1, the failure load was 54kN and the ratio J_p/J_e was 30. Figure 3 shows an a/W = 0.04 specimen which has been deformed without failure to a total J of 0.27MN/m.

The values of η_p used were those derived from limit load solutions for the 3PB specimen by Sumpter (11). More accurate values of η_p for a range of work hardening exponents have recently been presented by Kirk and Dodds (12), but, by coincidence, it appears that the accurate solution for n = 5, which is close to that of the mild steel tested here, is virtually identical to the slip line field solution, and the latter has been retained for this analysis.

J_p for the CCT specimens was calculated from equation [2] with η_p = 0.5 at all crack depths. Further details are given in Sumpter (8).

RESULTS

Trend of J_c with crack depth

Figure 4 shows J_c versus a/W for the new 3PB data (this report) and from Sumpter (8). The two sets of data are clearly from the same population, but the new results illustrate much more clearly the upswing in toughness at shallow crack depths (a/W <0.15).

Trend of J_c with T stress

Figure 5 shows all the 3PB data from this study, and the CCT and 3PB data from Sumpter (8), plotted against T stress normalised by yield stress, T/σ_y. The quantity T/σ_y is calculated individually for each specimen based on the failure load, but because all of the specimens failed near their plastic limit load, there is a fairly tight correlation between T/σ_y and a/W as shown in Fig. 6 (a/W for the CCT specimens is half crack length divided by half specimen width). Figure 5 shows that there is reasonable agreement between the CCT and 3PB data when plotted versus T/σ_y, but that, on average the CCT data show slightly higher toughness. What is clearly established is that both CCT and small a/W 3PB specimens are low constraint geometries which show elevated J_c compared with deeply cracked 3PB. This is in line with the trend predicted by T-stress analysis.

DISCUSSION

Predictions that the T-stress will be a useful parameter for indexing elastic-plastic crack tip constraint have arisen from finite element studies. Two other proposals for indexing constraint based on finite element studies have also been made recently - Anderson and Dodds (9) and Shih et al (10)*. The present data can be used to evaluate these proposals.

A common feature to emerge from all of the finite element studies is that work hardening is very influential in determining the extent to which J_c will vary with geometry. The analyses indicate that highly work hardened materials will show a less rapid variation of J_c with geometry than mildly work hardened materials. Low strength mild steel has fairly high work hardening. Using Fig. 12 from Kirk and Dodds (12) the work hardening exponent is around n = 6.

Anderson and Dodds (9) combine a critical tensile stress criterion with detailed finite element analysis of the crack tip region to make a direct prediction of how J_c will vary with a/W in the 3PB test. In Fig. 11 of their paper (9) they present predictions for elevation of J_c at three a/W ratios - 0.05, 0.15, and 0.5 - as a function of $a\sigma_y/J$. This dimensionless group is a measure of the extent of plastic deformation prior to attainment of J_c. It is implied that J_c is dependent on the level of prior plastic flow as well as geometry - a concept which is familiar through statements which appear in test standards related to the maintenance of J and K dominance. Table 1 compares averaged experimental data from Fig. 4 with the Anderson and Dodds (9) predictions. The agreement is excellent.

Shih et al (10) index constraint in terms of a parameter Q, which is an elastic-plastic equivalent of T. In its simplified form Q is the fraction of the yield stress by which the tensile stress ahead of the crack tip deviates from the Hutchinson, Rice, and Rosengren (HRR) prediction. Q has a similar sign convention to T: negative Q implies low constraint and high J_c, whilst zero or positive Q implies high constraint and lower bound J_c. In Figs. 7b and 7d of their paper Shih et al (10) present Q values for a deeply notched CCT specimen and for shallow cracked 3PB specimens as a function of plastic deformation. Table 2 shows the application of these results to the experimental data in this report. Both 3PB a/W = 0.1 and CCT a/W = 0.7 specimens have the same T-stress (T/σ_y = -0.6) which implies J_c should be the same. As noted previously this prediction appears to be slightly in error. Analysis based on Q, on the other hand, correctly predicts that the CCT specimen will be tougher.

---

*   It should be emphasised that the three constraint indexing procedures considered in this report are simply variants on the same theme. All the approaches stem from observation of crack tip tensile stress in elastic-plastic finite element analyses.

Table 1. Prediction of variation of J_c with a/W using Anderson and Dodds (9) approach

a/W	Experimental J_c MN/m [1]	$\dfrac{a\sigma_y}{J_c}$	$\dfrac{J}{J_{ssy}}$ [2]	J_c (a/W)/J_c (0.5) [3]	
				predicted	experimental
0.05	0.19	2	>>3	>>3	4.3
0.10	0.10	8	2.7	2.5	2.3
0.15	0.06	23	1.8	1.6	1.4
0.50	0.044	156	1.1	1	1

[1] Experimental J_c values are obtained by averaging all J_c values from Fig. 4 at the stated a/W ± 0.02.

[2] From Fig. 11 in Anderson and Dodds (9). J/J_{ssy} is the factor by which J in a low constraint geometry has to be elevated to obtain the stress distribution which exists at J_{ssy} in a high constraint geometry.

[3] J_c at a particular a/W divided by J_c at a/W = 0.5.

Table 2. Calculation of Q following Shih et al (10)

Geometry	a/W	Average J_c MN/m	$\dfrac{J}{a\sigma_y}$	$\log\dfrac{J}{a\sigma_y}$	Q [5]
3PB	0.1	0.10	0.123	-0.9	-0.95
CCT	0.7[1]	0.14[2]	0.0246[3]	-1.6[4]	-1.15

[1] Half crack length/half specimen width.

[2] CCT tests were actually at a/W 0.63 and 0.77. J_c = 0.14 MN/m is the average of all the tests at these two crack lengths.

[3] $J/b\sigma_y$ for CCT specimens. Ligament, b, taken as 21 mm (average of 16 and 26 mm, which were actual ligaments at a/W = 0.77 and 0.63).

[4] Log $J/b\sigma_y$ for CCT specimens.

[5] From Fig. 7 in Shih et al (10).

CONCLUSIONS

Geometry dependence of J_c has been a troublesome feature of fracture analysis for many years. Detailed finite element analyses now show that the observed trends are fully explicable in terms of variations of near crack tip stresses at a given J value in different geometries at different deformation levels. The problem is thereby reduced to one of deriving a framework to describe the observed geometry and deformation dependence of crack tip stresses in terms of easily calculated global parameters.

This report has presented data showing clear J_c geometry dependence in mild steel. Shallow notched three point bend (3PB) and deeply notched centre cracked tension (CCT) specimens both show elevated J_c compared to deeply notched 3PB specimens.

Three constraint indexing schemes have been considered to explain the observed experimental trends: T-stress analysis, following Betegón and Hancock (7); the scaling approach of Anderson and Dodds (9); and Q analysis, following Shih et al (10).

The T-stress approach is the easiest to apply and provides the best prospect for extension to design, since it requires only an elastic analysis to characterise constraint. Applied to the present data it correctly predicts that deeply notched CCT and shallow notch 3PB specimens will both show elevated toughness, but it fails to predict the slightly higher J_c in the CCT specimens compared to the shallow notched 3PB. Analysis based on Q does predict this trend. The Anderson and Dodds (9) procedure is currently available only for the 3PB geometry. It very closely predicts the precise trend of increasing J_c with decreasing a/W observed for the mild steel data in this report.

The experimental data presented in this report is thought to be the most comprehensive yet obtained showing trends in geometry dependence of J_c. The ability of recent constraint based fracture analysis methods to quantify the observed trends is highly encouraging. This success suggests that radical improvements will very soon be possible in critical defect size prediction procedures for structures.

REFERENCES

(1)   Larsson, S.G. and Carlsson, A.J., 'Influence of Non-singular Stress Terms on Small Scale Yielding at Crack Tips in Elastic-plastic Materials', Journal of Mechanics and Physics of solids, 21, 1973, pp. 263-77.

(2)   Rice, J.R., 'Limitations to the Small Scale Yielding Approximation for Crack Tip Plasticity', Journal of Mechanics and Physics of Solids, 22, 1974, pp. 17-27.

(3)   Leevers, P.S. and Radon, J.C., 'Inherent Stress Biaxiality in Various Fracture Specimen Geometries', International Journal of Fracture, Vol 19, 1982, pp. 311-325.

(4)   Bilby, B.A., Cardew, G.E., Goldthorpe, M.R. and Howard, I.C., 'A Finite Element Investigation of the Effect of Specimen Geometry on the Fields of Stress and Strain at the Tips of Stationary Cracks', Size and Effects in Fracture, Conference Proceedings, Farnborough. Published by Institution of Mechanical Engineers, 1986.

(5)   Harlin, G. and Willis, J.R., 'The Influence of Crack Size on the Ductile to Brittle Transition', Proceedings of the Royal Society of London, A415, 1988, pp. 197-226.

(6)   Al-Ani, A.A. and Hancock, J.W., 'J Dominance of Short Cracks in Tension and Bending' Journal of Mechanics and Physics of Solids, Vol 39, No 1, 1991, pp. 23-43.

(7)   Betegón, C. and Hancock, J.W., 'Two Parameter Characterisation of Elastic-plastic Crack Tip Fields' ASME Journal of Applied Mechanics, Vol 58, March 1991, pp. 104-110.

(8)   Sumpter, J.D.G., 'An Experimental Investigation of the T-Stress Approach' Constraint Effects in Fracture, ASTM Symposium, Indianapolis, May 8-9, 1991.

(9)   Anderson, T.L. and Dodds, R.H., 'Specimen Size Requirements for Fracture Toughness Testing in the Transition Region' Journal of Testing and Evaluation, March 1991, pp. 123-134.

(10)   Shih, C.F., O'Dowd, N.P., and Kirk, M.T., 'A Framework for Quantifying Crack Tip Constraint' Constraint Effects in Fracture, ASTM Symposium, Indianapolis, May 8-9, 1991.

(11)   Sumpter, J.D.G., 'J_c Determination for Shallow Notch Welded Bend Specimens' <u>Fatigue and Fracture of Engineering Materials and Structures</u>, Vol 10, No 6, 1987, pp. 479-493.

(12)   Kirk, M.T. and Dodds, R.H., 'J and CTOD Estimation Equations for Shallow Cracks in Single Edge Notch Bend Specimens', University of Illinois, Civil Engineering Studies, Structural Research Series No 565, 1992.

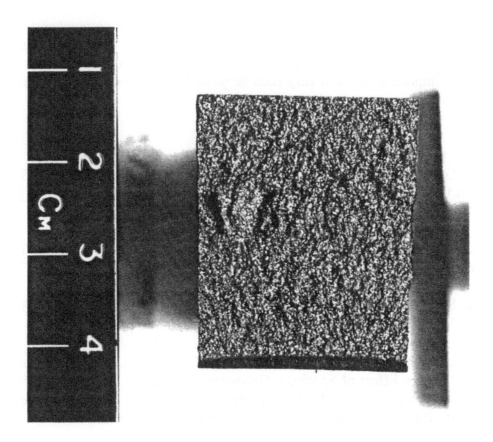

Figure 1: Shallow cracked 3PB specimen (a/W = 0.03) which failed at J_c = 0.23MN/m

Figure 2: Deeply cracked CCT specimen (a/W = 0.77) which failed at J_c = 0.15 MN/m

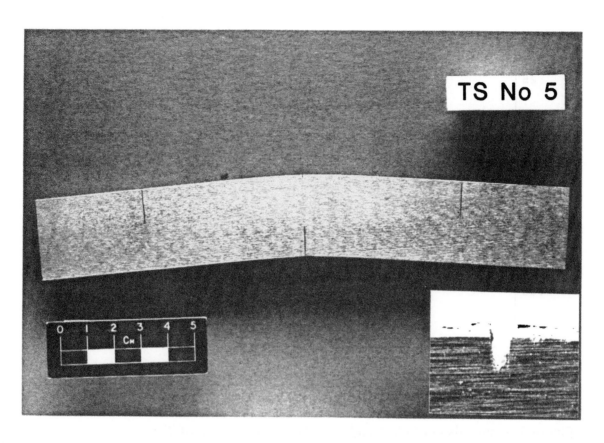

Figure 3: Shallow cracked 3PB specimen (a/W = 0.04) deformed without failure to a J of 0.2MN/m ($a\sigma_y/J = 1.5$). The insert shows a close up of the deformed crack. The depth of the crack is 1.15mm.

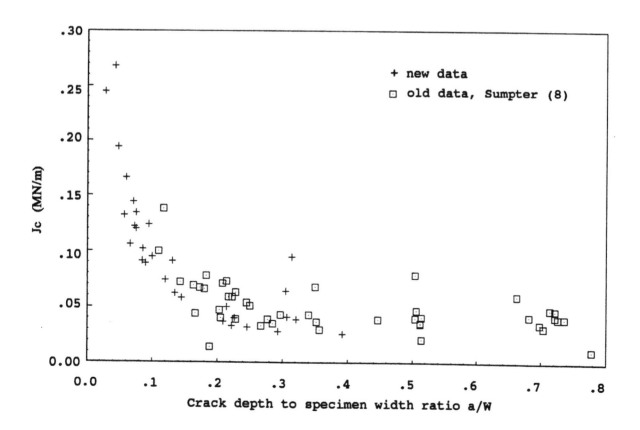

Figure 4: J_c as a function of crack depth for 3PB data

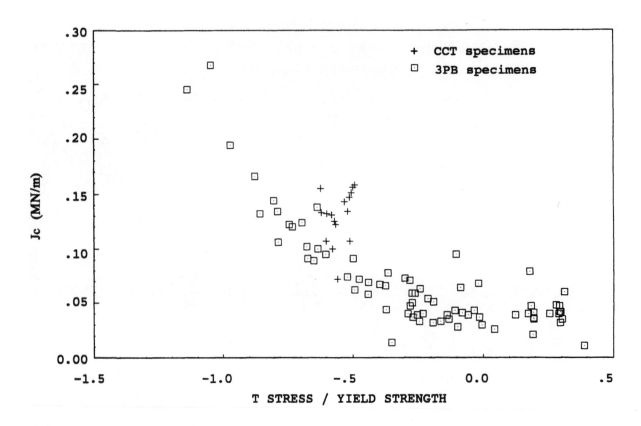

Figure 5: Comparison of CCT and 3PB data as a function of T/σ_y

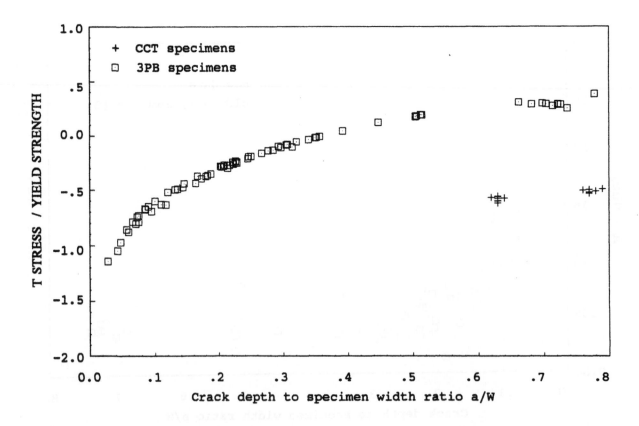

Figure 6: Relationship between a/W and T/σ_y at failure for all specimens

TWI

PAPER 27

Interpretation of shallow crack fracture mechanics tests with a local approach of fracture

M Di Fant, (IRSID, Saint Germain en Laye, France) and F Mudry, (Usinor-Sacilor, Paris, France)

SUMMARY

There are many situations where cleavage fracture is initiated at shallow surface cracks. This is of particular importance for structures containing welded joints where fracture starts at small defects in the heat affected zone. In this case the true fracture toughness measured on specimens with long cracks can hardly be used.

The local approaches of fracture (2-4) are felt useful for describing such difficult situations.They are based on the evaluation of a local damage criterion after a finite element calculation of strains and stresses . In that way, they can take into account the accurate elastoplastic behaviour of a fracture mechanics specimen which is quantitatively different for a short or a long crack. From a practical point of view it is possible to relate the local approach criterion to the classical fracture mechanics parameters (CTOD, K_{1c}, J_{1c}) and to predict the changes which are observed experimentaly. EMP (Ecole des Mines de Paris) and IRSID (Institut de Recherche de la Sidérurgie) have been involved in a project organised, with this objective, by TWI.

The first part of this paper is devoted to the experimental measurement of the local approach parameters on the steel M1 which is briefly presented. Then numerical simulations are described and are used in a third part to predict the behaviour of specimens tested at TWI. As a last part, conclusions will be drawn.

MEASUREMENT OF THE PARAMETERS OF THE LOCAL APPROACH

The steel investigated is designed M1 in the program. It was produced as a 25 mm thick plate corresponding to the grade 50 EE of the standard BS 4360. It is a normalised ferritic pearlitic steel. The chemical composition and stress-strain laws in tension given by TWI (1) and measured at IRSID are given in Table 1 and in Fig. 1. The yield strength and tensile strength of the material tested seem low compared to the figures given in the reports issued by TWI. From Fig. 1, it can be seen that the measured yield stress by IRSID is 8% lower or, conversely, for a given yield strength, the temperature shift is 30°C. This discrepancy cannot be explained and may be due the material variability.

Table 1. Chemical composition of the steel M1 measured at IRSID and given by TWI (weight %)

	C	Mn	Si	S	P	Ni	Cu	Cr	Mo	V	Nb	Ti	Al	O	N
IRSID	0,125	0,163	0,364	0,003	0,0166	0,161	0,154	0,028	0,011	<0,006	0,029	0,003	0,060	0,0005	0,0065
TWI	0,120	0,1510	0,400	0,004	0,015	0,140	0,150	0,020	0,005	0,002	0,025	0,002	0,054	0,0006	0,0054

In the local approach two parameters are used :

- σ_u which is a measure of the cleavage stress for cleavage fracture

- m which is the exponent of the Weibull statistic's law (Weibull's modulus)

σ_u has not the same definition than that given by KNOTT (5) because size effects are included.

It could be defined as the cleavage stress for an arbitrary volume Vo, here taken as 100 µm x 100 µm x 100 µm. The cleavage stress for a larger volume can be evaluated using the Weibull's statistics, ie :

$$\sigma_u{}^m \, Vo = \sigma_c{}^m V,$$

where σ_c is the cleavage stress for a volume V corresponding to a particular specimen geometry.

The measurement of σ_u and m has been done with 18 tensile experiments on round-notched specimens shown in Fig. 2. The procedure followed for this determination is described in reference (2) The results of the tests as well as their interpretation are given elsewhere (6). The final result is :

$$\sigma_u = 2525 \text{ MPa, m} = 30$$

NUMERICAL SIMULATIONS

Two dimensional numerical simulations have been carried out for four different tests :

- SENBL : 3 - point bend specimen with long crack : $a/w = 0,5$

- SENBS : 3 - point bend specimen with short crack : $a/w = 0,05$

- SENTL : single edge notched tension with short crack : $a/w = 0,5$

- SENTS : single edge notched tension with short crack : $a/w = 0,05$

The meshes and loading conditions are given in Fig. 3. The square elements at the crack tip are 100 µm x 100 µm. The stress-strain curve used corresponds to the one measured at IRSID at -125°C (6). Beyond necking, the stress-strain curve has been extrapolated linearly in a Ln-Ln plot.

The calculations have been performed in plane strain with the numerical code ZEBULON which has been developed during the first studies on local approach (2D calculations).

In order to use the local approach, the following parameters have been computed :

d, load line displacement,

CMOD, crack mouth opening displacement,

CTOD, crack tip opening displacement,

J, J integral,

σ_w, Weibull stress.

CMOD is measured 3,25 mm above the upper flank as in the experiments. CTOD has been estimated using a linear extrapolation of the flat part of the crack face Fig. 4. J has been estimated from the computed load-load line displacement diagrams using the same formulas as the one used in the experiments. σ_w is computed by the program in order to use Weibull statistics. Its definition corresponds to the following integral which is evaluated numerically for each loading step.

$$\sigma_w^m = \int_{PZ} \sigma_I^m \frac{dV}{V_0}$$
[1]

where m is the Weibull's modulus, here m = 30. Vo is the arbitrary unit volume.
σ_I is the maximum principal stress and the integral is extented to the plastic zone (PZ). This value is used to estimate the fracture probability :

$$P_F = 1 - \exp\left(-\left(\frac{\sigma_w}{\sigma_u}\right)^m\right)$$
[2]

where P_F stands for the fracture probability and σ_u is the cleavage stress, measured previously, ie : 2525 MPa. For more details see references (2, 3, 6).

For the four geometries, SENBL, SENBS, SENTL, SENTS the evolution of σ_w/σ_0 as a function of CMOD, CTOD, load line displacement, J/σ_0 have been computed, where σ_0 is the yield strength used, ie : 534 MPa. Examples are given in Fig. 5a. and 5b., where σ_w/σ_0 is plotted as a function of CMOD for the SENBS (a/W = 0,05) and SENBL (a/W = 0,5) geometries.

INTERPRETATION OF RESULTS

Figure 5 can be used as follow ; given an experimental result (CMOD in this case) at a given temperature, we use the yield strength at that temperature as a new σ_0. The yield strength used is the one given by TWI because the bent specimens and the tensile tests were taken from the same part of the plate. Then, using the diagram starting from the value of CMOD along the abcissa, we get two different values of σ_w/σ_0 depending on the a/w ratio. σ_w can be computed since σ_0 is known.

Using this technique the experimental points given by TWI have been evaluated. The results are given in Table 2. Some points could not be evaluated because the deformation at fracture was far beyond our numerical results.

Table 2 shows that σ_w is fairly stable. However, the average value in this table is approximately 100 MPa lower than the value evaluated from notched tensile tests. This is because the cleavage stress σ_u is higher than the average value, see [2]. This is a rather small discrepancy which could arise from :

- variability of the material already discussed,

- simplification of 3D problems to 2D plane strain simulation,

- use of the yield strength for σ_0 not taking into account the variation of the strain hardening behaviour with temperature.

It is noted that the SENT results are systematically lower than the SENB ones.

Table 2. Values of the Weibull stress, σw, corresponding to experimental points

	Temperature (°C)	Yield stress (MPa)	CMOD (mm)	d (mm)	σw (MPa)
SENB a/w = 0.05	- 150	702.3	0.07		1840
	- 140	676.4	0.25		2235
	- 135	663.9	0.55		2307
	- 130	651.2	0.98		2359
	- 125	638.4	1.05		2327
	- 125	638.4	1.45		2400
	- 120	625.6	0.92		2274
	- 120	625.6	1.22		2302
	- 120	625.6	1.56		2366
SENB a/w = 0.5	- 120	625.6	0.5		2530
	- 115	612.8	0.9		2499
	- 110	600	0.6		2434
	- 110	600	2.3		2645
	- 105	587.2	2.2		2576
	- 100	556.1	4		2648
	- 95	525	4.9		2596
	- 95	525	1.5		2213
SENT a/w = 0.05	- 160	782.2		0.28	1975
	- 150	702.3		0.55	2186
	- 140	676.4		0.46	2049
SENT a/w = 0.5	- 160	728.2		0.08	1924
	- 140	676.4		0.22	2393

Using σu = 2450 MPa instead of 2525 MPa allows us to predict the scatter band (PF = 10 %, PF = 90 %) of the experimental results. The results are shown in Fig. 6a and 6b. For the SENB geometries the temperature shift between the two crack lengthes (a/w = 0,05 and a/w = 0,5) is rather well predicted. However, in the lowest temperatures region the experimental points are below the predicted scatter bands. The same observation can be made for the SENT geometries where only few experimental values are available. Since the predictions of the BEREMIN model (2) are sensitive to the yield stress, the discrepancy between the TWI and IRSID values, as noted earlier, can be explained by the differences in yield strength seen in Fig. 1. This would induce a temperature shift of 30°C which would bring the experimental results back into the predicted scatter band of Fig. 6b. It is difficult to proceed further without a good variation of the yield strength with temperatue. However, the temperature shift is correctly predicted.

CONCLUSION

Some experimental results of the "Interntional project to develop shallow fracture mechanics tests" (2) have been interpreted with the BEREMIN model based on a local approach of fracture.

- The local criteria parameters of the material M1 have been measured (σu = 2525 MPa, m = 30).

- Numerical simulations of the SENB and SENT tests for two crack lengthes (a/w = 0,05 and a/w = 0,5) have been carried out. The CMOD, CTOD, J integral and Weibull stress were computed.

- For the SENB and SENT geometries the experimental temperature shift of the transition curves between short and long cracks is well predicted. However the quantitative predictions for the SENT specimens overestimate the displacement at fracture.

The yield strength and tensile strength of the material M1 measured at IRSID are 8% lower than the values given in the TWI reports. This point should be elucidated to conclude definitely. This discrepancy could explain the different values of the Weibull stress at fracture between the SENB and SENT specimens.

REFERENCES

(1) M.G. DAWES, R.H. LEGATT and G. SLATER,
"An international research projet to develop shallow crack fracture mechanics tests - European contribution", *First progress report, The Welding Institute,*5574/3/88, (March 1988).

(2) F.M. BEREMIN,
"A local criterion for cleavage fracture of a nuclear pressure vessel steel", *Met. Trans. A,* (1983), Vol 14 A, pp. 2277 - 2287.

(3) F. MUDRY, F. DI RIENZO and A.PINEAU,
"Numerical comparison of global and local criteria in compact tension and center-crack panel specimens".
Nonliniar in Fracture Mechanics : vol. 2, American Society for Testing and Materials 995, Eds J.D. LANDES, A. SAXENA, J.G. MAKLE (1989), pp. 29 - 49.

(4) A. FONTAINE, E. MAAS and J. TULOU,
"Prevision of the cleavage fracture properties using a local approach : application to the welded joint in a structural C-Mn steel", Proceeding, *International seminar on local approach of fracture,* Moret-sur-Loing, France, Ed, J. MASSON, Pub : Electricité de France, (1986), pp. 197 - 206.

(5) R.D. RITCHIE, J.F. KNOTT and J.R. RICE,
"On the relationship between critical tensile stress and fracture toughness in mild steel", *J. Mech. Phys. Solids,* (1973), Vol 21, pp.395 - 410.

(6)M. DI FANT, F. MUDRY,
"Interpretation of shallow crack fracture mechanics tests with a local approach of fracture", *IRSID report,* RI 91350, (1991).

(7) M.G. DAWES, R.H. LEGATT and G. SLATER,
"An international research project to develop shallow crack fracture mechanics tests - European contribution", *Second progress report,* The Welding Institute, 5574/6/88 (1988).

(8) M.G. DAWES, R.H. LEGATT and G. SLATER,
"An international research project to develop shallow crack fracture mechanics tests - European contribution", *Fourth progress report,* The Welding Institute, 5574/12/89 (1989).

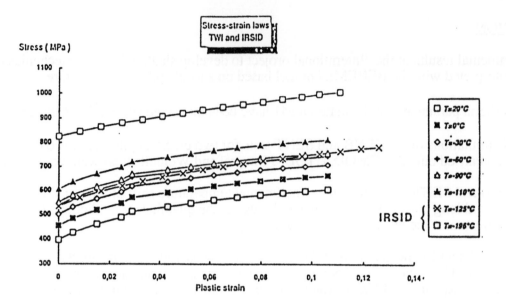

Figure 1   Comparaison of stress-strain low measured at IRSID and given by TWI

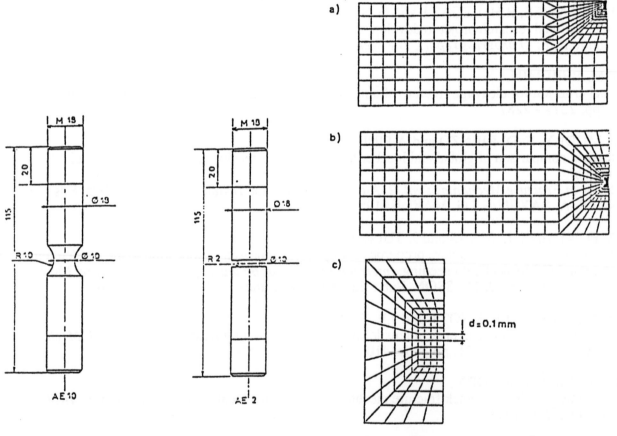

Figure 2 Geometries of the notched
Specimens used for measuring
the Parameters of the Beremin
model

Figure 3a   Meshes of the SENB Specimens
a) a/w = 0,05
b) a/w = 0,5
c) detail of the crack tip zone

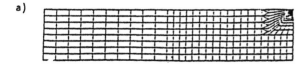

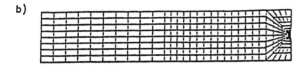

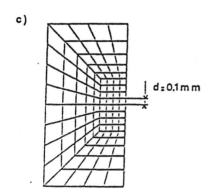

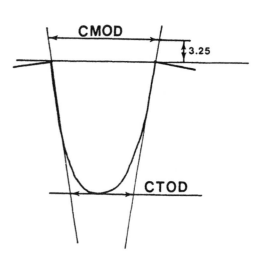

Figure 3b   Meshes ot the SENT Specimens
a) a/w = 0,05
b) a/w = 0,5
c) Detail of the Crack Tip Zone

Figure 4   Definition of CTOD and
CMOD

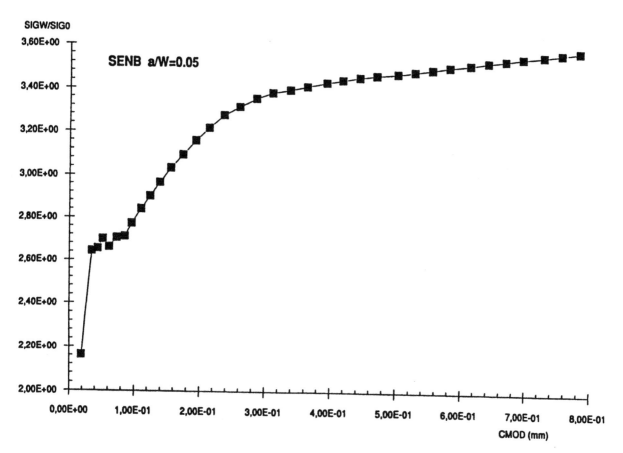

Figure 5a   Evolution of $\sigma w/\sigma o$ as a Function of CMOD (SENB, a/w = 0,05)

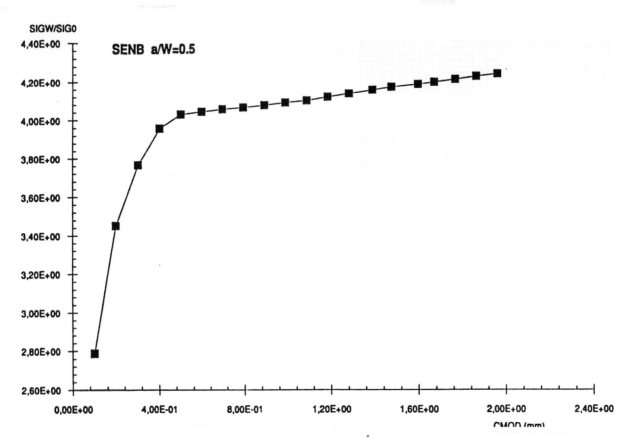

Figure 5b   Evolution of σw/σo as a Function of CMOD (SENB, a/w = 0,5)

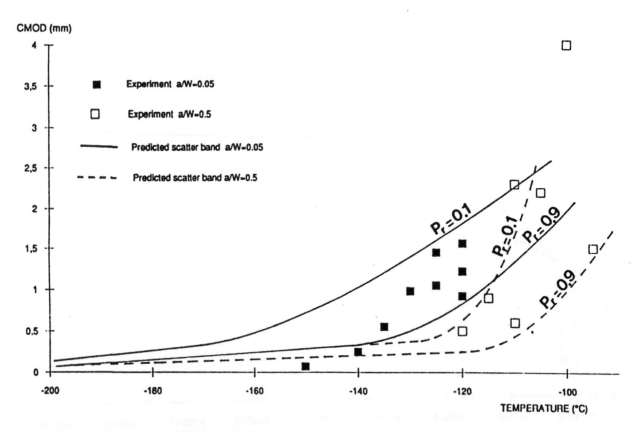

Figure 6a   Variation of CMOD with Temperature for the SENB Geometry
Comparaison of Experimental Results, Fig. 13 of reference (7) with scatter-bands
predicted by the Beremin Model

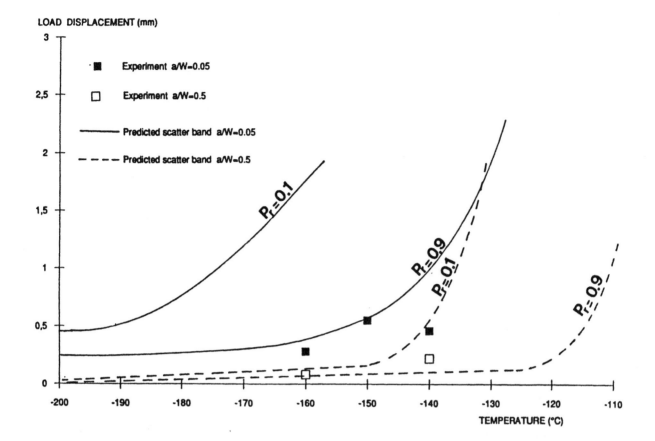

Figure 6b Variation of Critical d (Load Line Displacement at Fracture) with Temperature for the
SENT Geometrs
Comparaison of Experimental Results, Fig. 8 of reference (8) and scatter-bands
predicted by Beremin Model

TWI

PAPER 21

The effect of constraint on the ductile-brittle transition

I J MacLennan, BEng and J W Hancock, BSc, PhD (University of Glasgow, UK)

SUMMARY.

Betegón and Hancock (1) have described the elastic-plastic behaviour of short and deeply cracked bend bars by reference to modified boundary layer formulations based on the first two terms of the Williams (2) expansion. A local cleavage criterion has been applied to these fields to indicate the effect of the loss of constraint on lower shelf toughness of shallow cracked bend bars. The work models the maximum temperature at which cleavage is possible in these geometries to show the effect of constraint and the a/W ratio of cracked bend bars on the ductile-brittle transition temperature.

INTRODUCTION.

Temperature transitions in the fracture behaviour of steels are important to the integrity of engineering structures. Three principal forms of fracture can be identified in plain carbon structural steels. At low temperatures cleavage occurs by the direct separation of low index crystallographic planes. At higher temperatures the fracture mode passes through a transition in which initial crack extension by void growth and coalescence is followed by cleavage failure. At higher temperatures the toughness reaches the upper shelf level where crack extension occurs entirely by ductile mechanisms. The transition between brittle and ductile modes has particular significance for materials and structures operating near the transition temperature. In the present work interest is centered on the effect of geometry, and size, on the lower shelf toughness and transition temperature.

Ludwik and Scheu (3) first postulated brittle failure at a critical stress. The critical stress level is found to be independent or very weakly dependent of temperature and strain, (4, 5), unless twinning is involved, (6, 7). The application of the critical stress criterion in the stress and strain gradients ahead of a crack requires the introduction of microstructural size scale. In the work of Ritchie, Knott and Rice (8), (henceforth RKR), cleavage fracture is associated with the cracking of grain boundary carbides, which necessitates that the critical stress is achieved over distances of at least one grain diameter, (9). More sophisticated models have been developed by Beremin (10), Mudry (11) and Lin, Evans and Ritchie (12), who address the statistical problem associated with the application of weakest link criteria to the finite volumes of material in the crack tip field.

Until the recent work of Anderson and Dodds (13) most descriptions of cleavage failure have concentrated on the behaviour of deeply cracked bend bars, in which crack tip deformation is very similar to that of small scale yielding. This has enabled the HRR fields to be used as small strain descriptions of crack tip deformation. However the deformation mode of fully plastic shallow cracked bend bars differs from that of deeply cracked bars as discussed by Ewing (14). Similarly fully plastic shallow cracked bars are known to exhibit enhanced toughness values due to the loss of crack tip constraint. Hancock and co-workers (1, 15, 16) have shown that the lack of constraint originates from the nature of the elastic stress field associated with short cracks and in particular the non-singular stress in the Williams expansion, that Rice (17) has denoted the T stress. Betegón and Hancock (1) and Al-Ani and Hancock (15) have matched full field elastic-plastic solutions of a range of geometries including shallow edge cracked bars in tension and bending with boundary layer formulations of crack tip deformation in which the boundary conditions comprise the first two terms of the Williams expansion, K and T. Subsequently, Shih and co-workers (22, 23, 24) have adopted a related approach describing the crack tip constraint by a parameter Q which they claim to have more general application than the J - T approach. Betegón and Hancock (18), Sumpter and

Hancock (19), Sumpter (20), and Hancock, Reuter and Parks (21) have expressed the geometry dependence in the form of a failure locus in J-T space. A failure criterion is applied by matching the constraint of the cracked structure to the failure locus established by testing geometries such as shallow edge cracked bend bars through the T stress.

The present work adopts a complementary approach, in which local failure criteria are applied to elastic-plastic crack tip fields. The work is based on the ability to describe the crack tip deformation fields of shallow cracks with modified boundary layer formulations. The present authors prefer characterisation based on J and T although an identical approach is possible using Q as a second parameter as advocated by Shih and co-workers (22, 23, 24).

The work expresses the ratio of the critical values of J for shallow cracked bend bars to that of deeply cracked or small scale yielding, as a function of T. This is interpreted in terms of the effect of specimen size and temperature on the lower shelf toughness. Finally the shift in transition temperature associated with the change in geometry from deep to shallow cracked bars is predicted.

NUMERICAL METHODS.

Crack tip deformation under small scale yielding has been modelled by boundary layer formulations which make use of elastic displacements derived from the first two terms of the Williams expansion, K and T, as boundary conditions for the outer elastic field. The details of this technique have been described by Betegón and Hancock (1) following the work of Tracey (25) and Bilby et al (26). Crack tip plasticity has been described by a Ramberg-Osgood stress-strain relation, with a power hardening exponent, $n = 13$. Given the framework of small strain theory, the stress field within the plastic zone, and directly ahead of the crack can be described by the family of small strain solutions shown in Figure 1. The same Figure also shows a range of large strain solutions which explicitly recognise the large geometry change solutions and the the shape change associated with the blunting crack tip. Small strain solutions such as the HRR field exhibit stress singularities at the crack tip, in contrast the large strain solution exhibits finite maximum stress at a distance which is approximately two crack tip openings from the tip. As T becomes more negative the maximum achievable stress reduces and the distance at which it is achieved moves closer to the tip.

In the RKR analysis, brittle failure is postulated to occur when a critical cleavage stress, σ_f, is attained over a microstructurally significant distance, r^*, from the the crack tip. However by comparing the family of fields with the reference $T = 0$ field, at a given stress level, it is possible to remove the distance term, r^*.

$$\frac{J_T}{J_{T=0}} = \frac{\dfrac{r^*\sigma_0}{J_{T=0}}}{\dfrac{r^*\sigma_0}{J_T}} \qquad [1]$$

The T=0 field is the limiting field which applies at infinitesimally small applied loads, and is hence identified as the limiting small scale yielding field (SSY). The ratio $\dfrac{J_T}{J_{T=0}}$ is shown in Figure 2 as a function of $\dfrac{T}{\sigma_0}$ for a range critical stress levels, σ_f. This may be compared with data of Betegón (29), (17). The form of the failure locus is clearly captured, while an exact fit depends on the value of σ_f. More detailed analysis of this data is given by Parks (27), and Wang (28) using the Lin, Evans and Ritchie (12) approach and T as the parameter which quantifies constraint. The same procedure has been described by, Shih, O'Dowd, and Kirk (24) using the RKR approach and Q.

Supplying dimensional values of σ_f and r^*, combined with the experimentally determined temperature dependence of the yield stress (28), allows the temperature dependence of J to be modelled, as shown in Figures.3 and 4. In these Figures r^* was taken as 100 μm combined with a critical cleavage stresses of σ_f of 1400 MPa, which preliminary studies indicate to be appropriate to a plain carbon steel described by BS 4360. For the purposes of a sensitivity study the results of analysis with a fracture stress of 1800 MPa are also shown. The data are truncated at the maximum temperature at which cleavage initiation can occur

The J - T loci shown in Figures 5 and 6 unify the geometry dependence of toughness by quantifying the associated constraint of the crack tip fields by T. It is now convenient to recast this information in terms of a range of specific geometries. In the present case interest has been arbitrarily restricted to edge cracked three point bend bars. J is related to the load by relations given by Kumar and Shih (30). For a Ramberg-Osgood power hardening law J is decomposed into elastic and plastic components:

$$J^{total} = J^{elastic} + J^{plastic} \qquad [2]$$

$$J^{elastic} = \frac{K^2}{E'} \qquad [3]$$

$$J^{plastic} = \alpha \, \sigma_0 \, \varepsilon_0 \, (w-a) \, h_1 \left(\frac{a}{w}, n\right) \left(\frac{P}{P_0}\right)^{n+1} \qquad [4]$$

Here P is the load per unit thickness, P_0 is the limit load per unit thickness and h_1 is a function of geometry and strain hardening rate tabulated by Kumar and Shih (30). P_0 is a limit load given by Green and Hundy (32) as:

$$P_0 = \frac{0.728 \, \sigma_0 \, (w-a)^2}{L} \qquad [5]$$

The corresponding T value can be calculated from the calculations of Sham (33) who gives the T stress for a wide range of specific geometries in the form of a biaxiality parameter β :

$$\beta = \frac{T\sqrt{\pi a}}{K} \qquad [6]$$

For any given specimen, geometry, and size, the J - T history can be determined, allowing failure to be predicted at the intersection with the J-T failure locus. On this basis it is possible to determine the effect of temperature, size, and geometry on the lower shelf toughness as shown in Figure 7.

Figures 8 and 9 show the maximum temperatures, ϕ_c, at which cleavage initiation can occur as a function of geometry and size.

DISCUSSION.

The RKR analysis and its refinements provide a foundation for modelling the temperature cleavage in deeply cracked bend bars. The ability to describe the crack tip fields of unconstrained specimens, such as edge cracked bend or tension bars with short cracks allows the method to be extended to a wide range of geometries. A particular difficulty with the RKR analysis, and a source of much discussion has been the size of micostructural distance, r^*. Fortunately this can be removed by comparing the effect of constraint on cleavage toughness, as shown in Figure (2), and presenting data in the non-dimensional form $\frac{J_T}{J_{T=0}}$. Dimensional values can be supplied by using the RKR method to determine $J_{T=0}$. The experimentally determined temperature dependence of the yield stress (31) now allows the temperature dependence of J to be determined as shown in Figures 3, 4 Such results can be compared with those of Anderson and Dodds (13) who use an amplitude parameter to scale the HRR field to fit the fields of bend bars. This also leads to the ratio of the toughness of short and deeply cracked bars, and like the present method the ratio is independent of the micro-structural distance. However unlike the present analysis of weakly hardening boundary layer formulations, Anderson and Dodds (13) analysis appears to be independent of the critical stress, σ_f.

It is now appropriate to turn to the conditions under which cleavage cannot initiate ahead of a crack tip. Here it is necessary to recognise that the large strain solutions indicate that there is a maximum stress achievable at finite distances ahead of a blunting crack tip, as shown in Figure 1. In small scale yielding (T = 0) this is approximately 3.96 times the yield stress. At elevated temperatures cleavage initiation is prevented by the fact that the yield stress is too low to allow the local stress to reach the critical cleavage stress. The maximum temperature at which cleavage is possible is given by the temperature at which the maximum principal stress can just reach σ_f. The maximum stress that can be achieved ahead of the crack decreases with constraint, as shown in Figure 1. This implies that the highest temperature at which cleavage can initiate failure is lower in short crack geometries than in deeply cracked configurations as shown in Figures 7 and 8. This is in accord with experimental observations (34, 35, 36) that the transition temperature decreases with (a/W) ratio.

The effect of size on the transition temperature, ϕ_c, for specific geometries is shown in Figure 9. Size independence is exhibited by the deep cracked geometries, for which T is positive or close to zero, given (W-a $\geq$ 25J/σ_0). The transition temperature of the shallow cracked configurations is however is geometry dependent until a critical size is reached. Figure 9 shows the effect of specimen size on the transition temperature for a range of geometries. Geometry and size independence occur only in very large specimens.

CONCLUSIONS.

A J-T description of elastic plastic crack tip fields has been combined with local fracture criteria to indicate the effect of constraint on toughness, as measured by a J-T failure locus. An identical approach is possible in terms of Q. Such failure loci can be interpreted in terms of the effect of size and geometry on both the lower shelf toughness and the ductile-brittle transition temperature as measured by the maximum temperature at which cleavage initiation can occur. The analysis indicates that the transition temperature decreases with crack tip constraint in accord with experimental data. The model also predicts the size dependence of the transition temperature for short and deeply cracked geometries.

ACKNOWLEDGEMENTS.

Acknowledgements are due to the support and encouragement of Dr J Sumpter through MoD grant D/ERI/9/4/2048/59/ARE(D). Thanks are due to Hibbitt Karlsson and Sorensen Inc for access to ABAQUS under academic license.

REFERENCES.

1. Betegón C. and Hancock J.W., (1991) J Appl Mech, Vol 58,104
2. Williams, M. L.,(1957), J Appl Mech, Vol. 24, 111.
3. Ludwik, P. and Scheu, R.,(1923), Stahl und Eisen, 43, 999.
4. Orowan, E., (1948), Rep. Prog. Phys., 12, 185.
5. Knott, J. F.,(1966), J. Iron Steel Inst., 204, 104
6. Oates, G., (1968), J. Iron Steel Inst., 206, 930.
7. Oates, G., (1969), J. Iron Steel Inst., 207, 353.
8. Ritchie, R. O., Knott, J. F., and Rice, J. R., (1973), J. Mech. Phys. Slds.,21,395.
9. Curry, D. A., and Knott, J. F., (1976) , Metal Science, 10, 1.
10. Beremin, F. M., (1983), Metal. Trans. A., 14A, 2277.
11. Mudry, F., in "The Assessment of Cracked Components by Fracture Mechanics", ed

L .H.Larson, 133, EGF4, MEP London

12. Lin, T., Evans, A. G. and Ritchie, R.O., (1986), J. Mech. Phys. Slds.,5, 477.

13. Anderson,T.L. and Dodds R.H.1991 J. Testing and Evaluation, 123

14. Ewing D.(1968),J. Mech Phys Slds, Vol 16, 205

15. Al-Ani A. M. and Hancock, J. W.,(1991) J. Mech Phys Slds, 39, 23.

16. Du, Z.-Z. and Hancock, (1991)J. Mech Phys Slds, Vol. 22,.17.

17. Rice, J. R., (1974), J. Mech Phys Slds, Vol. 22, 17.

18. Betegón C. and Hancock J.W.,(1990), Fracture Behaviour and Design of Materials and Structures" 999, ECF8 Turin (ed D. Firrao), EMAS, Warley , U.K. .

19. Sumpter, J. D. G., and Hancock, J. W., (1991), Int. J. Press Ves. and Piping., 45, 207.

20. Sumpter, J. D. G.(1991), ASTM Symposium on Constraint Effects in Fracture, Indianapolis

21. Hancock, J. W., Reuter, W. A. and Parks, D. M.,(1991), ASTM Symposium on Constraint Effects in Fracture, Indianapolis

22. O'Dowd N.P. and Shih C.F.1991 J. Mech Phys SldsVol 39,989

23. O'Dowd N.P. and Shih C.F.1991 J. Mech Phys Slds in press

24. Shih C.F, O'Dowd N.P.and Kirk M.T. (1991), ASTM Symposium on Constraint Effects in Fracture, Indianapolis.

25. Tracey D. 1976 ASME J. Eng. Mats and Tech.98,146.

26. Bilby, B. A., Cardew, G. E., Goldthorpe, M. R. and Howard, I. C., (1986), in "Size effects in Fracture", I. Mech. E., London, pp. 37-46

27. Parks D.M.1991 in "Some Topics in Deformation and Fracture" ed A.Argon Springer-Verlag

28. Wang Y.Y.PhD Thesis Department of Mechanical engineering M.I.T. Cambridge Mass

29. Betegón, C.(1990) Ph.D Thesis , University of Oviedo.Spain

30. Bennet, P. E. and Sinclair, G. M., (1966), J. Basic. Eng, Trans, ASME Series D, 88,518.

31. Kumar V.and Shih C.F 1980 in "Fracture Mechanics"ASTM STP 700, 406

32. Green A.P. and Hundy B.B. 1956J. Mech Phys Slds Vol 4,128

33. Sham, T.- L. (1991), Int. J. Fracture., 48, 81.

34. Li Q.F. (1985) Engineering Fracture Mechanics Vol 22, 9

35. Li.Q.F.,Zhou L. and Li S. (1986) Engineering Fracture Mechanics Vol 23, 925

36. Al-Ani A.M.(1991) PhD Thesis Department of Mechanical Engineering University of Glasgow

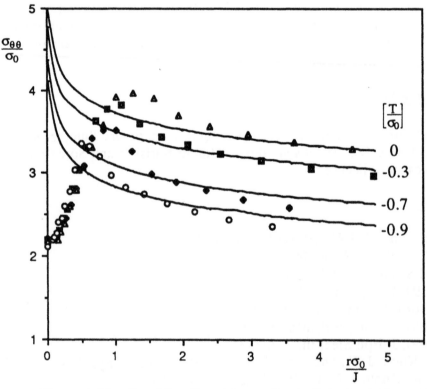

Figure 1: Family of Small and Large Strain Solutions for n=13.

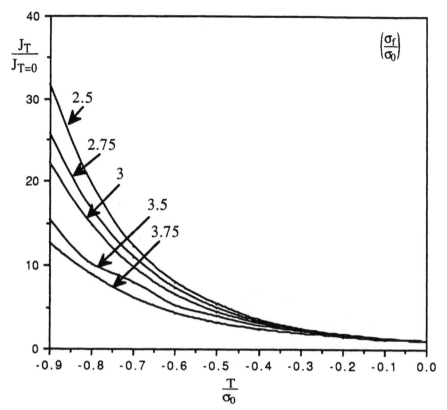

Figure 2: Non-dimensionalised J-T loci for Cleavage at a Range of Fracture Stresses for n=13.

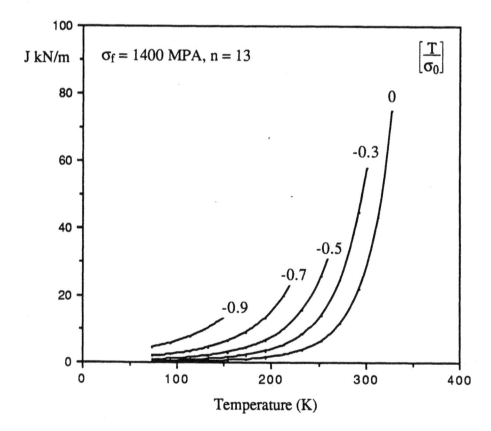

Figure 3: Effect of Temperature on Toughness with varying degrees of Constraint.

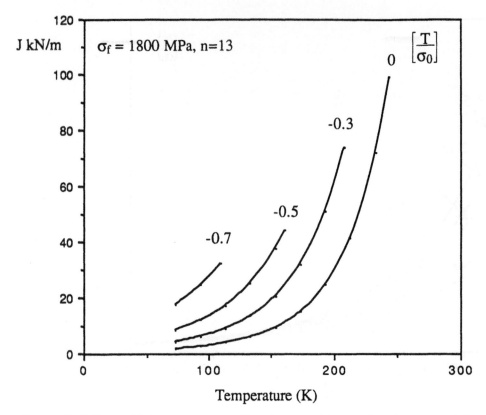

Figure 4: Effect of Temperature on Toughness with varying degrees of Constraint.

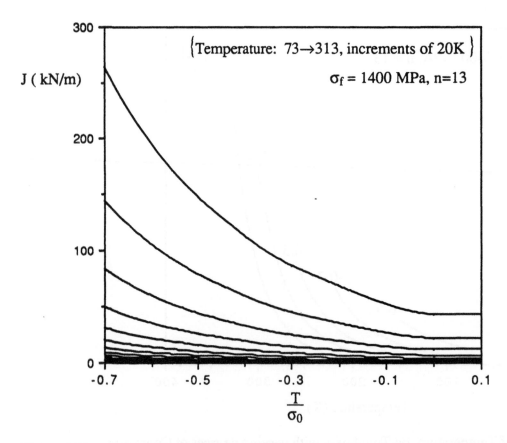

Figure 5:Temperature Dependence of J-T Loci for a Critical Cleavage Stress of 1400MPa.

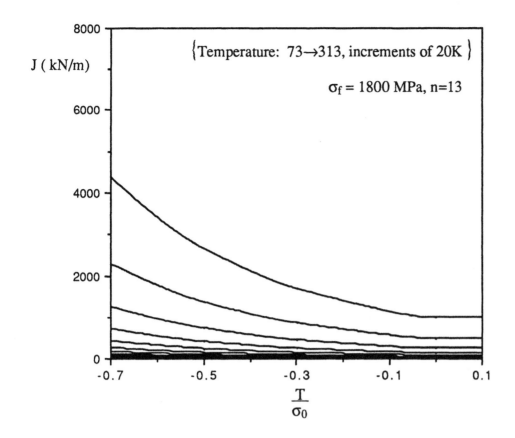

Figure 6: Temperature Dependence of J-T Loci for a Critical Cleavage Stress of 1800MPa.

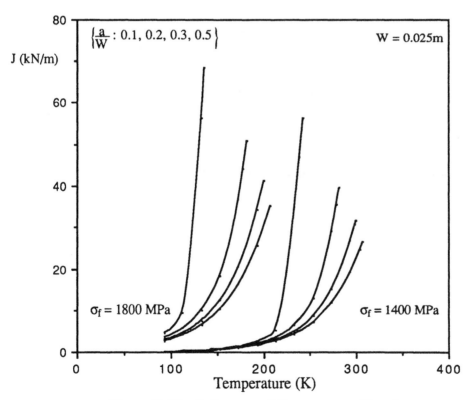

Figure 7: The Influence of Geometry on Toughness.

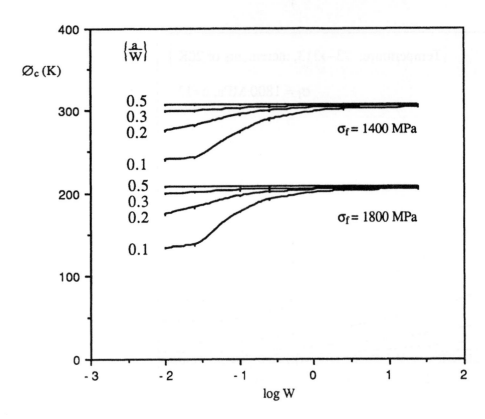

Figure 8: The Effect of Size on the Transition Temperature for a Range of Geometries.

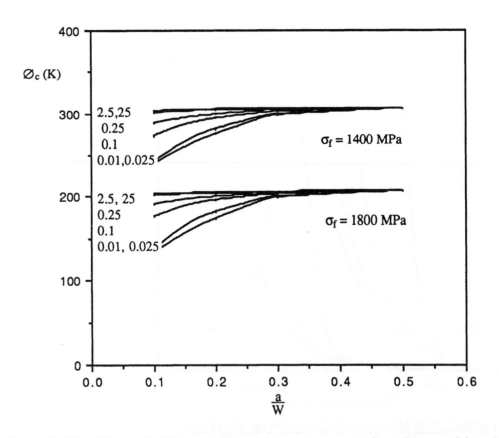

Figure 9: The Effect of a/W on the Transition Temperature for a Range of Specimen Widths.

The effect of constraint on fracture of carbon and low alloy steel

Brian Cotterell (National University of Singapore) and Shang-Xian Wu, and Yiu-Wing Mai (University of Sydney, Australia)

SUMMARY

The constraint at a crack tip has an effect on the fracture of all but the most brittle materials unless the specimen size is very large. Standard fracture toughness tests are designed so that constraint is high and the fracture parameters obtained from such tests can be conservative when applied to shallow cracks of low constraint in carbon and low alloy steels. Low constraint has two effects. It can greatly increase the critical CTOD or J_{Ic} for ductile tearing or cleavage fracture and also, because low constraint limits the stress, it can cause a change in the mode of fracture from cleavage to ductile tearing. For ductile tearing it is suggested that the effects of constraint can be predicted if the stress-strain relationship for the fracture process zone is known. For cleavage fracture and fracture transition it is shown that the critical stress over a characteristic distance is applicable.

INTRODUCTION

Classic fracture mechanics is based on the idea of identifying an autonomous region or fracture process zone (FPZ) at a crack tip where the criterion of fracture can be based on a single parameter. This concept required modification when it was realized that in general resistance to crack propagation increases with crack growth. However, a single fracture parameter was retained with the modification that this parameter was a function of the crack extension. It is now realized that for ductile metals and many other materials, such a simple description of fracture is insufficient in many circumstances. Because all the present fracture toughness test methods derive from tests for high strength brittle metals, the specimens have long cracks which give high constraint. Under these circumstances, crack initiation in low strength steels can be defined by a single parameter such as crack tip opening displacement (CTOD) or J-integral. However, even in the specimens with high constraint, there is the realization that the crack growth resistance is not necessarily well described by a single parameter that depends solely on crack extension (1,2). If the constraint is not high, the critical condition for the FPZ cannot be described by a single parameter (3-10).

The mode of fracture initiation in low strength steels can be either ductile tearing or cleavage and also it is possible for a ductile tear after a short propagation to initiate a cleavage fracture. The two modes of fracture are governed by different factors.

DUCTILE TEARING

Ductile tearing is the result of void initiation and growth. Voids of two different sizes form at the tip of a notch. The larger voids form at non-metallic inclusions. It is the growth of these voids that govern the early stages of tearing (11). Very much smaller voids form at carbides and create a porosity that makes the continuum criterion of plastic flow of the material between the large voids sensitive to hydrostatic stress and cause strain localization in the last stages of forming a ductile tear (12). Many inclusions are poorly bonded and the growth of the larger voids is more important than their initiation. Thompson and Hancock (13) found

25% of the total void population to be pre-existing. There is a large variation of the strain necessary to initiate the well bonded inclusions, though there is a tendency for the critical strain to decrease with increasing triaxiality (13). The growth of isolated voids is strongly dependent on the hydrostatic stress (14). However there is a difference in behaviour of isolated voids and those that form near a notch tip where there is interaction between the voids themselves and the notch tip. Theoretical analyses of ductile rupture by void growth and strain localisation such as that of Needleman and Tvergaard (15) have had a crack tip embedded in stress fields dominated by crack tip stress singularities and while these provide a much needed insight into the mechanism of ductile fracture they do not help our understanding of the effects of constraint. In modelling the mechanics of ductile tearing so that stable tearing and the onset of unstable fracture can be predicted it is not necessary in our view to model the precise behaviour within the FPZ. For ductile tearing the FPZ can be defined roughly by that region where the microvoid porosity volume fraction becomes significant so that the constitutive equations for continuum plastic deformation do not apply, as shown schematically in Fig. 1.

Examination of finite element calculations (12,16) for Gurson type materials where the nucleation of microporosity depends either on strain or stress suggest that the maximum nominal stress is reached at about the same moment that the volume fraction of the microvoids starts to increase. It therefore seems reasonable to identify the FPZ in ductile tearing with the region of strain softening. In the strain-softening region the stress decreases. For low constraint the decrease in stress is caused by simply area reduction that can be calculated from the normal continuum equations. However, for high constraint the nucleation of microporosity becomes important as is demonstrated in Fig. 2 where the maximum traction stresses calculated by Tvergaard (12) for a Gurson material with an array of large voids are compared with the values obtained for a dense material simply on the basis of area reduction (allowance has been made in these latter calculation for the reduction in area caused by the large void array). The stress necessary to initiate the formation of a FPZ obviously depends on the degree of constraint. The same conclusion stated in some what different terms is made by Al-Ani and Hancock (10).

The stress-displacement relationship for the FPZ of a ductile tear is shown schematically in Fig. 3. The maximum stress σ_m depends upon the degree of constraint which can be quantified by the ratio of the hydrostatic stress to the flow stress. This ratio is not constant along the FPZ but varies from a maximum at its tip to near unity at the notch tip (15). The most important parameters of the stress-displacement curve are the maximum stress σ_m and the CTOD δ_t at which the voids link up to form a ductile tear. If it is assumed that the shape of the stress-displacement is self similar so that

$$\sigma / \sigma_m = f(\delta / \delta_t) \tag{1}$$

then the essential work of fracture w_e is given by

$$w_e = \sigma_m \delta_t \int_0^1 f(\delta / \delta_m) \, d(\delta / \delta_m) . \tag{2}$$

Outside the FPZ the material obeys the normal constitutive equations thus it can be argued that J_{Ic} is the energy flowing to the FPZ and identical with w_e. The equality between J_{Ic} and w_e can also be seen from the definition of the J-integral, if the FPZ is assumed to be thin and the contour integral is performed round the edge of the FPZ. Hence

$$\int_0^1 f(\delta/\delta_t)\, d(\delta/\delta_t) = J_{Ic}/\sigma_m \delta_t. \qquad (3)$$

Compare Eq. 3 with the usual equation relating J_{Ic} with the CTOD

$$J_{Ic} = m\sigma_Y \delta_t, \qquad (4)$$

where σ_Y is the uniaxial yield strength and m is a constant in the range $1 \leq m \leq 3$ where the lower values come from tensile tests and the higher from bending tests. If σ_Y is replaced by the stress σ_m, where for tension $\sigma_m \approx \sigma_Y$ and for bending (deep notches) $\sigma_m \approx 2.5\sigma_Y$, then it becomes apparent that there may be a universal constant connecting J_{Ic} and δ_t. The CTOD is an easily measured parameter (3-6,8,9), but the estimation of σ_m is more problematical. It is suggested that, in fully yielded specimens, the stress calculated from slip line theory using a flow stress of $\sigma_f = (\sigma_Y + \sigma_{ult})/2$ should be a reasonable estimate.

Estimation of the stress-displacement relationship

Three-point-notch-bend test specimens were made from a free-cutting steel in the normalized condition and after straining to 5% and 15%. The mechanical properties of the steel are given in Table 1.

Table 1. Mechanical properties of free cutting steel

Condition	0.2% Proof stress (MPa)	Ultimate tensile strength (MPa)	Elongation (%)	Reduction in area (%)
Normalized	202	338	29	35
5% Prestrain	325	385	19	27
10% Prestrain	421	424	7	22

The specimens were 14mm thick and 25mm deep; the loading span was 100mm. Two notch depths were used: a/W=0.1 and a/W=0.5. In both cases the machined notches were sharpened by fatigue cracks. CTOD and J_{Ic} tests were performed using multiple specimens to obtain the initiation values by extrapolation. The CTOD and J_{Ic} for the deep notches were obtained according to BS5762 and ASTM E813-83 respectively, except that the rotation factor was taken to be 0.45. For the shallow notches, the CTOD was determined by a section metallography technique, whereby the specimens were sectioned at 1/6, 1/3, and 1/2 of the thickness from one surface of the specimen and the crack length, crack growth, and CTOD measured with an optical microscope. J_{Ic} for the shallow notches was calculated from the plastic η factor given by Wu et al. (19,20). Further experimental details can be found in (9,21). The experimental results are given in Table 2.

Table 2. Experimental results for a free cutting steel

Condition	Inclusion/ void size (µm)	a/W	σ_m/σ_f	δt (mm)	J_{Ic}(kJ/mm²)	$J_{Ic}/\sigma_m\delta_t$
Normalized	4.0	0.1	1.95	0.192	81.0	0.80
		0.5	2.51	0.082	30.3	0.55
5% Prestrain	4.7	0.1	1.95	0.086	41.4	0.70
		0.5	2.51	0.043	19.7	0.51
10% Prestrain	6.0	0.1	1.95	0.043	24.5	0.69
		0.5	2.51	0.013	9.5	0.69

The main effect of prestraining is to increase the flow stress of the material. The inclusions are not well bonded and, though they open during prestraining, it is assumed that there is little damage done, so that the material entering the FPZ is essentially unchanged by the prestraining operation. What the prestraining does is to increase the stress at the tip of the FPZ. Using the estimates of σ_m/σ_f obtained from slip-line theory (22) the "m" factors in column 7 of Table 2 have been calculated. There is surprisingly little variation in these values which leads one to suggest that the stress-displacement curves for the FPZ are self similar. Assuming the power law

$$\sigma/\sigma_m = f(\delta/\delta_t) = 1 - (\delta/\delta_t)^n, \tag{5}$$

adequately represents the stress-displacement relationship the value of the exponent that gives an average "m" factor of 0.66 is n=1.9. The value of the CTOD to cause ductile tearing is plotted against the value of the maximum stress at the tip of the FPZ estimated from slip-line theory in Fig. 4. The CTOD non-dimensionalized by the inclusion spacing is presented as a function of the void diameter/spacing ratio in Fig. 5, the results of Thompson and Knott (23) for a similar free-cutting steel are included. An attempt (9) was made to estimate δ_t from the void spacing/size ratio using the theory of Rice and Johnson (24), modified to take account of the differences in the slip-line fields for deep and shallow notches, but it was not very successful. The experimental results indicate a much stronger dependence on void spacing/ size than theory.

The experiments described clearly show the effect of constraint on ductile tearing. By closer examination of the relationship between J_{Ic} and δ_t it is shown that the relationship is independent of constraint provided the effects of constraint on the stress in the FPZ are taken into account. Unfortunately as yet it is not possible to predict the effect of constraint on δ_t.

CLEAVAGE FRACTURE

Cleavage fracture is controlled by stress. Provided the cleavage is slip-induced, the stress necessary to produce fracture is relatively independent of temperature (25). The work of Ritchie et al. (26) has shown that it is also necessary that the critical stress be exceeded over a microscopically significant distance at the crack tip (26). If the stresses over the critical distance are J-dominated, then this simple criterion of cleavage fracture translates into a critical J-integral J_{Ic}, or CTOD. For specimens of high constraint J-dominance is ensured when

$$b > 25 \sim 50 \, J_{Ic}/\sigma_{Y'} \qquad\qquad (6)$$

where b is the remaining ligament. As shown recently, the constant in this inequality can be underestimated by more than a factor of ten for specimens with low constraint (10). If the constraint is reduced, then not only will J_{Ic} increase, but also it is possible for the mode of fracture to change to ductile tearing as the specimen needs to be strained further to achieve the critical cleavage stress over its characteristic distance. The effect of reducing the constraint is analogous to increasing the temperature.

Ductile-cleavage transition

Three and four point bend tests were made on a 0.3% carbon steel (CS1030) in the as received condition at room temperature. The width of the specimens was 25mm and the notched ligaments were 12.5 mm. The relative notch depths varied from 0.1 to 0.5. The three-point-bend specimens were tested with a span that was 4W. The specimens with (a/W)>0.2 failed in a brittle manner and the critical CTOD, δ_c, was obtained according to BS5762, but the rotation factor of the plastic hinge was taken as 0.45 for the three-point-bend tests and 0.37 for the four-point-bend specimens. In the case of the shallow notches, where failure was by ductile tearing, the CTOD at initiation, δ_i, was obtained by the multiple specimen technique and the plastic component of the CTOD was measured from two indentations on the surfaces of the specimen. Further experimental details can be obtained from references (8,21). The CTOD values for the bend tests are shown as a function of the relative notch depth a/W in Fig. 6.

The fracture propagation in the shallow notched three and four point bend specimens was quite different. In the three-point-bend tests the ductile tear initiated along the centre line, but propagated along the slip-line direction at approximately 45° to the initial tear. In these specimens the whole fracture was a ductile tear. However, under four-point bending the ductile tear continued to propagate along the centre line until an unstable cleavage fracture was initiated. A series of experiments was run to determine the relative crack depth at which transition from ductile to cleavage propagation occurred. These results are shown in Table 3

Table 3 Relative crack depth a_f/W for transition to cleavage fracture in four-point-bend specimens as a function of initial a_o/W

a_o/W	0.10	0.09	0.18	0.08	0.14	0.14
a_f/W	0.20	0.21	0.30	0.41	0.43	0.44

In general the relative crack depth for transition from a ductile tear to a cleavage fracture during propagation is larger than at initiation. A possible cause of the delayed transition is crack tip blunting during ductile tearing.

The critical fracture stress σ_f for cleavage fracture was estimated to be 1520 MPa by performing four-point-bend tests at -196°C and -100°C on bars with 45° V-notches containing a small root radius in a similar manner to Ritchie et al. (27).

Since the deep notched specimens did not yield completely before fracture slip-line field solutions are inappropriate and finite element analyses were undertaken. These analyses used eight-noded isoparametric elements and were based on small strain theory with J_2 flow theory. The actual stress-strain relationship for the experimental steel was used in the analysis. The normalized stress ahead of the crack tip for four-point-bend specimens obtained by this means is shown in Fig. 7. These are very similar to those obtained by Al-Ani and Hancock (10) for a hypothetical material with a similar strain hardening exponent. Because the finite element analyses were based on small strain theory there is a singularity at the crack tip. Slip-line theory (21,22) was used to estimate the position of the maximum stress. The accuracy of the estimates of the maximum stresses obtained by this process is confirmed by the results of Al-Ani and Hancock (10) who present both small and large geometry change solutions. The critical cleavage fracture stress of 1520 MPa is superimposed on Fig. 7.

Using The four-point-bend specimen with a/W=0.5 as the master result, it is seen that the critical fracture stress is exceeded over a distance of $X/(J/\sigma_Y)=4.7$ from the crack tip which corresponds to a distance of $X \approx 9.4/\delta_c$ if m is assumed to be 2. Since for this specimen $\delta_c=0.025$mm the characteristic distance is about 0.23mm or 9 times the mean ferrite grain size. This characteristic distance has been used to predict the critical CTODs for the other specimens that failed by cleavage fracture. For a/W<0.2 the cleavage stress is not exceeded and hence the fracture is by ductile tearing. These predictions, which do not depend on the value chosen for m provided the same value is used for all a/W, are indicated in Fig. 6.

The results indicate that neither the CTOD nor J-integral alone can be used as criteria of cleavage fracture in these specimens. The analysis of Al-Ani and Hancock (10) indicates that for J-dominance a four-point-bend specimen would have to have a crack length larger than $350\delta_c$ for a/W=0.2. With the present material this would require a crack length of about 30mm and a specimen depth of 150mm such a specimen be approaching the size requirements of LEFM. The RKR model of cleavage fracture is shown to be capable of predicting the effect of constraint on the critical CTOD δ_c necessary for cleavage fracture. The same model can also predict the transition in fracture mode to ductile tearing at low constraint. The present method cannot be used to predict the transition in fracture propagation

CONCLUSIONS

It has been demonstrated that in laboratory sized specimens constraint has a large effect on the critical CTOD or J_{Ic} for the initiation of either ductile tearing or cleavage fracture. Because of this effect the fracture of low strength metals like carbon and low alloy steels cannot be defined by a single parameter.

The RKR critical stress over a critical distance criterion of cleavage fracture (26) is shown to be capable of predicting the variation of δ_c with specimen constraint and also predicting when ductile tearing occurs in preference to cleavage.

For ductile fracture it is possible to predict the effect of constraint on the critical CTOD or J_{Ic} if the stress-strain relationship for the FPZ is known. It has been shown that the relationship between δ_t and J_{Ic} can be expressed in terms of the maximum stress at the tip of the FPZ. In specimens that have fully yielded an estimate of this maximum tensile stress can be found from slip-line theory using a flow stress which is the mean of the yield and ultimate strengths. The stress-displacement relationship in the FPZ for the free-cutting steel studied is roughly parabolic. Further work is needed to be able to predict the stress-displacement relationship, since the model based on the simple theory of void growth and coalescence originally proposed by Rice and Johnson (24) does not give good predictions.

REFERENCES

(1.) C.E. Turner, *Fracture Behaviour and Design of Materials and Structures,* Proc. ECF8 (Ed. D. Firrao) pp. 933-968 (1990).

(2.) M.R. Etemad and C.E. Turner, *Fracture Mechanics: Twenty-First Symposium, ASTM STP 1074* (Eds. J.P. Gudas, J.A. Joyce, and E.M. Hackett) pp. 289-306 (1990).

(3.) J.W. Hancock, and M.J. Cowling, *Metal Sci.,* 293-304 (1980).

(4.) C.G. Chipperfield, *Proc. Specialists' Meeting on Elasto-Plastic Fracture Mechanics,* Vol. 2, paper 15, OECP Nuclear Energy Agency, Daresbury (1978).

(5.) B. Cotterell, Q.F. Li, D.Z. Zhang, and Y.W. Mai, *Engng. Fract. Mech.* **21**, 239-244 (1985).

(6.) G. Matsoukas, B. Cotterell, and Y.W. Mai, *Engng. Fract. Mech.* **24**, 837-842 (1986).

(7.) H.W. Liu and T. Zhuang, *Theor. Appl. Fract. Mech.,* **7**, 149-168 (1987).

(8.) S.X. Wu, Y.W. Mai, B. Cotterell, and C.V. Le, *Acta Metall. Mater.,* **39**, 2527-2532 (1991).

(9.) S.X. Wu, B. Cotterell, and Y.W. Mai, *Int. J. Fract.,* **51**, 207-218 (1991).

(10.) A.M. Al-Ani and J.W. Hancock, *J. Mech. Phys. Solids,* **39**, 23-43 (1991).

(11.) S.H. Goods and L.M. Brown, *Acta Metall.,* **27**, 1-15 (1979).

(12.) V. Tvergaard, *J. Mech. Phys. Solids,* **30**, 265-286 (1982).

(13.) R.D. Thomson and J.W. Hancock, *Int. J. Fract.,* **26**, 99-112 (1984).

(14.) J.R. Rice and Tracey, *J. Mech. Phys. Solids,* **17**, 201-217 (1969).

(15.) A. Needleman and V. Tvergaard, *J. Mech Phys. Solids,* **35**, 151-183 (1987).

(16.) R. Becker and A. Needleman, *J. Appl. Mech.,* **53**, 491-499 (1986).

(17.) C.L. Hom, and R.M. McMeaking, *J. Appl. Mech.,* **56**, 309-317 (1989).

(18.) C.E. Turner, *Post-Yield Fracture Mechanics,* D.G.H. Latzko, C.E. Turner, J.D. Landes, D.E. McCabe, and T.K. Hellen, 2nd. edition, Elsevier Applied Science, London pp.25-222 (1984).

(19.) S.X. Wu, Y.W. Mai, and B. Cotterell, *Int. J. Fract.,* **45**, 1-18, 1990.

(20.) S.X. Wu, *Advances in Fracture Research,* ICF-7, Eds. K. Salama, et al., Vol. 1, 517-524 (1989).

(21.) S.X. Wu, *Fracture analysis and toughness measurements of specimens with deep and shallow cracks,* Doctoral Thesis, University of Sydney (1990).

(22.) S.X. Wu, B. Cotterell, and Y.W. Mai, *Int. J. Fract.,* **37**, 13-29 (1988).

(23.) H.E. Thompson and J.K. Knott, *Fracture Control of Engineering Structures, Proc ECF6,* Vol.3 pp.1737-1749 (1986).

(24.) J.R. Rice, and M.A. Johnson, *Inelastic Behavior of Solids,* Eds. M.F. Kauninen et al., McGraw Hill, New York, 641-672 (1970).

(25.) J.F. Knott, *J. Iron Steel Inst.,* **204**, 104-111 (1966).

(26.) R.O. Ritchie, J.F. Knott, and J.R. Rice, *J. Mech. Phys. Solids,* **21**, 395-410 (1973).

(27.) R.O. Ritchie, W.L. Server, and R.A.Wullaert, *Metall. Trans.,* **10A**, 1557-1570 (1979).

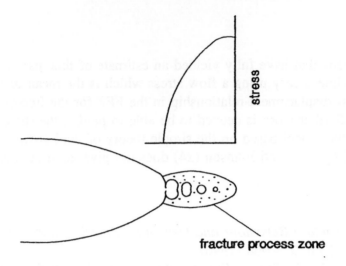

Figure 1 A Schematic illustration of a ductile FPZ

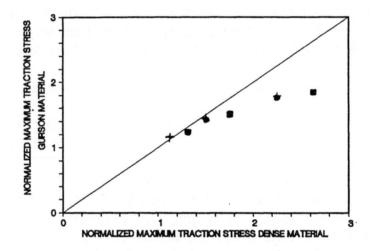

2 The maximum traction stress for a Gurston material with voids compared with a dense material (+, $\sigma_3 = \sigma_1$; ∎, $\varepsilon_3 = 0$; ●, $\varepsilon_3 = \varepsilon_1$)

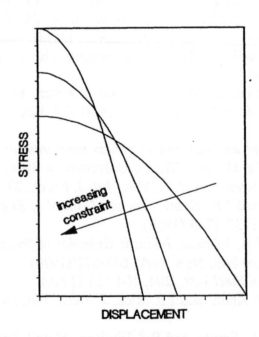

3 Schematic stress-displacement curves for a ductile FPZ

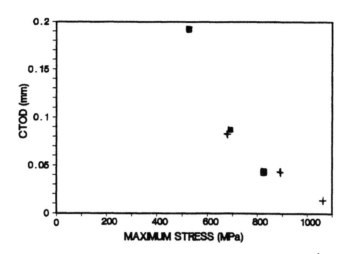

Figure 4 The critical CTOD, δ_t, for ductile tearing as a function of the maximum stress in the FPZ (■, $\sigma_m/\sigma_f = 1.95$; +, $\sigma_m/\sigma_f = 2.51$)

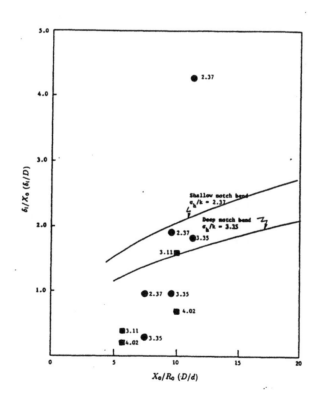

Figure 5 The critical CTOD, δ_t, for ductile tearing as a function of the void spacing/size ratio (numerals are values of σ_h/k; results this paper ●, Thompson & Knott (23) ■)

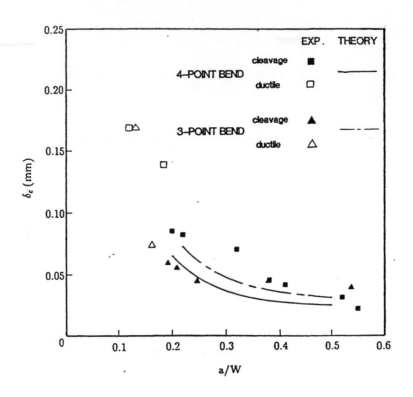

Figure 6 CTOD for cleavage initiation δ_c and the CTOD for ductile tearing δ_t as a function of a/W

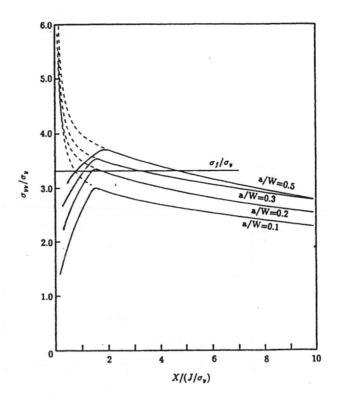

Figure 7 The normalized stress σ_{yy}/σ_Y ahead of a crack tip of four-point-bend specimens

SHALLOW CRACK FRACTURE MECHANICS,
TOUGHNESS TESTS AND APPLICATIONS

TWI

Cambridge, UK
23-24 September 1992

PAPER 16

A general discussion of short crack fracture

J R Matthews, BEng, MEng, PhD, PEng, J F Porter, BEng, PEng, C V Hyatt, BSc, PhD and
K J KarisAllen, BEng, MASc (DREA, Canada)

SUMMARY

This paper discusses experimental and analytical techniques for characterizing plastic strain around crack tips from deep to short crack depths. Numerically derived load deflections and plastic strain patterns for single edge notched bend bars, fabricated from ASTM A710 Grade A plate, are compared with experimental measurements. The shapes of the plastic zones are experimentally characterized employing visual techniques. Experimental trends in J and stretch zone width as crack depth is varied from the deep to short crack depth are described. Stretch zone width measurements between 0.08 a/W and 0.95 a/W have shown that there is no observable difference in stretch zone width between the deep and shallow notched specimens. This work indicates that a well defined plastic strain pattern develops around the crack, which precedes stable crack growth.

INTRODUCTION

In the past few years, fracture control technology has seen a new focus on short crack fracture because of a perception that cracks would either be detected when shallow, or that defects would not grow deep in a stable manner due to the nature of the service of the structure (1). This has caused a re-evaluation of material toughness determination methods and of the structural significance owing to the increased loss of constraint at the crack front. The trend has been to modify the existing elastic-plastic concepts and technology to compensate for the loss of J dominance (2-5).

Paramount in the application of fracture technology for the prediction of critical events is the guarantee of conservatism in the absence of knowledge. Applying deep crack toughness values into structural analysis models meets this goal by efficiently converting applied deflection into crack driving force owing to the existence of a maximum level of constraint. Figure 1 schematically indicates failure conditions for a material/geometry combination exhibiting elastic-plastic behavior in the deeply cracked configuration. For the short crack situation, deflection is no longer converted into constrained plasticity at the crack tip. Plastic yield is now partitioned between localized plastic damage at the crack tip and general yield in the surrounding area of the crack. As cracks tend toward zero a/W the failure is increasingly dominated by a fully plastic failure mechanism. At some point the plastic strain pattern changes dramatically and the yield path goes from the crack tip to the top surface rather than from the crack tip to the back surface. The point at which the plastic strain pattern changes should be used as a bench mark for defining the limiting value of a/W for short cracks. In cases where the plastic strain pattern is clearly from the crack tip to the top surface, ie. a short crack, strain exhaustion at the crack tip or strain to fracture should be the relevant material property defining the onset of tearing. This strain to fracture could equally well be described by the strain energy density at the crack tip. It remains to be determined also, how far ahead of the crack tip one should measure or calculate the strain or strain energy density as it has not yet been proven in tearing dominated by the growth and coalescence of holes whether the crack jump is from the hole ahead of the crack tip to the crack tip or from the crack tip to the hole.

Further to the above, sophisticated analytical models based on strain energy density have shown promise when applied to predicting ductile initiation events at short crack depth (6). The basic difficulty with this technique comes in trying to determine a critical strain energy density value. This aside, the computational concept is rigorous and in application should be relevant and meaningful not only for the short crack case, but across all levels of constraint, linear elastic, elastic plastic and plastic. All efforts in this area support strain domination of the initiation process for stable ductile extension for the short crack case.

The COD Design Curve gives additional insight into the size effect question and the role of strain in short crack fracture. Regardless of the material, there exists a crack depth below which strain is the applicable failure property. This factor is evident in the design curve. The design curve starts on the low end (elastic) with an allowable defect size, a_m, based on linear elastic fracture mechanics and K_I equal to .7 K_{Ic}.(7).

$$a_m = \frac{1}{\pi} \left(\frac{K_I}{\sigma_{design}}\right)^2 \frac{1}{f(geometry)} \qquad [1]$$

On the upper end of the curve, where plastic conditions exist, the design curve is defined by a quasi description of applied strain, ε, and failure defined by strain analogy formulas.

$$a_m = \frac{1}{2\pi} \left(\frac{\delta_{crit}}{\varepsilon_y}\right) \frac{1}{(\varepsilon/\varepsilon_y - 0.25)} \qquad \text{ferritic steel} \qquad [2a]$$

$$a_m = \frac{1}{2\pi} \left(\frac{\delta_{crit}}{\varepsilon_y}\right) \frac{1}{(\varepsilon/\varepsilon_y)^2} \qquad \text{other materials} \qquad [2b]$$

The central elastic plastic area, does not have any special definition to cover failure but simply is covered by a smooth running together of the linear elastic failure line and the line defined by strain. It implies that for cases of high plasticity (i.e. short cracks or ductile materials) strain to fracture defines separation and tearing. The design curve has been created by joining the curves for the two extremes of fracture, stress dominated and strain dominated, and therefore bounded accurately all subsequent predictions. The Engineering J design curve (8), proposed by Turner for extreme plasticity, is also based on a strain analogy.

In this paper, evidence is introduced to reinforce the role of strain at the crack tip and plastic strain pattern development around the crack in determining the onset of tearing in short crack fracture. Experimental and analytical methods for obtaining strain based fracture information are highlighted along with J based fracture information.

EVALUATION OF TOUGHNESS MEASUREMENT TECHNIQUES

This section is divided in three parts: J - based techniques, strain to fracture techniques, and stretch zone techniques.

J - Based

To overcome the size limitation of plane strain fracture toughness testing, the fracture criterion based on the J integral, J_{Ic}, has proven to be the necessary link between elastic and elastic-plastic fracture mechanics. It is assumed that the onset of fracture is caused by a critical level of stress or strain and the associated critical level of potential energy being reached at or near the crack tip. This condition is measured with the J integral, a two-dimensional energy line integral defined by Rice (9,10).

The J integral was originally used as an analytical tool for stress and strain determination at the crack tip. As such, the development of a J_{Ic} criterion was a logical extension of the original concept. Fracture starts from a sharp fatigue crack when the specimen or structure containing that crack is loaded beyond the prior cyclic loading level. At that point the crack tip becomes blunted. This blunting increases with the loading until at some critical stress the crack advances ahead of the original blunted flaw. This point is marked by a change in slope of the J versus crack extension curve, since crack advance caused by tearing ahead of the blunted crack develops at a much faster rate than the blunting process. J_{Ic} can then be determined by developing a curve of J versus crack advance (using multiple specimens) and marking the point of deviation from the blunting line, or by using one of a variety of single specimen techniques (11). When using the J integral as a failure criterion, specimens can be subject to large scale or general yielding provided plane-strain conditions exist at the crack tip. J is independent of geometry provided the thickness, crack depth and remaining ligament are large enough for J to characterize the crack tip field. The minimum size requirement is expressed as:

$$a, b, B \geq \alpha \, (J_{Ic} / \sigma_{flow}) \qquad [3]$$

where, a is the crack size, b is the remaining ligament, B is the thickness and α is a nondimensional constant on the order of 25 to 50.

Figure 2 shows the load deflection response of three 25 mm by 41 mm specimens of A710 Grade A steel with crack depths of 0.07 a/W, 0.2 a/W and 0.51 a/W. Load-displacement records were analyzed by normalization methods (12) to develop the corresponding resistance curves. Center thickness J_c was determined using near field direct current potential difference techniques. Figure 3 shows the resistance curves for the 0.2 a/W and 0.51 a/W tests indicating only a slight elevation in the J - Δa relationship for the shorter crack. For the 0.07 a/W crack depth, a resistance curve could not be obtained using normalization owing to lack of plastic containment in the remaining ligament; however, applied J at initiation was obtained from the potential drop signal and the load displacement record. Mid plane initiation J values for the 0.07 a/W, 0.2 a/W, and 0.51 a/W were 662 kJ/m^2, 347 kJ/m^2, and 215 kJ/m^2 respectively. Crack front profiles for the specimens tested are shown in Figure 4. While the 0.07 a/W and 0.5 a/W specimens maintained relatively straight crack fronts, the 0.2 a/W specimen experienced considerable tunneling in the center of the specimen owing to the lack of through thickness constraint. The tension surface of the 0.07 a/W specimen revealed visual evidence of plastic deformation which will be characterised further in the next section.

Strain to Fracture

In this section we examine experimentally and analytically the strain developed around the crack tip and throughout the specimen as a function of crack size. In the short crack case, which we define as crack sizes for which yielding breaks through to the top surface, strain exhaustion at the crack tip (strain to fracture) should best define the point of tearing.

Surface Plastic Strain Measurements

To accurately quantify the effect of notch depth on the development of surface plastic strain, FAMOSS, a system developed at DREA (13), was employed. Small circles with an original diameter of 2.5 mm were electro-etched on the surfaces of the 0.07 a/W and 0.2 a/W specimens. The samples were loaded to a point where crack extension was visually apparent. The deformed samples with distorted circles are shown in Figures 5a and 5b. The dimensions of each distorted circle were measured using the computer controlled, video based system and the maximum principal plastic strains calculated. Figures 6a and 6b present the resultant strain contours for one quadrant of the samples for 0.2 a/W and 0.07 a/W, respectively.

As can be seen in Figure 6b, the onset of crack advance for the 0.07 a/W specimen, is preceded by extensive plastic flow of the material on the tension surface of the sample. Measured strain levels on this surface reached a maximum of 16 percent at a distance of 1.6 cm from the edge of the notch. In the 0.2 a/W specimen no plastic deformation on the tension surface was detected.

On the side of the 0.2 a/W specimen, the zone of plastic strain exceeding 4 percent is continuous from the crack tip to the bottom surface, whereas for the 0.07 a/W sample, there are two separate regions of plastic strain exceeding 4 percent. The region of plastic strain exceeding 13 percent is approximately three times as extensive on the 0.07 a/W specimen surface compared to the 0.2 a/W sample.

Finite Element Results

The influence of a/W on plastic strain pattern development was numerically investigated using the finite element program ABAQUS (14). Plastic strain patterns were determined for two crack depths to provide results for comparison with the experimentally measured surface strains. Single edge notch specimens under three point bending, 0.1 a/W and 0.5 a/W, were modelled in a plane strain two dimensional nonlinear analysis. Collapsed eight noded quarter point elements were employed around the crack tips, while eight and six noded quadratic isoparametric were employed for the rest of the model. With the exception of the crack tip elements, reduced integration was employed to enhance the efficiency of the calculations. In both analyses, the level of mesh refinement around the crack tip region was similar. The level of mesh refinement between this crack tip region and the free surfaces was higher for the 0.5 a/W case. This was done in order to accurately model the development and extension of the plastic zone to the point at which this zone interacts with the free surface. Earlier three dimensional analysis results showed good correlation with two dimensional plane strain results for similar geometries. Since the aim of this current numerical effort was to define trends in plastic zone development as a function of crack depth, it was decided that a two dimensional analysis was sufficient.

For these elastic plastic analyses, the equivalent plastic strain variable, which defines the total accumulated strain for the isotropic hardening plasticity assumption, was used to define the plastic zone. Figures 7a and 7b present a summary of the numerical results to date for the of 0.5 a/W and 0.1 a/W models, respectively. The darkened regions correspond to the zones in which the equivalent plastic strain exceeds 0.2 percent. Figure 8 shows the calculated load deflection results for the two models along with a reproduction of the curves in Figure 2. The tick marks on the curves denote the J_c points as determined by potential drop. For the two models, the extent of the plastic zone development is strikingly different. As was noted in the surface set strain measurement work described above, the top surface of the shallow notched sample experiences extensive plasticity, whereas for the deeper notched case, the plastic zone is confined to the region below the crack tip.

Plastic Zone Size and Diffraction Techniques

In an effort to devise experimental procedures to characterize plastic zone sizes in test samples as well as in structural applications, DREA has attempted to extend well established diffraction technologies for measuring elastic residual strain to include a means of quantifying plastic strains around cracks. X-ray diffraction techniques are being applied to determine plastic strain on the surfaces of specimens and neutron diffraction to characterize the interior plastic strains ahead of a crack. A special three point loading apparatus, as shown in Figure 9 was constructed which permits controlled loading of standard single edge notch bend specimens. At the present time (March'92), neutron diffraction measurements were being conducted by Atomic Energy of Canada on specimens with a/W's of 0.1 and 0.5, to study the influence of crack depth on plastic zone development as a function of the applied deflection.

Stretch Zone

Determination of the initiation fracture toughness by stretch zone width (SZW) measurement is well known and accepted, at least in principle (15). It involves determining the pre-tearing relationship between J and SZW and then measuring the stretch zone width on a specimen which exhibits stable tearing. Since this SZW is on the pre-tearing line determined in the first step, the value of J at the initiation of tearing can be simply calculated. Apriori understanding is that the pre-tearing relationship between SZW and J is roughly linear and is determined by the stress strain behavior of the material (16,17). As well, the critical stretch zone width appears to be nearly independent of constraint changes produced by changing specimen size and type (18).

The relationship between the pre-tearing line determined by stretch zone width and the blunting line defined in resistance curve terminology comes from the following equation.

$$\Delta a = \Delta a_e + SZW - L_{sz} \qquad [4]$$

where Δa_e is the elastic component of blunting and L_{sz} is the original length of stretched surface. Because Δa_e and L_{sz} are small (on the order of 20 μ and 70 μ for a 20 mm crack), the nominal assumption that $\Delta a = SZW$ has resulted in a reasonable first approximation. Clearly, however, more work is needed to develop a better formulation for the relationship between J (or perhaps only the plastic component of J) and the SZW at a point.

Since data showing how pre-crack depth affects stretch zone were lacking and are relevant to the current discussion, we measured the critical stretch zone widths formed on seven prefatigued samples between 0.08 a/W and 0.95 a/W. These specimens were dynamically loaded at room temperature in a drop weight test machine using an impact velocity of 2.4 m/s. After testing, and the specimens were fatigued apart, the pre-crack depth was measured according to the standard 9 point averaging scheme (19) and the stretch zones were measured using techniques developed at DREA (20). Briefly, this technique involves: recording stereo-pairs of scanning electron microscope images of the blunted fatigue crack at 3/8, 4/8 and 5/8 B; examining three dimensional re-creations of the crack tip with a stereoviewer to identify the limits of the stretch zone - the points where yielding and tearing initiate; and measuring the SZW at 10 equally spaced points on each set of images for a total of 30 measurements per specimen.

Examples of the results of the stretch zones measured are shown in Figures 10a, 10b and 10c. They include stretch zones from specimens with crack depths of 3.5 mm corresponding to 0.08 a/W, 37.7 mm corresponding to 0.95 a/W, and 17.1 mm corresponding to 0.43 a/W. The similarity in SZW for specimens of different crack depths illustrated by these extreme specimens was exhibited by all the stretch zones examined, as shown in Figure 11. The data points shown in Figure 11 are the average of the 30 measurement points and the error bars are the standard deviation of the measurement points used in the average. Particularly for the shallowest and deepest cracks, the uncertainty in determining where the stretch zone started and ended also approached the size of the error bars shown. Thus, if there is a change in stretch zone width with crack depth, it is smaller than the measurement errors.

DISCUSSION

The tunnelling present on the 0.2 a/W fracture surface, Figure 4, indicates a loss of through thickness constraint in this specimen. The absence of significant tunneling on the 0.07 a/W specimen would indicate that the major loss of constraint in this specimen is from the tension surface. This conclusion is supported by the inclined sides and the elevated J_c value calculated for this crack depth.

The strain based results in Figures 6 through 8 show that plastic zone development takes on two distinct patterns, one for deep cracks and one for short cracks. The deep crack plastic zone development is always toward the back surface while the short crack pattern is toward the top or tension surface. This latter strain development is referred to as a pattern because it is not characteristic of normal plastic zone development. When the plastic pattern breaks to the top surface, the surface yields and two hinges develop, one on either side of the crack. In the limit of very short cracks the majority of deformation is at these two hinges and the deformation in the specimen almost becomes blind to the crack. Tearing at the crack tip finally occurs at strain exhaustion.

In examining the stretch zone width results an important feature of fracture is seen. The stretch zone width is the same for the short crack case, the elastic plastic case, and final shear of the specimen. This suggests that the stretch zone should be evaluated as a universal fracture criterion and it also suggests that conditions at the crack tip or slightly ahead of the crack tip may exert an influence over a longer range than previously accepted. Alternatively the stretch zone may simply reflect the natural development of the plastic zone ahead of the crack tip governed by constraint and the basic laws of mechanics (such as equations of equilibrium, constitutive relations and conservation of volume). Regardless of this, there is a clear variation of stretch zone with transition behavior (20), and stretch zone provides a bench mark for advanced finite element solutions.

The application of J based fracture to short crack fracture seems unwarranted, based on the role of strain and the loss of J dominance, as dramatically characterized by the plastic strain pattern breaking to the top surface. J based fracture may appear to be overly conservative but only if deep cracks never exist. Such questions as "do we assume large cracks are present or small ones", can only be assured by NDT and a rationale of fabrication practice. When short cracks are proven to be the only relevant crack depths to a given structure, strain to fracture preceded by limit load failure of the structure would be suitable for predicting tearing. Finally, it is suggested that the point at which the plastic zone breaks to the top surface be used to define the region between true short crack (strain dominated) fracture and deep crack fracture (J - dominant).

CONCLUSIONS

Plastic zone development for three point bend specimens exhibits two distinct patterns, one for deep cracks and one for short cracks. Deep crack development is always toward the back surface while the short crack pattern is toward the top or tension surface.

For true short crack fracture, tearing at the crack tip occurs at strain exhaustion.

Stretch zone width measurements between 0.08 a/W and 0.95 a/W have shown that there is no observable difference in the stretch zone width between deep and shallow notched specimens.

Research in the future must focus on identifying the true links between the various degrees of constraint and plasticity. Research is also recommended on the direct application of finite element codes to the prediction of fracture response.

ACKNOWLEDGEMENTS

Special thanks to Dr Nader Zamani of the University of Windsor for the 2D finite element results with ABAQUS. Thanks to technologists Gary Dease, Dan Morehouse and Bob Armstrong and co-op student Paul Bishop.

REFERENCES

(1) L.N. Gifford, J.R. Carlberg, A.J. Wiggs and J.B. Sickles, "Explosive Testing of Full Thickness Precracked Weldments", DTRC-SSDD-88-172-42, May 1988.

(2) A. Al-Ani and J.W. Hancock, in Proceedings of the 7th International Conference for Fracture, Houston, Texas, March, 1989.

(3) R. H. Dodds, Jr., T.L. Anderson and M. T. Kirk, "A Framework to Correlate a/W Ratio Effects on Elastic-Plastic Fracture Toughness J_c", Inter J of Fracture, Vol 48, 1991.

(4) C.V. Shih, N.P. O'Dowd and M.T. Kirk, "A Framework for Quantifying Crack Tip Constraint", presented at ASTM Symposium on Constraint Effects in Fracture, Indianapolis, Indiana, USA, May 8-9, 1991.

(5) J.D.G. Sumpter, "An Experimental Investigation of the T Stress Approach", ibid., 1991.

(6) V.G. DeGiorgi, "Prediction of Fracture Initiation in Surface Flaws", CANCAM 91, Winnipeg, Manitoba, Canada, June 1991.

(7) "Guidance on some methods for the derivation of acceptance levels for defects in fusion welded joints", British Standards Institution, PD6493, 1980.

(8) C.E. Turner, "Further Development of a J based Design Curve and its Relationship to Other Procedures", ASTM STP 803, Philadelphia, 1981.

(9) J.R. Rice, in Journal of Applied Mechanics, Series E of the Transactions of the ASME, Vol 90, June 1968.

(10) J.R. Rice, "Mathematical Analysis in the Mechanics of Fracture:, FRACTURE, H. Liebowitz ed., Vol. II, Academic Press, NY, 1968.

(11) J.A. Joyce and J.P. Gudas, " Computer Interactive J_{Ic} Testing of Navy Alloys", Elastic-Plastic Fracture, ASTM STP 668, J.D. Landes, J.A.Begley and G.A. Clarke, eds., ASTM, 1979.

(12) J.D. Landes, Z. Zhou, K. Lee and R. Herrera, "Normalization Method for Developing J - R Curves with the LMN Function", Journal of Testing and Evaluation, Vol 19, July 1991.

(13) J.F. Porter and J.R. Matthews, "The Role of Explosion Bulge Trials in Submarine Materials Research at DREA" , TTCP Workshop on Approval of Materials for Use in Submarine Pressure Hulls, DTRC, Annapolis, Maryland, May 1991.

(14) Hibbitt, Karlsson and Sorensen, Inc., "ABAQUS Theory Manuel", and "ABAQUS User's Manual", Hibbitt, Karlsson and Sorensen, Inc., Providence, RI, 1987.

(15) A. Halim, W. Dahl, and K.E. Hagendorn, *Engng Fracture Mech.* 31, 857-866 (1988).

(16) J. Heerens, K.H. Schwalbe, and A. Cornec., "Modifications of ASTM E 813-81 Standard Test Method for an Improved Definition of J_{IC} Using New Blunting-Line Equation," in Fracture Mechanics: 18th Symposium, ASTM STP 945, D.T. Read and R.P. Reed eds., 1988.

(17) O. Kolendnik and H.P. Stüwe, "A Proposal for Estimating the Slope of the Blunting Line." Inter J of Fracture, 33, 1985.

(18) G.P. Gibson and S.G. Druce, "Some Observations on J-R Curves", ASTM STP 856, E.T. Wessel and F.J. Loss, eds., ASTM, 1985.

(19) ASTM standard E813-1987, Annual Book of ASTM Standards, 3.01, ASTM, 1987.

(20) C.V. Hyatt and J.R. Matthews, "The Use of Stereoscopic Imaging in a Scanning Electron Microscope for Stretch Zone Width Determination on High Strain Rate Fracture Mechanics Samples of ASTM A710 Grade A Steel", in proceedings of *Canadian Fracture Conference 21*, J.R. Matthews, ed., Defence Research Establishment Atlantic, April 1990.

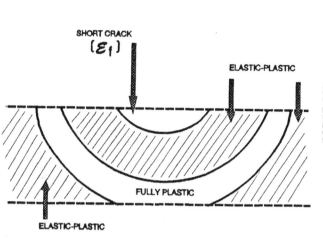

Figure 1. Schematic representation of zones of applicable failure conditions.

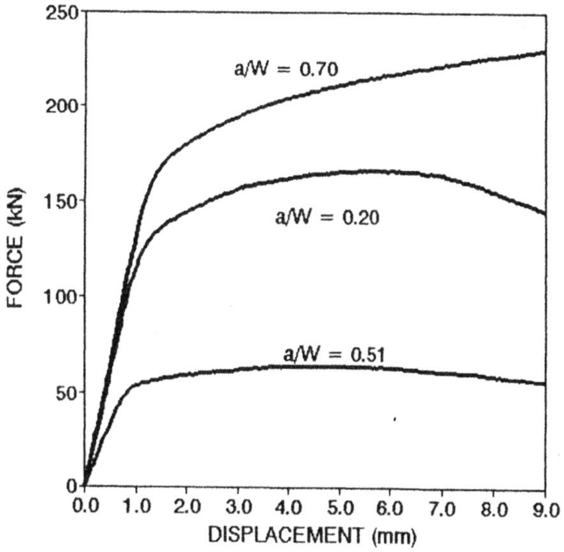

Figure 2. Load Deflection

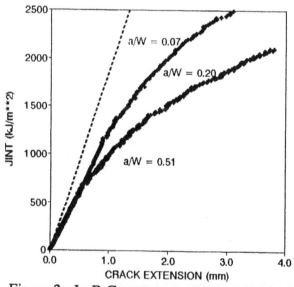

Figure 3. J - R Curves

Figure 4. Fracture surfaces.

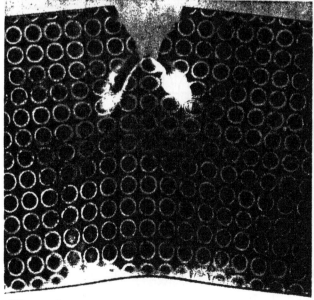

Figure 5a. Photograph of deformed specimen, a/W = 0.2

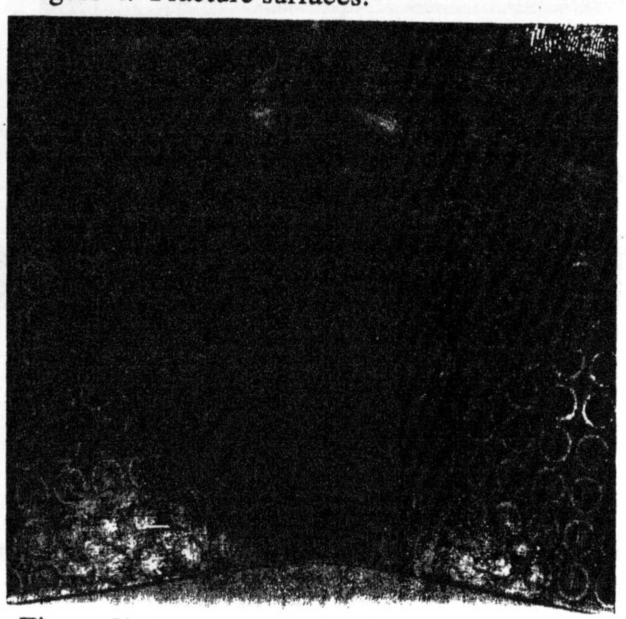

Figure 5b. Photograph of deformed specimen, a/W = 0.07

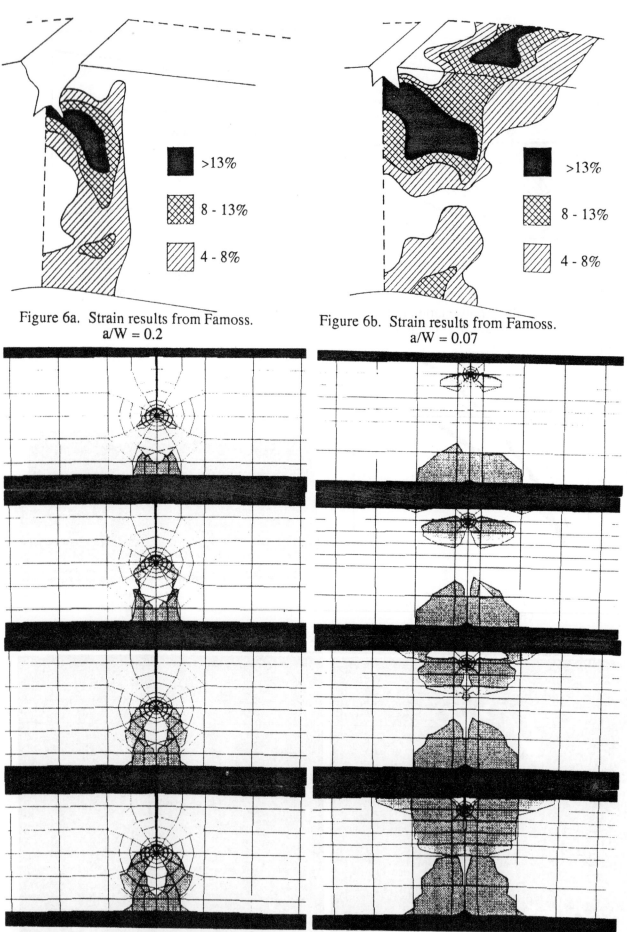

Figure 6a. Strain results from Famoss.
a/W = 0.2

Figure 6b. Strain results from Famoss.
a/W = 0.07

Figure 7a. Finite element results for a/W of 0.5

Figure 7b. Finite element results for a/W of 0.1

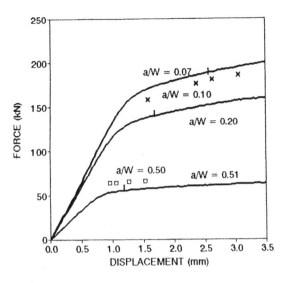

Figure 8. Analytical and Experimental load deflection results.

Figure 9. Special loading jig for neutron diffraction studies

Figure 10a. Stretch Zone (dashed line) for pre-crack depth of 3.5 mm.

Figure 10b. Stretch Zone (dashed line) for pre-crack depth of 37.7 mm.

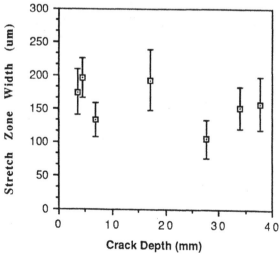

Figure 10c. Stretch Zone (dashed line) for pre-crack depth of 17.1 mm.

Figure 11. Stretch zone width as a function of crack depth.

A crack stability approach to inelastic dynamic shallow-crack fracture

L N Gifford, BSc, MSc (Naval Surface Warfare Center, USA), J W Dally, BSc, MSc, PhD (University of Maryland) and
A J Wiggs, BSc, MSc (Naval Surface Warfare Center, USA)

SUMMARY

This paper describes a simple engineering approach for establishing the shallow
crack depth limit at which the dynamic failure mode of a structure transitions
from premature fracture to plastic exhaustion of the remaining ligament. A
crack within this limit is said to be "stable", which (by definition) means that
the structure can reach its limit state (allowing for loss of cross-sectional
area due to the crack) without fracture occurring first. Therefore, a stable
crack cannot cause premature fracture. For stable cracks, a more global limit
analysis, rather than nonlinear crack tip analysis, may be more suitable for
evaluating energy-absorbing capacity and margin of safety of a structure.

The crack stability limit in a structure depends on inherent material
properties, the mode of loading (relative amount of tension vs. bending), the
structural thickness, and crack-structure interaction, i.e. load re-distribution
capacity at the crack site due to global plastic deformation. This paper
addresses experimental and analytical steps necessary to estimate the stability
of a shallow structural crack. The concept is useful for defining acceptable
defects in welded structures which must resist dynamic (explosive) loading to
high inelastic strain level. The approach is especially suitable for situations
in which the load severity cannot be accurately anticipated, since the response
of a crack (stable or unstable) is not a function of load amplitude.

INTRODUCTION

In critical welded naval structures upon which life and mission depend, tough
materials are employed and known weld defects are normally repaired. Small weld
flaws (assumed to be planar cracks) will nonetheless be inadvertently
introduced, or may develop later due to fatigue. Safety assurance requires that
nondestructive evaluation reveal all flaws which are potentially dangerous,
while in the interests of economy, false alarms should not be raised at flaws
which can be safely ignored. The fundamental question is how to distinguish
between these two cases, thereby allowing an engineer to assure that fracture
will not cause premature structural failure. This problem is compounded for
ship structures which may be subject to unusual plastic overloads (explosions,
collisions, etc.), for which the load amplitude cannot be readily quantified in
advance.

Accordingly, the concern here is with fracture under the following conditions:
1. The material is tough and cracks are shallow (a/W < 0.2). The situation
   is far beyond the current valid limits of nonlinear fracture mechanics.
2. Transient loading is severe, sufficient to induce net section yielding
   at the crack site, but is otherwise unquantified.
How can one establish that a crack is "safe" under these conditions?

This question was initially approached experimentally, using explosive loading
against simple (but redundant, structure-like) full-thickness "structural
elements" designed to simulate shallow crack fracture response in welded
structures. The system developed to explosively load structural elements in

bending is shown schematically in Fig. 1. Two sizes of structural element can be used, the smaller to test plate thicknesses in the 19 to 38 mm (0.75 to 1.5 in.) range and the larger to test plate thicknesses in the 38 to 64 mm (1.5 to 2.5 in.) range. The use of a water column (as opposed to air) as a shock transmission medium allows high inelastic strain to be achieved using small sheet charges (on the order of 500 g or 1 lb.) at a 203 mm (8 in.) standoff. Tests at minimum service temperature are achieved by pre-chilling the structural element in an electric freezer. Details of this test system have been described by Gifford et al. in an earlier publication (1). Typical strain versus time histories (and corresponding charge weights) for uncracked HY-100 calibration elements (with yield strength around 760 Mpa or 110 ksi) are shown in Fig. 2.

Exploratory tests of 32 shallow-cracked 51 mm (2 in.) thick HY-80 and HY-100 GMA butt weldments loaded past net section yield (nominal uncracked geometry peak strain of about 2.5 percent) have been described by Gifford and Dally (2). These tests revealed a rather curious crack growth response. This is illustrated in Fig. 3, where the amount of crack growth is plotted against the initial crack depth. It is obvious that without exception, crack growth was either severe, in excess of 25 mm (1 in.), or else very small, less than 5 mm (0.2 in.). In no case did crack extension fall between these extremes.

Scatter in Fig. 3 is evidenced by the fact that there is a wide range of initial crack depths for which crack growth was either zero (or insignificant) or large (25 mm or more). Such scatter is to be expected when testing inherently heterogeneous multi-pass welds, even when fabricated from a single welding consumable. In effect, many different materials with differing fracture resistance and differing initial crack depth were evaluated. There is additional reason for this apparent scatter which will be explained later.

What was not initially expected of these tests was the all-or-nothing (binary) crack growth response, with no cracks extending to intermediate depths. This behavior obeys an analogy suggested by Rolfe and Barsom (3) between column instability and crack instability. The analogy suggests that as a column (crack) becomes shorter and shorter (shallower and shallower), a point is reached at which the failure transitions from premature column buckling (premature crack extension) to failure by yielding of the column cross-section (plastic exhaustion of the crack ligament). This implies that a crack stability theory might be useful as an engineering method for interpreting (and quantifying) experimental dynamic crack growth data at high inelastic loading level. Subsequent analysis has suggested that a crack stability approach can also be useful as an engineering tool for predicting shallow-crack fracture response in structures dynamically loaded to (and past) net-section yielding. This paper examines these possibilities.

A SIMPLE CRACK STABILITY ANALYSIS

Potential failure by net section yielding

Consider an edge-cracked structure with rectangular cross-section, subjected to both tensile load P and bending moment M as shown in Fig. 4. If the ratio of M to P remains constant (M = kP) as load increases, and if the material is modelled as elastic-perfectly plastic as shown in Fig. 5, then the stress distribution in the crack ligament at the limit state will be as shown in Fig. 4. The moment produced by this stress distribution approximates the limit moment M_L:

$$M_L = B\alpha\,\sigma_{ult}[\frac{(W-a)^2}{4} - e^2] \qquad [1]$$

Here, B is the element width, W is the plate thickness, and a is the crack depth. e (given in terms of k below) is the offset of the neutral axis from the ligament midpoint under combined bending and tension load as shown in Fig. 4 (center). σ_{ult} is the uniaxial limit state stress shown in Fig. 5. The α factor is used to account for the fact that the axial stress shown in Fig. 4 must be large enough to bring the von Mises (effective) stress in the ligament to σ_{ult}. For plane stress, α is unity. For plane strain (an approximation appropriate for the structural element shown in Fig. 1), $\alpha = 1/(1-\nu+\nu^2)^{\frac{1}{2}}$, where ν is Poisson's ratio. When $\nu = 0.3$, $\alpha = 1.125$.

The tensile component of load at the limit state is similarly approximated as

$$P_L = 2B\alpha\sigma_{ult}e \qquad [2]$$

By substituting $M_L = kP_L$ into Eq. [1] and using Eq. [2], the unknown parameter e can be expressed in terms of the known parameter k:

$$e = -k + [k^2 + (W-a)^2/4]^{\frac{1}{2}} \quad \text{(for combined tension and bending)}$$
$$e = 0 \quad \text{(for } k = \infty, \text{ pure bending)} \qquad [3]$$
$$e = (W-a)/2 \quad \text{(for } k = 0, \text{ pure tension)}$$

Equations [1] through [3] turn out to be the same as the limit load solution given by Merkle and Corten (4) for a cracked compact tension specimen loaded in tension as well as in bending.

Potential failure by fracture

A valid crack tip fracture criterion which characterises fracture propensity over a range of material response-- from elastic to fully plastic-- does not exist. Fortunately, such a criterion is unnecessary for the development of crack stability theory. As will be shown subsequently, the theory leads to a crack stability measure "F" which is an integrated (not separable) measure of the ratio of the material fracture resistance (however quantified) to the material flow stress (quantified herein in terms of σ_{ult}). This ratio is measured experimentally for a given material and thickness. It will become apparent later that in this process, crack tip constraint conditions appropriate for this measure are automatically achieved.

To describe the basic idea in the simplest way, the (mode I) stress intensity factor K_I can be used to quantify the crack tip "loading". This is a matter of convenience; any other crack tip characterising parameter will lead to a similar (but more complicated) analysis. The use of K_I does not imply the assumption that the elastic crack tip stress field governs, or even exists. It does imply that for analysis purposes, a measure of fracture propensity which is directly proportional to the load and to the square root of the crack depth is assumed.

Under the combined loading shown at the center of Fig.4, K_I is usually expressed in terms of applied bending and tensile stress components[*] σ_b and σ_t:

$$K_I = \sqrt{\pi a} \ [\ \sigma_b f_b(a/W) + \sigma_t f_t(a/W) \ ] \qquad [4]$$

where $f_b(a/W)$ and $f_t(a/W)$ are crack "shape functions" for bending and tension as

---

[*] In elastic fracture mechanics, the applied stress components (acting on the remote thickness W) are assumed to have an elastic distribution. At high load, this may not be so. For pure bending and $a/W \leq 0.1835$, for example, the remote stress distribution will be elastic-plastic at the ligament limit state. The K_I solution is unaffected, however, since it depends linearly on M and P.

given, for example, by Tada (5). The nominal (uncracked geometry) applied stresses used in this formula are related to the loading M and P by:

$$\sigma_b = 6M/BW^2 \text{ and } \sigma_t = P/BW$$

Using this and $P = M/k$, Eq. [4] can be written in terms of the applied moment:

$$K_I = \frac{M\sqrt{\pi a}}{B\,W}\left[\frac{6f_b(a/W)}{W} + \frac{f_t(a/W)}{k}\right] \qquad [5]$$

Since the applied moment reaches its critical value M_c when K_I reaches its critical value K_c, Eq. [5] can be solved for the cracking moment M_c:

$$M_c = \frac{K_c BW}{\sqrt{\pi a}}\;\frac{1}{\dfrac{6f_b(a/W)}{W} + \dfrac{f_t(a/W)}{k}} \qquad [6]$$

It is important to reiterate that the use of K_c here is conceptual, for analysis purposes only. It is in no way intended to represent a "material property". It is merely the supposed value of K_I at fracture, and it thus would depend on structural geometry and crack tip constraint as well as material. As will be shown below, an explicit value for K_c is neither determined nor used in crack stability theory.

Transition from stable to unstable crack response

Depending on material, crack depth, and mode of loading, either Eq. [1] (for a stable crack) or Eq. [6] (for an unstable crack) will govern the response of the crack ligament in the structure of Fig. 4**. This is illustrated schematically in Fig. 6 for the case of pure bending. At shallow crack depth, M_L is less than M_c, so the crack is "stable" (M_L can be reached before fracture). At deeper crack depth, the opposite is true and the crack is unstable. The transition crack depth (from stable to unstable response) shown in Fig. 6 depends on the values chosen for K_c and σ_{ult}. Such a transition point theoretically exists for all materials, since the limit moment is bounded whereas the moment to initiate cracking tends to infinity as the crack depth tends toward zero.

At the transition from stable to unstable crack response, $M_L = M_c$. Equating Eqs. [1] and [6] and solving for K_c/σ_{ult} gives the material requirement for crack stability in terms of crack depth, loading mode, and plate thickness:

$$F \equiv \left(\frac{K_c}{\sigma_{ult}}\right) = \alpha\sqrt{\pi a}\left[\frac{(W-a)^2}{4} - e^2\right]\left[\frac{6f_b(a/W)}{W^2} + \frac{f_t(a/W)}{kW}\right] \qquad [7]$$

The parameter defined in Eq. [7] as F is a measure of the material fracture resistance necessary for a crack of given depth to respond in a stable manner, i.e. $M_L < M_c$. F should not be taken literally as the value one would obtain after conducting independent tests for K_c and σ_{ult}. It will be shown later that F for material/thickness combinations (F_{MAT}) can be measured directly and used to estimate stable or unstable crack response in structures.

_____

** The special case for pure tension in terms of P_L is omitted for brevity.

If Eq. [7] is used to plot F (required for crack stability) vs. crack depth, the resultant "crack stability curves" are material-independent and influenced only by plate thickness and the ratio of tensile to bending load. The importance of material thickness is clear from the cases shown in Fig. 7 for pure bending. For example, the F value required for a 5 mm (0.2 in.) deep crack to remain stable is much higher for 50 mm (2 in.) thick plate (where a/W = 0.1) than for 25 mm (1 in.) thick plate (where a/W = 0.2). Note also that for a given F value less than the unconditionally stable limit, there are two crack depths which are stable, one shallow and one deep. The deeper stable crack depth is not of interest here, but may be of future utility in interpreting results of crack arrest tests.

Figure 8 shows the entirely different crack stability curves which result for the special case of pure tension. The effect of material thickness is not as pronounced, and for this mode of loading there is only one stable crack depth for given F. For shallow cracks, the required value of F for stability is less in tension than in bending (compare Figs. 7 and 8), while the reverse is true at deeper crack depths. Unlike the case for bending, there is no theoretical value of F for tensile loading for which all crack depths are stable.

Figure 9 shows cases for combined tension and bending for a 51 mm (2 in.) plate thickness. Clearly, the mode of loading (ratio of nominal tensile stress to nominal bending stress, σ_t/σ_b) is an extremely important factor influencing the stability of a crack. A relatively small tensile component of applied stress (say σ_t/σ_b = 0.2) superimposed on a bending field can dramatically increase the F value required for stability of a shallow crack. But this is true only up to a point. If the ratio σ_t/σ_b is increased to the point that tension dominates (say σ_t/σ_b = 10 as shown in Fig. 9), the required F value for stability of a shallow crack decreases to less than that required for pure bending. This up-and-down behavior is illustrated in Fig. 10, where the F value required for crack stability is plotted against σ_t/σ_b for several shallow crack depths. As σ_t/σ_b increases from zero, the required F first increases, then decreases. The value of σ_t/σ_b at which the required F becomes a maximum depends on crack depth. Figs. 9 and 10 together suggest interesting scenarios, not at all intuitive, in which the stable crack depth is deeper for (mostly) tension than (mostly) bending, and vice-versa-- all depending on material (i.e. the F value measured experimentally for a material, F_{MAT}).

INTERPRETATION OF STRUCTURAL ELEMENT TESTS

The theoretical crack stability curves for pure bending shown in Fig. 7 can be used to interpret (quantify) the results of explosive tests against shallow-cracked structural elements of the type shown in Fig. 1. This is done by establishing the stable crack depth experimentally, and then using the crack stability curve to read off the F value, F_{MAT}, corresponding to the stable crack depth. The experimental procedure is one of trial and error as illustrated schematically in Fig. 11. Once the stable crack depth is known approximately, statistical analyses can be used to estimate the number of tests necessary to establish F_{MAT} to any required level of confidence. The resulting F_{MAT} is thickness dependent, i.e. in general, one cannot directly use F_{MAT} determined at one thickness to predict the stable crack depth in the same material but with a different thickness.

It is worth repeating that a crack is stable or unstable regardless of the magnitude of the applied load. However, in <u>determining</u> experimentally whether a crack is stable or unstable, it is essential that the explosive loading be severe enough to bring the cracked section to the limit state (in bending), since otherwise a lack of crack extension would not necessarily imply that the crack is stable. In other words, a non-test results unless the load is sufficient to bring the cracked section to its limit state. The tests shown in Fig. 3 were conducted at strain levels well above this requirement, which explains the small amounts of ductile tearing observed for some cracks which nonetheless behaved in a stable manner. The extent of ductile tearing, on the order of 2.5 mm (0.1 in.), is consistent with a tearing velocity of about 2.5 m/s (100 in./s) and a time at load (Fig. 2) of about 1 ms. For comparison, the speed of propagation of unstable cracks in steel is on the order of 500 times as great (6).

While the test load severity must meet the above minimum, it is unnecessary to choose a load severity much greater than the maximum anticipated under worst-case service conditions. One can always load so severely that the test section will rupture—— regardless of whether the crack is stable or not, or even if there is no crack. Short of this extreme, overloading beyond service conditions could produce a needless excess of ductile tearing at stable cracks. This excess tearing could afford opportunity for transition to unstable cleavage (initiating at some brittle inhomogeneity) that would not occur under service conditions, leading to a false "unstable" test result.

It should also be noted that the structural element does not respond precisely in pure bending. A small tensile (membrane) component of loading develops due to finite deflection of the test section. As shown clearly in Fig. 9, even a relatively small component of tensile loading can change the crack depth at which the stable/unstable transition occurs. Fortunately, the inadvertent introduction of a small tensile component of loading will have a conservative effect on the measurement of F_{MAT}, i.e. the apparent value of F_{MAT} will be less than the actual. The structural element crack stability test, interpreted in terms of pure bending, is thus inherently conservative.

A CAVEAT FOR UNDERMATCHING YIELD STRENGTH WELDS

The analysis described above for interpreting crack stability experiments applies to homogeneous materials or to welds which closely match the baseplate in yield strength. For cracks in undermatching yield strength welds, the analysis will not hold if the geometry of the weld is such that the inelastic stress field which would develop in an all-weld-metal structural element is changed by the higher-yield-stress (unyielded) surrounding baseplate (i.e. for contact-strengthening undermatching welds). For such welds, the analysis does not account for two factors: First, the limit load for the contact-strengthening weld will be higher than predicted by Eq. [1] (using the all-weld-metal σ_{ult}). This strength increase is developed from constraint (afforded by the unyielding baseplate) of the strain-hardening phase of plastic weld deformation, and this behavior is not incorporated into the elastic-perfectly plastic material model assumed in the theory. Second, after weld yielding becomes significant, the crack tip constraint in a contact-strengthening undermatching weld will differ from that in a weld of the same material at the same load, but surrounded by matching yield stress baseplate.

Because of these differences, crack stability tests of the type shown schematically in Fig. 11 will measure F_{JOINT} (rather than F_{MAT}) for contact-strengthening welds. F_{JOINT} is defined as

$$F_{JOINT} \equiv (\beta/\gamma) K_c/\sigma_{ult} = (\beta/\gamma) F_{MAT} \qquad [8]$$

where $\beta \equiv M_{L(joint)}/M_{L(all-weld-metal)}$ and $\gamma \equiv M_{c(joint)}/M_{c(all-weld-metal)}$.

It is not straightforward to obtain F_{MAT} from F_{JOINT}. The ratio β can be calculated approximately using finite element analysis with suitable stress-strain properties (and strain-hardening material models) for the weld metal and base plate. The γ ratio is much more difficult to estimate. Without this quantification, the extrapolation of crack stability test results for a contact-strengthening undermatching weld to predict response of an undermatching weld with different geometry cannot be made. By the same token, results for F_{MAT} obtained for the all-weld-metal condition cannot be used to predict the crack stability limit in structural welds which exhibit contact strengthening. In short, for contact-strengthening undermatching welds, the geometry of the weld in the structural element must match the geometry of the weld in the structure if crack stability testing (as currently quantified) is to be applicable.

EXTRAPOLATION TO ACTUAL STRUCTURES

Although F_{MAT} is presently determined using a shallow-cracked structural element explosively loaded in bending, any cracked structure (of the proper thickness) and any mode of tensile/bending load could be used. The resultant differences would be in the shape of the theoretical crack stability curve used to evaluate F_{MAT} from the experimental stable crack depth, and (quite likely) the stable crack depth itself. But F_{MAT} should be the same. This is the basis upon which laboratory data can be extrapolated (using F_{MAT}) to predict crack stability (or instability) in general structures.

The crack stability curve for the structural element is known in closed form as given in Eq. [7]. For general structures, no such closed form solution can be developed. Nevertheless, it is possible to calculate a single point (corresponding to a particular crack depth of interest) on the crack stability curve of any structure using numerical methods. To remain consistent, the same assumptions used to develop the crack stability curve for the structural element are used to treat the structure. To illustrate this, assume that we have some general structure with a generalized load Q. The following two-step approach would then be used:

1. Using elastic (perhaps finite element) analysis, calculate the load Q_C necessary for crack extension in terms of K_c, i.e

$$K_I = \frac{Q}{C_1} \quad which \ gives \quad Q_c = C_1 K_c \quad\quad [9]$$

This equation is analogous to Eq. [6], but evaluated at a specific crack depth. C_1 is a function of the structural geometry, mode of loading (tension/bending), and the crack geometry and location. Because Q_c and K_c are directly proportional, C_1 is a known constant and it is not necessary to know or assume a value for K_c. By using an elastic analysis, the calculated K_I is consistent with that used to develop the crack stability curve for the structural element.

2. Using elastic-plastic (finite element) analysis with an elastic-perfectly plastic material model, calculate the limit load Q_L for the cracked structure in terms of σ_{ult}. This gives

$$Q_L = (a \ number) = C_2 \sigma_{ult} \quad\quad [10]$$

This equation is analogous to Eq. [1] or [2] evaluated at a specific crack depth. C_2 is a function only of structural and crack geometry, and loading mode (tension/bending). If geometric nonlinearities (large deflections) are not significant, Q_L and σ_{ult} are directly proportional. In this event, it is not necessary to know σ_{ult} to calculate C_2.

Now, just as for the simple structural element, the loads Q_c and Q_L must be equal at the transition from stable to unstable crack response. Equating Eqs. [9] and [10] yields:

$$C_1 K_c = C_2 \sigma_{ult}, \quad which\ gives \quad F_{REQ} \equiv (\frac{K_c}{\sigma_{ult}})_{REQ} = \frac{C_2}{C_1} \qquad [11]$$

Since C_1 and C_2 are known, F_{REQ} is known. F_{REQ} is the value of F required for the crack under analysis to be stable. As in the analysis of the simple structural element, it does not depend on material or on the actual load amplitude applied to the structure. If $F_{REQ} \leq F_{MAT}$ (from structural element tests of the actual material at structural thickness), then the crack will be stable. Otherwise, the crack is unstable.

It is of course possible to estimate the entire crack stability curve for any complex structure subject to any given mode of loading. One would simply repeat the process outlined here for a range of discrete crack depths. Such an undertaking can be computationally demanding.

DISCUSSION

The test results illustrated in Fig. 3 serve as experimental confirmation of the stable/unstable response of shallow cracks subjected to severe (beyond limit state) transient overload. The scatter in Fig. 3 (manifested in the fact that there is a wide range of crack depths for which the response was both stable and unstable) is attributable to two factors. First, there is the high local variation in material properties associated with multi-pass welds. Second, the welds tested were all of unusually high quality, leading to high values for F_{MAT}. Referring to Fig. 11, it is evident that for F_{MAT} above about 5 mm$^{1/2}$ (1.1 in.$^{1/2}$), small changes in F_{MAT} lead to relatively larger changes in the crack depth corresponding to the stable/unstable transition. In this sense, crack stability testing as formulated here is not very sensitive in distinguishing good materials from even better materials. Experience with more marginal materials, in baseplate form, is desirable but currently lacking.

In analyzing crack stability, K_c and σ_{ult} were used as parameters governing the load at which cracking or plastic exhaustion of the crack ligament occurs. It is important to re-emphasize that these parameters are neither individually measured nor directly used in crack stability theory. Instead, it is the stable crack depth which is measured experimentally, and this measure is _interpreted_ in terms of the ratio $F_{MAT} \equiv K_c/\sigma_{ult}$. In effect, K_c and σ_{ult} are used only in an engineering sense to assign a numerical value (F_{MAT}) to the crack stability limit. Other parameters could have been used, provided they characterize the stable to unstable transition in crack response (cross-over point) shown schematically in Fig. 6. This implies that consistency in analysis (of both the test element and the structure) is necessary to extrapolate laboratory performance to structural performance. With the choice of K_c as the conceptual fracture-governing parameter, this extrapolation requires only a linear fracture analysis, followed by a limit load analysis which requires no special treatment of the crack tip region. On the negative side, the resultant analysis is not well suited to crack stability testing and extrapolation for undermatching yield strength welds which develop contact strengthening.

The approach suggested here is motivated by the need to prevent premature structural failure due to fracture-- without prior knowledge of the actual dynamic loads to be applied to the structure. The approach allows limiting shallow crack depths to be determined for a structure which theoretically assure that premature crack extension will not be the cause of failure, regardless of load severity. Experimental verification of stable/unstable predictions made in this manner, however, have not yet been obtained. The testing side of the approach requires a dynamic loading history and plate thickness which are representative of structural conditions. This is achieved for many naval structures of interest by means of the simple structural element of Fig. 1. In addition, the experimental loading must be sufficiently severe to bring the cracked test section to at least its limit state. (Otherwise, a non-test results.)

A shallow crack in a structural element (and structure) has been defined as "stable" if the element (structure) can reach its nominal limit state without (premature) fracture occurring first. This does not mean that the structure will not fail, nor does it mean that structures which contain only stable cracks are the best structures. For example, if a given crack depth must be tolerated in a structure, choosing a low-strength material for which that crack depth is safely stable may not optimise load resistance. Another material, with higher yield stress, may be able to safely resist a greater dynamic load, even though the presumed crack is no longer "stable". (Picture the performance of cracked lead vs. cracked steel.) Indeed, adopting a crack stability approach does not mean that one can get away from fracture mechanics. Instead, the approach is just another tool, perhaps most useful as an engineering device for assuring fracture safety once a material system is in place.

Note also that the crack stability approach tells nothing about the maximum deformation that a stable crack can withstand beyond the deformation associated with the limit state. Post-limit behavior of stable cracks is an open question. This means that even a stable crack can serve as a potential initiation site for ductile rupture (as opposed to unstable fracture) at deformation beyond the limit state. This failure mode, however, is more related to fully plastic separation of a notched structure than to fracture. Thus if a crack is known to be stable, there appears to be no need for fracture mechanics analysis, but rather for a more global post-limit analysis of some kind, perhaps a "hydrocode" analysis. For unstable cracks, on the other hand, the principles of nonlinear fracture mechanics can be invoked to estimate an allowable load or deformation level-- which will be less than for the limit state.

It appears that the simple crack stability theory developed here can provide an estimate of the natural envelope within which nonlinear (crack-tip) fracture mechanics is useful. This envelope depends explicitly on crack depth, structure, structural thickness, and loading mode (tension/bending). It also depends implicitly on inherent material properties (measures of flow stress and fracture toughness). For example, regions within the fracture mechanics envelope (for the simple structure shown in Fig. 4) are marked "UNSTABLE" in Figs. 7 through 9, i.e. fracture analysis is necessary. Outside these regions (i.e. for "STABLE" cracks), the structure of Fig. 4 can reach its fully plastic limit state without premature fracture, and is beyond the useful range of conventional nonlinear fracture analysis.

REFERENCES

(1) Gifford, L.N., J.R. Carlberg, A.J. Wiggs, and J.B. Sickles, "Explosive Testing of Full Thickness Precracked Weldments", in Fracture Mechanics: Twenty-First Symposium, ASTM STP 1024, J.P. Gudas and E.M. Hackett, Eds., American Society for Testing and Materials, Philadelphia, 1990, pp. 157-177.

(2) Gifford, L.N. and J.W. Dally, "Dynamic Fracture Resistance of Metal Structures Loaded into the Plastic Regime", in Advances in Marine Structures, C.S. Smith and R.S. Row, Eds., Elsevier Applied Science, 1991, pp. 23-41.

(3) Rolfe, S.T. and J.M. Barsom, Fracture and Fatigue Control in Structures, Prentice-Hall, Inc., Englewood Cliffs, 1977, pp. 19-20.

(4) Merkle, J.G. and H.T. Corten, "A J-Integral Analysis for the Compact Specimen, Considering Axial Force as well as Bending Effects", Trans. ASME, J. of Pressure Vessel Technology, Vol. 96, 1974, pp.286-292.

(5)   Tada, H., <u>The Stress Analysis of Cracks Handbook</u>, Del Research Corporation, Hellertown, Pa., 1973.

(6)   Kanninen, M.F. and C.H. Popelar, <u>Advanced Fracture Mechanics</u>, Oxford University Press, New York, 1985, pp. 196-202.

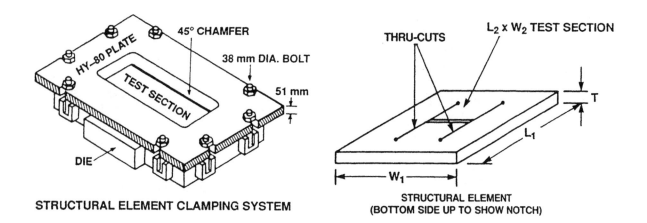

STRUCTURAL ELEMENT CLAMPING SYSTEM

STRUCTURAL ELEMENT
(BOTTOM SIDE UP TO SHOW NOTCH)

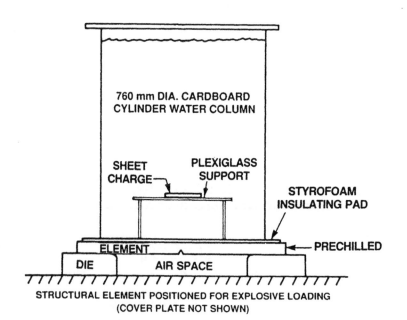

STRUCTURAL ELEMENT POSITIONED FOR EXPLOSIVE LOADING
(COVER PLATE NOT SHOWN)

	STRUCTURAL ELEMENT DIMENSIONS, mm (in.)	
	SMALL	LARGE
T	19 TO 38 (0.75 TO 1.5)	38 TO 64 (1.5 TO 2.5)
L_1	660 (26)	965 (38)
L_2	406 (16)	610 (24)
W_1	406 (16)	559 (22)
W_2	152 (6)	203 (8)

Figure 1   Schematic of System for Explosive Testing of Shallow-Crack Welds

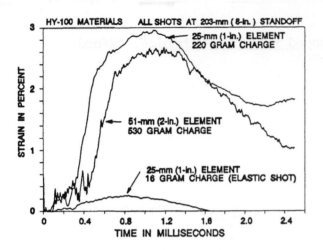

Figure 2 Typical Strain-Time Histories at Center of Explosively Loaded (Uncracked) Structural Elements

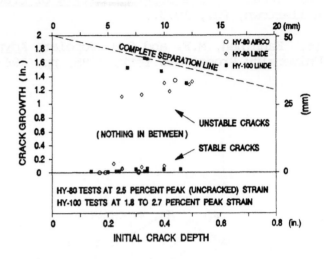

Figure 3 Crack Growth Results for 51-mm (2-in.) Thick GMA Butt Welds

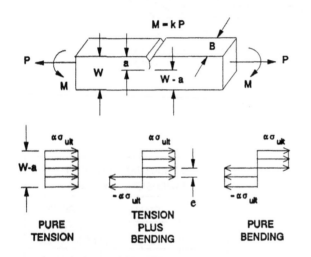

Figure 4 Possible Stress Distributions in the Crack Ligament at the Limit State

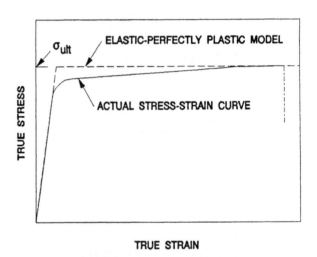

Figure 5 Elastic-Perfectly Plastic Model of the Stress-Strain Curve

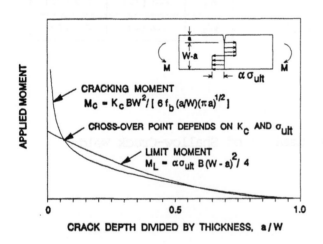

Figure 6 Transition in Pure Bending from Premature Cracking to Limit Response

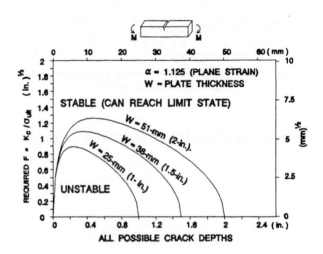

Figure 7 Crack Stability Curves for Pure Bending Showing Thickness Effect

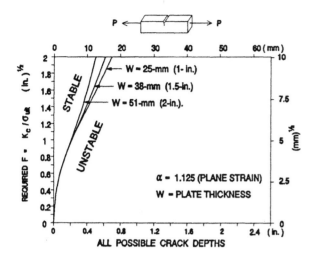

Figure 8 Crack Stability Curves for Pure Tension

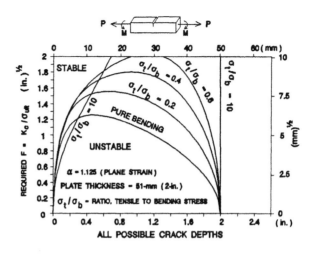

Figure 9 Crack Stability Curves for Combined Bending and Tension

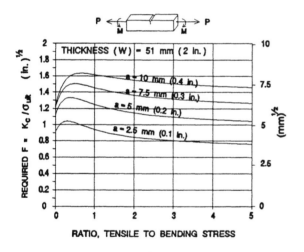

Figure 10 Effect of Tensile Stress to Bending Stress Ratio on F Required for Crack Stability

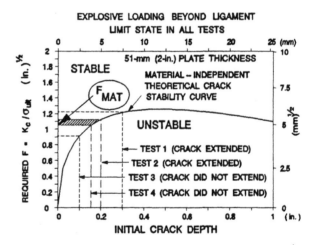

Figure 11 Schematic of Trial and Error Structural Element Tests for F_{MAT}

SESSION B: Fracture toughness

Shallow crack test methods for the determination of K_{Ic}, CTOD and J fracture toughness (Paper 15)
M G DAWES, G SLATER, J R GORDON and T H McGAUGHY

Shallow surface notch — experience from bead in groove testing (Paper 19)
C THAULOW and A J PAAUW

CTOD toughness evaluation of hyperbaric repair welds with shallow and deep notched specimens (Paper 20)
I RAK, M KOÇAK, M GOLESORKH and J HEERENS

Shallow-crack toughness results for reactor pressure vessel steel* (Paper 37)
T J THEISS, S T ROLFE and D K M SHUM

TWI

PAPER 15

Shallow crack test methods for the determination of K_{Ic}, CTOD and J fracture toughness

M G Dawes, PhD, CEng, FWeldI and G Slater, MA, AMIMechE, SenMWeldI (TWI, UK),
J R Gordon, BSc, PhD, CEng, SenMWeldI and T H McGaughy, BSc, ME (EWI, USA)

SUMMARY

The paper describes the development of shallow crack fracture mechanics toughness test specimens, appropriate fatigue precracking conditions, and instrumentation for measurement of crack mouth opening and load–line displacements. These are complementary to those in the standard deep notch fracture mechanics test methods for the determination of fracture toughness in terms of K_{Ic}, CTOD and J. A wide range of single edge notched bend, single edge notched tension and single edge notched arc bend specimen geometries were investigated. These contained crack length to specimen width ratios ranging from 0.5 to 0.05, and absolute crack lengths as small as 1.25mm.

INTRODUCTION

There are many structural situations involving shallow surface cracks, for which it is not possible to make direct measurements of the relevant fracture toughness using existing British Standards Institution (BSI) and American Society for Testing and Materials (ASTM) deep notch (a/W >0.45) fracture mechanics tests (1–6). These situations often involve cracks of the order of 2mm in depth in regions where mechanical properties vary considerably with position. Although these regions are commonly associated with welded joints, they can also be associated with shallow cracks in flame cut edges, coated, clad, locally heat treated or corroded materials, and sometimes even cast, forged and wrought materials.

When a shallow surface crack is sited in a region containing a brittle microstructure, the true fracture toughness may be **dangerously overestimated** by a standard test specimen having a relatively deep notch in a more ductile region. At the opposite extreme, a standard deeply notched test specimen may **grossly underestimate** the true (higher) fracture toughness associated with a shallow surface crack in a similar or relatively ductile microstructure.

The above situation, and its serious implications for structural integrity, safety assessments and the requirements for non–destructive testing and repair, led to a joint TWI/Edison Welding Institute (EWI) proposal (7) to develop appropriate shallow crack fracture mechanics tests.

The major objective of the project was to develop shallow–crack test methods to cover:

* Bending and tension loading
* Testing of both flat and curved (e.g. pipe) material
* Characterisation of fracture toughness in terms of K, J and CTOD.

The project, which started on 1st October 1987, involved the development of test specimens and testing procedures, numerical analyses for the determination of CTOD and J estimation formulae, and fracture toughness tests on four very different steels over a wide range of temperatures, i.e. to cover a full range of ductile and brittle fracture behaviour.

This paper summarises the development of shallow crack fracture mechanics test specimens, appropriate fatigue precracking procedures, instrumentation, and experience gained in tests at temperatures ranging from 25 to −160°C.

Another paper at this conference describes the numerical work (8), and a future paper will describe the results of the fracture tests.

MATERIALS

Two European steels were used for the tests at TWI, and two North American steels were used for the tests at EWI. These were designated M1 to M4 as follows:

M1 – BS4360:1986 Grade 50EE normalised steel plate in 25mm thickness (European)
M2 – AISI 4145H quenched and tempered drill pipe casing steel, 127mm outside diameter x 25mm wall thickness (European)
M3 – ASTM A36–84a steel in 25mm thickness (North American)
M4 – API 5CT L80 drill pipe steel, 270mm outside diameter x 21mm wall thickness (North American)

The average tensile properties for the four steels are summarised in Table 1.

Table 1. Tensile test results (tests at 20 to 25°C)

Steel	Yield strength σ_{YS} N/mm^2	Tensile strength σ_{TS} N/mm^2	Elongation, %	Reduction of area %
M1	388	530	28	74
M2	850	980	16	34
M3	283	491	44	68
M4	590	705	–	–

TEST SPECIMENS

Three specimen geometries and modes of loading were selected for the development of shallow crack test methods. These are shown in the figures:

1.   Single edge notched bend (SENB)
2.   Single edge notched tension (SENT)
3.   Single edge notched arc bend (SENAB)

The SENB and SENT specimens were used for the experiments on plate materials M1 and M3, and the SENAB specimens for experiments on the pipe materials M2 and M4.

Throughout this paper the SENB and SENT specimens are described as being through thickness (TT) or surface (S) notched, as illustrated in Fig. 4. Similarly, the SENAB specimens are referred to as containing internal surface radial (ISR) notches, as indicated in Fig. 3.

Table 2 gives the full range of specimen type/material combinations and crack–plane dimensions investigated.

Table 2. Specimen geometries investigated

Specimen type	Material	Notch orientation *	Section B x W	B, mm	a/W	Min. a, mm
SENB	M1	TT	B x B	25	0.05, 0.1, 0.15 0.2, 0.3, 0.5	1.25
	M3	TT	B x B	25	0.05, 0.1, 0.2 0.3, 0.5	1.25
	M1, M3	TT	B x 2B	25	0.5	12.5
	M1, M3	S	B x B	25	0.1, 0.5	2.5
	M1, M3	S	2B x B	25	0.1, 0.5	2.5
	M1, M3	S	3B x B	25	0.1, 0.5	2.5
SENT	M1	TT	B x B	25	0.05, 0.1, 0.15, 0.2, 0.3, 0.5	1.25
	M3	TT	B x B	25	0.05, 0.1, 0.2, 0.3, 0.5	1.25
	M3	S	B x B	18	0.1, 0.5	1.8
	M3	S	2B x B	18	0.1, 0.5	1.8
	M3	S	3B x B	18	0.1, 0.5	1.8
SENAB	M2	ISR	B x B	25	0.1, 0.15, 0.20, 0.3, 0.5	2.5
	M3	ISR	B x B	21	0.1, 0.15, 0.5	2.1

* TT = through thickness, S = surface (see Fig. 4) and ISR = internal surface radial, see Fig. 3.

Further details of the three types of specimen are given below.

SENB specimens

All the SENB specimens were designed for testing in three–point bending using a span, S = 4W, which is consistent with the standard test methods (1–6). However, all except the most deeply notched (a/W = 0.5) specimens had crack depths that were too shallow to meet the requirements of the present test methods (1–6).

SENT specimens

The SENT specimen geometry shown in Fig. 2a was used for all the through–thickness notched specimens, and some of the B x B section surface–notched specimens. However, most of the surface-notched specimens were tested using the geometry in Fig. 2b, which was more convenient for the specimens of greater breadth, i.e. those having the 2B x B and 3B x B Sections (Table 2). None of these SENT specimen geometries are covered in the latest standards for fracture mechanics tests (1–6).

SENAB specimens

The pipe specimen designs in Fig. 3 are similar to those in the latest version of the ASTM E399 test method (3), except that it was necessary to machine flats on the inner radii. As will be apparent later, the flats were necessary for the attachment of the shallow–crack clip gauge instrumentation. The design of the flats was based on the studies of Underwood et al (9), and Dodds et al (10), the latter showing that the machined flats have a negligible effect on the stress intensity functions of the SENAB specimens.

DEVELOPMENT OF FATIGUE PRECRACKING PROCEDURES

Fatigue precracking procedures for 'deep notch' fracture toughness specimens having a/W ratios of about 0.5 are well established and documented in the national test methods (1–6). However, procedures for shallow cracks with a/W ratios as low as 0.05 are not, and it was therefore necessary to develop these for particular application to the specimen geometries covered in this project.

Crack–starter notches

For the deep notch tests (a/W = 0.50), the notches were machined using a standard 3.2mm wide milling cutter with a 60 degree included angle V nose. For specimens with smaller a/W ratios the notches were machined using a 0.15mm wide slitting disc. TWI also investigated the possibility of using a hardened tool steel knife edge having a 60 degree included angle to plastically form a notch in a manner similar to a hardness indenter. The slitting disc produced a squarish notch tip with corner radii approximately equal to 0.04mm, compared to the pressed notch tip radius of approximately 0.02mm. However, it was concluded that the pressed notch offered no advantage over the slit notch, and it was not as versatile in terms of the notch depth that could be achieved. It was also established that there was no evidence of multiplanar crack initiation from the slit notch. Therefore, the slit notch method was adopted for all subsequent starter notches for shallow cracks.

Precracking SENB specimens

Most fracture toughness testing procedures include requirements on fatigue precracking. These either limit the load or the stress intensity factor that can be applied during fatigue precracking. The objective of these requirements is to limit the size of the plastic zone at the crack tip. The most common restriction is to limit the maximum permissible load to a certain fraction of the general yield load of the specimen.

Since there was no clear justification for using any one of the various fatigue precracking procedures for deeply–cracked specimens (1–6), a purely experimental approach was adopted. This attempted to define the lowest value of stress intensity factor for fatigue precracking (K_f) that would give uniform fatigue crack growth across the full thickness of shallow notched specimens (a/W ≥ 0.05). At the same time, it was required to produce an acceptable crack growth rate for crack length monitoring and control at a cyclic rate up to approximately 150Hz.

Using approximately 0.15mm wide by 0.5mm deep starter notches, TWI obtained acceptable shallow crack (a/W = 0.05) shapes in all the SENB specimens manufactured from material M1 (σ_{YS} = 388N/mm^2) with final K_f values of approximately 1200Nmm$^{-3/2}$. Subsequently, this nominal value of K_f was used to prepare a set of a/W = 0.05 SENB specimens that was tested to fracture over a wide range of temperatures to produce a ductile/brittle transition curve. After the tests, the final K_f values were compared with the maximum stress intensity factors (K_{max}) encountered in each test.

It was found that the ratio K_f/K_{max} was in all cases less than 0.57. Since this satisfies the K_{Ic} test requirements (1,3), the fatigue precracking procedure was considered acceptable.

In the case of the lower yield strength material M3, (σ_{YS} = 283N/mm^2) it was not possible to apply the fatigue precracking procedure developed by TWI, as the loads required to produce a K_f value of 1200Nmm$^{-3/2}$ would have resulted in gross plastic deformation of the specimens. An estimate of the general yield load (P_{YS}) of an SENB specimen can be obtained from the following expression, see Fig. 1:

$$P_{YS} = \frac{4B(W-a)^2\ \sigma_{YS}}{3S} \tag{1}$$

A study was undertaken to compare the ratio of fatigue load (P_F) to general yield load (P_{YS}) for various levels of K_f and a/W ratio. The study was performed for plate materials M1 and M3 and assumed a square section 25 x 25mm (B x B) SENB specimen geometry. The results are presented in Fig. 5 and 6.

For material M1 it is evident from Fig. 5 that an applied stress intensity factor of 1200Nmm$^{-3/2}$ results in a P_F/P_{YS} ratio of less than 1.0 over the entire a/W range studied, i.e., $0.05 \leq a/W \leq 0.6$. In particular, for an a/W ratio of 0.05, the P_F/P_{YS} ratio is approximately 0.8.

However, for material M3, Fig. 6 shows that an applied stress intensity factor of 1200Nmm$^{-3/2}$ results in a P_F/P_{YS} ratio larger than 1.0 for a/W ratios less than approximately 0.1. Therefore to prevent gross plastic deformation of material M3, the applied stress intensity factor needed to be reduced to at least 1000Nmm$^{-3/2}$.

Based on the results in Fig. 5 and 6, the a/W = 0.05 SENB specimens from material M3 were fatigue precracked using a maximum K_f of 700Nmm$^{-3/2}$. For a/W values $\geq$ 0.1 the maximum K_f was limited to 800Nmm$^{-3/2}$.

Precracking SENT specimens

The SENT specimens were easily fatigue cracked in three point bending using the same procedures and maximum K_f values developed for the corresponding SENB specimens.

Precracking SENAB specimens

Since the yield strengths of the pipe materials M2 and M4 were significantly higher than those of the plate materials M1 and M3, see Table 1, larger K_f values could have been employed for the SENAB specimens without the risk of gross plastic deformation. Nevertheless, it was decided to use the earlier more conservative K_f values. Thus, for fatigue precracking material M2 at TWI, a maximum K_f of 1200Nmm$^{-3/2}$ was used, and for precracking material M4 at EWI, a maximum K_f of 800Nmm$^{-3/2}$ was used.

Precracking – general experience

The above fatigue precracking procedures resulted in acceptable fatigue precrack front shapes for all the specimen types, materials and crack-plane dimensions given in Table 2. Furthermore, in all the subsequent fracture tests on materials M1 – M4, the K_f/K_{max} ratios were less than 0.6, which is consistent with the required K_f/K_{Ic} ratio in the BSI and ASTM K_{Ic} test methods (1, 3).

DEVELOPMENT OF MEASUREMENT TECHNIQUES

This aspect of the project concentrated on methods of attaching clip gauges, for the measurement of crack mouth opening displacement (CMOD), and also the measurement of load line displacement (LLD); both of these displacements being important to determinations of fracture toughness in terms K, CTOD and J (1–6).

Measurement of CMOD

When a shallow crack specimen is tested beyond general yield, significant plastic deformation and curvature occurs on the notched surfaces of the specimen. This can occur to within a 30° included angle of the crack tip, and can lead to errors in the measurement of CMOD when the standard methods of attaching clip gauges (1–6) are applied to specimens having small absolute crack sizes, such as those investigated in this project (Table 2). To overcome this problem, a double clip gauge technique was developed.

The knife edges for the double clip gauge measurements were mounted on steel shims, which were micro – TIG welded to the notch mouth. This is indicated in Fig. 7, which shows the shim and knife edge arrangement on a plastically deformed shallow crack SENB specimen after testing. Figure 7 also demonstrates how the curvature of the notched surface of a shallow crack specimen results in displacement between the knife edges and the specimen. Clearly, the displacement between the lower and upper knife edges does not represent the rotation of the crack flanks during testing. Also, the displacement between the lower knife edges does not represent the true CMOD. Nevertheless, CMOD can still be calculated with good accuracy from the double clip gauge data. Referring to Fig. 8, and regardless of the sources of knife edge rotation,

$$CMOD = V_1 - z_1\left(\frac{V_2 - V_1}{z_2 - z_1}\right) - 2x \cos\left[\sin^{-1} 0.5\left(\frac{V_2 - V_1}{z_2 - z_1}\right)\right] + 2x \qquad [2]$$

Where x, z_1 and z_2 are known dimensions, and V_1 and V_2 are the lower and upper clip gauge displacements, respectively. For small displacements, the last two terms tend to cancel out.

The shim and double knife edge/clip gauge arrangement was used successfully with Eq. [2] for tests on all the designs of SENB, SENT and SENAB specimens (Fig. 9–11).

To illustrate the difference between the lower clip gauge displacement (V_1) and CMOD, a range of data for material M1 were analysed using Eq. [2]. For shallow notched bend specimens (a/W = 0.05) the ratio CMOD/V_1 was in the range 0.76 – 0.83; whereas for the deep notch specimens (a/W = 0.5) the ratio was in the range 0.85 – 0.87. For the shallow notch tensile specimens (a/W = 0.05), in which there was very little bending, the ratio CMOD/V_1 was in the range 0.93 to 0.98. However, for the deep notch tensile specimens (a/W = 0.5), the ratio CMOD/V_1 fell to the range 0.8 to 0.82. Hence, significant errors will arise if the CMOD is assumed to be equal to V_1 measured at some distance from the notch mouth.

Although the double clip gauge technique can provide good estimates of the CMOD, it has the disadvantage of requiring more instrumentation than is currently necessary in the standard deep notch test methods (1–6). An attractive alternative for both deep and shallow notch tests, would be measurement techniques that were capable of measuring CMOD directly. One such technique, which was not studied in this project, is non–contact laser extensometry.

Measurement of LLD

Successful measurements of load line displacements (LLD) in all the SENB and SENAB specimens were obtained using a standard comparator bar technique (1), which is also shown with the double clip gauge instrumentation in Fig. 9 and 10.

For the SENT specimens, the LLD was measured using a telescopic arrangement and a standard clip gauge. This enabled the linear displacement between reference points 100mm apart on the sides of the specimen to be measured regardless of specimen bending during the test. The LLD and double clip gauge instrumentation (for CMOD determination) are shown in Fig. 11.

Measurements at low temperatures

Measurements of CMOD and LLD on SENB and SENAB specimens at temperatures down to −160°C were straightforward, since the instrumentation was above the cooling medium. However, SENT specimens are normally tested vertically, and this makes it difficult to cool the whole specimen uniformly without exposing the instrumentation to the cooling medium. To avoid this problem, the SENT specimens in this project were tested horizontally. This required the preparation of special equipment, which made the SENT tests relatively complicated and expensive compared to the tests on SENB and SENAB specimens.

CONCLUSIONS

1. For steels having yield strengths greater than approximately $300N/mm^2$, successful fatigue precracking, instrumentation and testing procedures have been developed for single edge notched bend (SENB), single edge notched tension (SENT) and single edge notched arc bend (SENAB) specimens having crack length to specimens width ratios (a/W) down to 0.05, and absolute crack lengths down to 1.25mm. These procedures are complementary to those in the BSI and ASTM test methods (1–6) for specimens having a/W ≥ 0.45.

2. The most practical shallow crack test methods involve SENB specimens for full thickness tests on flat material (1), and SENAB specimens for tests associated with internal longitudinal surface cracks in pipes.

ACKNOWLEDGEMENTS

The authors are grateful to the sponsors of this project for their financial support and technical advice. The authors also wish to thank their colleagues at TWI and EWI for helpful suggestions and for conducting the experimental work.

REFERENCES

(1) BS7448: Part 1:1991: 'Fracture Mechanics Toughness Tests, Part 1. Method for Determination of K_{Ic}, Critical CTOD and Critical J Values of Metallic Materials'. (This Standard replaced BS5447:1977: K_{Ic} Tests, and BS5762:1979: CTOD Tests).

(2)   BS6729:1987: 'BS6729:1987: 'British Standard Method for Determination of the Dynamic Fracture Toughness of Metallic Materials'.

(3)   ASTM E399-90: 'Standard Test Method for Plane-Strain Fracture Toughness of Metallic Materials'.

(4)   ASTM E813-89: 'Standard Test for J_{Ic}, a Measure of Fracture Toughness'.

(5)   ASTM E1152-87: 'Standard Test Method for Determining J-R Curves'.

(6)   ASTM E1290-89: 'Standard Test Method for Crack-tip Opening Displacement (CTOD) Fracture Toughness Measurement'.

(7)   Contract Proposal TWI Ref: CP/MAN/3299-1, EWI Ref: A6098: 'An International Research Project to develop Shallow Crack Fracture Mechanics Tests'. July, 1987.

(8)   Gordon J R and Leggatt R H: '3D Elastic-Plastic Finite Element Analyses for CTOD and J in SENB, SENAB and SENT specimen geometries'. Paper 14, 1st Int. Conf. on shallow Crack Fracture Mechanics Toughness Tests and Applications, TWI, Cambridge, Sept, 1992.

(9)   Underwood J H, Kapp J A and Witherell M D: 'Fracture Testing with Arc Bend Specimens'. ASTM STP 905, 1984, pp.279-296.

(10)   Dodds R H, Vargas P M and Keppel M: 'Stress-Intensity Factors for Slotted SENAB specimens'. University of Illinois, Urbana, IL 61801, May 1989.

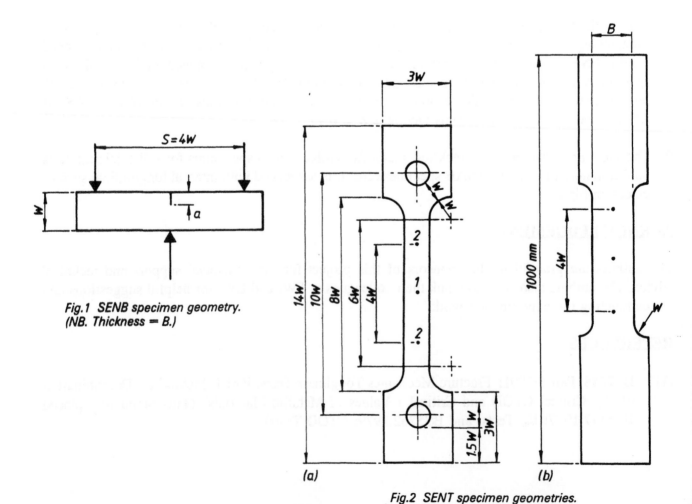

Fig.1 SENB specimen geometry.
(NB. Thickness = B.)

(a)   (b)

Fig.2 SENT specimen geometries.

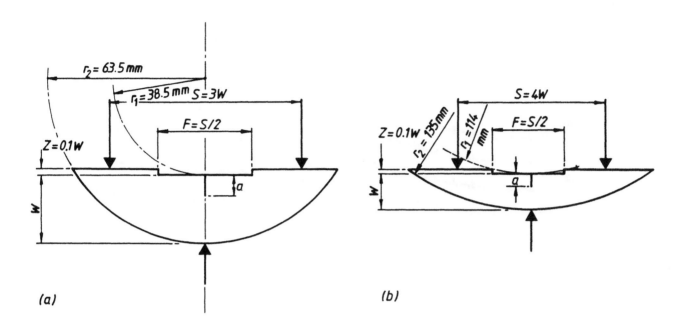

Fig.3 SENAB specimen geometries:
a) Used for pipe material M2;   b) Used for pipe material M4.

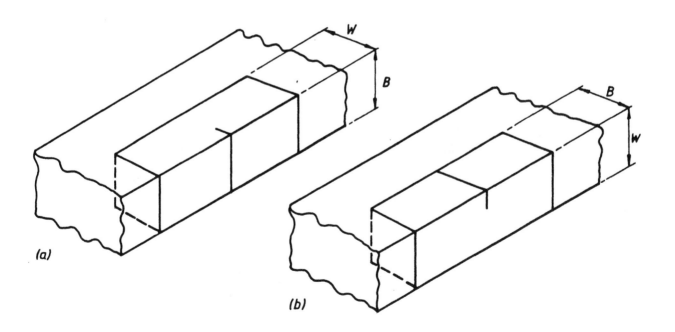

Fig.4 Specimen orientations:
a) Through-thickness notch;   b) Surface notch.

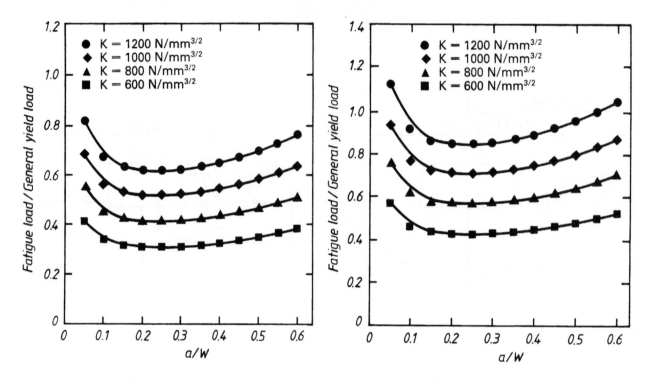

Fig.5 Fatigue loads for material M1.

Fig.6 Fatigue loads for material M3.

Fig.7 Deformed SENB specimen showing
shim and knife edge assembly.

58489/3

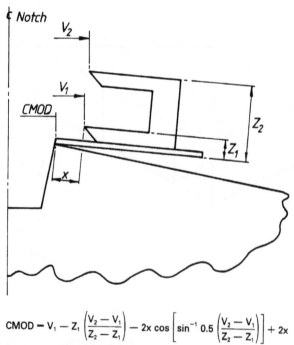

$$\text{CMOD} = V_1 - Z_1\left(\frac{V_2 - V_1}{Z_2 - Z_1}\right) - 2x \cos\left[\sin^{-1} 0.5\left(\frac{V_2 - V_1}{Z_2 - Z_1}\right)\right] + 2x$$

Fig.8 Effect of knife edge rotation on
measurement of CMOD.

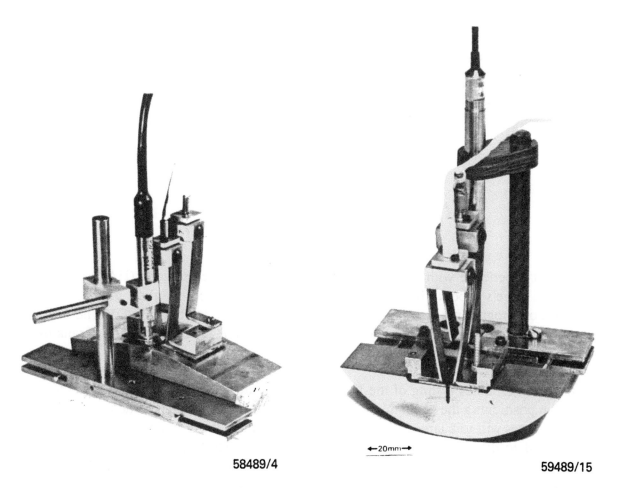

58489/4

59489/15

Fig.9 Instrumented SENB specimen.

Fig.10 Instrumented SENAB specimen.

←20mm→

59488/1

Fig.11 Instrumented SENT specimen in test rig.

TWI

PAPER 19

Shallow surface notch — experience from bead in groove testing

Prof C Thaulow (NTH/SINTEF) and A J Paauw (Welding Centre, SINTEF, Norway)

SUMMARY

After the first experience with low CTOD values in the heat affected zone (HAZ) of low carbon micro-alloyed steels about 10 years ago, SINTEF proposed a cost effective materials characterisation programme for the Norwegian Petroleum Dirtectorate (NPD). The aim of this NPD programme was to identify the microstructures most prone to brittleness and to rank the HAZ toughness of steel qualities in a quantitative way.

In the fracture mechanics testing of the HAZ, a large scatter is normally experienced because of difficulties in positioning the notch tip in the desired microstructure. The geometry and distribution of the brittle zones will to a large extent depend on the detailed welding procedure applied. Hence, there is a need to distinguish between the microstructural dependent toughness and the lower bound value that can be obtained for a specific weldment. The HAZ of the cap layer has been selected as the most critical and representative area for fracture testing.

On this backgroud the NPD programme includes two independent toughness tests, weld thermal Charpy simulation testing and shallow surface notch CTOD bead-in-groove testing. The bead-in-groove weldments are prepared in 50 mm thick plates, and CTOD specimens are used with an electro discharged machined (EDM) notch and a depth of a/W = 0.06.

Up to now around 30 steels have been tested and the state-of-the-art was presented in the TWI Conference on WELD FAILURES in 1988. In the present paper recent experience from high strength steels are presented including a TiO-steel. Attention is also drawn to the effect of weld metal yield strength, and special tests have been performed to document the effect of yield strength overmatch. The effect of substituting the EDM notch with a fatigue pre-crack will also be presented.

INTRODUCTION

In the fracture mechanics testing of weldments there is a need to distinguish between materials and welding procedure dependent test results. The toughness will, to a large degree, depend upon the detailed welding procedure applied, the resulting geometry and distribution of microstructures along the fusion line and the notch positioning procedure. (1) .

The Norwegian Petroleum Directorate (NPD) test programme, (3), (4) and (5) , aims at ranking the HAZ toughness of steel qualities in a quantitative way, so that different steels can be compared. The test can then serve as the background for an estimate of the toughness level to be expected in real weldments. The test programme includes shallow surface notched bead-in-groove CTOD tests and impact toughness testing of weld thermal simulated specimens.

The shallow surface notch was selected because it was thought that the HAZ of the cap layer will be the most realistic notch position with respect to practical cases. The cap layer of a weldment will have the largest brittle zones due to lack of heat treatment from a subsequent layer. This will also be a preferred position for weld defects and represents also the area with the highest tensile residual stresses. In addition, the local stress concentration factors are high, because of surface irregularities, favouring the initiation of fatigue cracks.

The weld thermal simulation testing will be independent of the notch location, but the significance of the toughness with respect to real HAZ weldments can be questioned. The link to the real weldment is then introduced with the CTOD bead-in-groove testing. The microstructures and the toughness of the two test methods are compared, and hence the confidence in the results increased.

TEST PROGRAMME

The NPD materials characterisation programme consists of the following tasks:

CTOD bead-in-groove test

Bead-in-groove weldments are prepared in 50 mm thick plates and CTOD specimens are machined with a shallow surface electro discharged notch (a/W = 0.06), Fig. 1. The discharged notch was chosen in order to optimise the notch positioning at the fusion line. Both the shallow notch depth and the electro discharging will favour higher CTOD values than will be experienced with a deep notch and a fatigue pre-crack. The calculation of the CTOD values is still based on BS 5762:1979. In this standard, however, the a/W values are restricted to 0.15 < a/W < 0.7 and so the CTOD values were calibrated using a feeler gauge inserted directly close to the crack tip. For CTOD values lower than about 0.5 mm direct correlations were obtained between values calculated from BS 5762:1979 and the calibration curve, Fig. 2. For larger CTOD values, the calibration curve gave smaller values. The results from the feeler gauge calibration curve have later been confirmed with rubber replicas. (2) . A total of 9 duplicate specimens are tested at -10°C. After the test the specimens are sectioned in the area of brittle fracture initiation and the fracture path is correlated with the microstructure.

Weld thermal simulation test

The weld thermal simulation specimens are machined 2 mm below the plate surface with their axes in the transverse direction. A Smitweld TSC 1405 thermal simulator was used for the heat treatment cycles. The specimens were machined to Charpy V-notch specimens and the absorbed impact energy was determined at a testing temperature of -40°C. Three test series are performed:

1)   The influence of the cooling time from 800°C to 500°C on the impact toughness at a peak temperature of 1350°C.
2)   The influence of the peak temperature on the impact toughness for a cooling time from 800°C to 500°C of 12 seconds.
3)   The influence of the peak temperature on the impact toughness for a cooling time from 800°C to 500°C of 24 seconds.

Selected specimens are prepared for optical microscopic examination. The microstructures are correlated with the impact toughness results and compared with the HAZ bead-in-groove microstructures.

The results from the testing of 15 steels used in offshore structures have been published previously

(3), (4) and (5) . The present paper presents the NPD test results from additional steels, including a titanium oxide steel (see Table 1). Furthermore, results from EDM notching versus fatigue pre-cracking are presented and the effect of weld metal yield strength overmatch.

TMCP TITANIUM OXIDE STEEL

Five different philosophies in how to assess low fracture toughness values in weldments have been proposed (6) . This includes the manufacturing of "LBZ free steels". The latest development in this direction is the so called Particle Controlled Microstructures, where small particles with a high thermal resistivity will restrict grain growth in the heating cycle and serve as nucleating agent for tough microstructures, eg. acicular ferrite, on cooling. One example is the titanium oxide bearing steel, with Ti_2O_3 particles.

One commercial available titaniumoxide treated steel plate (Table 1) has been tested according to the NPD procedure. The low carbon, phosphorus and sulphur content is obvious. Of special interest is the low level of oxygen, aluminium and nitrogen, while titanium has been added. The silicon content is low because the steel has been partly deoxidized with titanium.

Notch toughness of weld thermal simulated specimens

The notch toughness for a peak temperature of 1350°C did not change significantly with cooling time, Fig. 3, and kept stable at a rather high toughness level.

The influence of the peak temperature on the notch toughness for a cooling time ($\Delta t_{8/5}$) of 12 s and 24 s is presented in Fig. 3. A high toughness level is found in the whole peak temperature range. The result is compared with the previous NPD test results in Fig. 4, where the high toughness level is evident.

Shallow surface bead-in-groove CTOD tests

Six specimens revealed maximum load after ductile crack growth through all the zones of the HAZ, and three specimens revealed pop-ins which were followed by ductile tearing. The pop-ins were located in the coarse grained zone and terminated in the fine grained zone. The results are compared with the previous NPD test results in Fig. 5.

Evaluation

Comparing the notch toughness of the weld thermal simulated specimens with the bead-in-groove CTOD test results, the evaluated fracture behaviour of the HAZ from the two test methods are in good agreement.

Comparing the obtained results with a previously investigated TMCP steel "Steel 11", the titanium oxide steel "Steel 18" revealed a lower notch toughness level for the weld thermal simulation specimens in the whole HAZ. This was rather remarkable, as the prior austenite grain size was larger for "Steel 11". Furthermore, all the shallow surface notched CTOD tests of the bead-in-groove weldments revealed a maximum load result for "Steel 11", while "Steel 18" showed for three specimens a pop-in in the coarse grained HAZ. Comparing the chemical compositions of the two steels, "Steel 11" has a lower titanium and sulphur content, Table 1, while the aluminium, nitrogen and oxygen contents are higher. More research has to be performed to clarify the significance of the different factors.

EDM NOTCH VERSUS FATIGUE PRE-CRACK

An accurate notch positioning is a critical point in CTOD testing of weldments. The EDM notch represents better conditions for reproducible notch positioning in the desired microstructures compared to a fatigue pre-crack, especially for shallow surface notches where fatigue crack-growth can be difficult to control.

At an early stange of the development of the NPD procedure, the effect of EDM notch and fatigue pre-crack in various zones of HAZ were examined by means of CTOD weld thermal simulation testing of 10 x 10 mm specimens /(1)/. In the coarse grained zone, the CTOD values for the EDM notches for a low carbon micro-alloyed steel were in the range of 0.12 - 0.19 mm, compared to 0.04 - 0.05 mm for the fatigue pre-cracked specimens.

The results have now been followed up by fatigue pre-cracking of CTOD bead-in-groove specimens. Two steels were chosen, a normalised steel with yield strength 356 MPa, "Steel 14", and a quenched and tempered high strength steel with yield strength 585 MPa, "Steel 15", see Table 1. Both steels had previously been tested in accordance with the NPD procedure.

Different approaches were examined in order to initiate and extend a straight fatigue crack. The best results were obtained with introducing a 2 mm deep EDM notch in the weld metal overlay, and then extend the notch with 3 mm fatigue crack growth. Afterwards, the weld metal overlay was machined away.

From a series of seven test specimens, brittle fracture or pop-in was initiated in five specimens. The results are compared with the corresponding NPD test in Fig.6.

The fatigue pre-crack has resulted in a substantial reduction of the fracture toughness for the normalised steel, while the quenched and tempered steel had the same lower bound level as for the EDM notching.

The welding consumables used in the NPD test will result in a significant weld metal yield strength overmatch for the normalised steel and a more evenmatch for the high strength steel. The trends are clearly indicated with hardness measurements, Fig. 7.

Hence, as the load level during three point bend testing is increased, the weld metal in the vicinity of the crack tip will suffer substantial yielding before the HAZ and base metal are affected, and the plastic zone will reach the surface at an early stage of deformation. Hence, regardless if there is an EDM notch or a fatigue pre-crack, the notch tip will deform until a particular strain level is built up at the fusion boundary. In order to verify this explanation, new experiments were performed on a series of bead-in-groove test specimens with weld metal overmatch.

EFFECT OF WELD METAL OVERMATCH

The titanium oxide TMCP steel "Steel 18" and the quenched and tempered steel "Steel 15" were selected to investigate the effect of weld metal overmatch. These steels had a yield strength of about 454 and 585 MPa respectively.

All bead-in-groove testing in the NPD programme has been carried out so far with an electrode which is classified as ASME E7018 with a minimum weld metal yield strength of 420 MPa and minimum tensile strength of 500 MPa. The hardness profile across the weldment shows that the weld metal overmatch has been considerably reduced compared with the normalised steel, Fig. 7.

In order to evaluate the importance of the matching/overmatching, the steels were welded with a high strength electrode, ASME E11018, with a minimum yield strength of about 750 MPa. The hardness profiles illustrates clearly the overmatch condition, Fig. 8. Three bead-in-groove specimens of the TMCP steel and nine specimens of the QT steel were tested.

All three CTOD specimens of the TMCP steel revealed pop-ins followed by ductile crack growth. Sectioning revealed that the local brittle fracture had initiated in the HAZ coarse grained zone, after limited ductile crack growth in the weld metal, and terminated in the HAZ fine grained zone. Ductile crack growth occurred through the remaining HAZ and into the base material.

The CTOD results from the two test series are compared in Fig 9 with their cumulative distibutions. It is obvious that the fracture toughness of the coarse grained HAZ is influenced by the weld metal yield strength.

All nine CTOD specimens of the QT steel "Steel 15", revealed a pop-in or brittle fracture at low CTOD values which was initiated in the HAZ coarse grained zone. The comparison between the two test series is made in Fig. 9 and clearly shows the large influence of the overmatch.

The hardness measurements show the increasing degree of weld metal / base metal overmatch for the E11018 weldment, Fig. 8. For the coarse grained HAZ, however, the situation is changed from a clear undermatch for the E7018 weldment to an even/overmatch for the E11018 weldment. Hence, the observed change in toughness behaviour can be affected by the differences in strength level of both the base material and the HAZ compared with the weld metal.

In the CTOD bead-in-groove tests, the notch tip is normally positioned in the weld metal close to the fusion line. During testing, the material in the vicinity of the crack tip will yield and gradually create a larger and larger plastic zone until the local strains and stresses reach such magnitudes that ductile crack growth or unstable fracture is initiated. In the case of the E7018 weldment, the weld metal has a lower yield strength compared to the coarse grained zone. Hence, during testing the material will start to yield in the weld metal and reduce the possibility to build up high local stresses at the fusion line. For the E11018 weldment, all stresses will be directed towards the HAZ where the coarse grained zone has the unfavourable combination of a low yield strength and a low fracture toughness compared to the weld metal. See also /(7)/ for further discussions. The higher stress level in the HAZ coarse grained zone in the case of weld metal overmatch has also been confirmed by FEM calculations (to be presented in the near future).

CONCLUSIONS

Shallow surface notch CTOD bead-in-groove testing, with a notch depth of 3 mm in 50 mm thick plates, has been used to quantify the HAZ fracture toughness of offshore steels. Some recent test results and investigations on the effect of electro discharged machined (EDM) notch versus a fatigue pre-crack and the effect of weld metal yield strength overmatch, has been performed.

-   Both measurements with feeler gauges and rubber replica have concluded that shallow crack CTOD values can be calculated in accordance with BS 5762:1979 for CTOD values up to about 0.5 mm.

-   Testing of a titanum oxide TMCP steel revealed a high toughness level, but the level was somewhat lower than an "ordinary" TMCP steel previously tested.

- It was found that the microstructures and toughness in the shallow surface notched bead-in-groove specimen were in good agreement with the results of the weld thermal simulation experiments.

- Fatigue precracking reduced the CTOD value substantially for a normalised steel, but had no effect on a high strength quenched and tempered steel. This was explained on the basis of weld metal overmatching for the normalised steel, and close to matching for the high strength steel. With matching, the material will start to yield in the weld metal, and eventually reach the surface, before a high strain and stress level can be built up in the HAZ. Hence a machined notch and a fatigue pre-crack will both develop to a blunted notch, before fracture can be initiated.

- The effect of weld metal overmatch was verified with experiments which showed a distinct reduction in HAZ fracture toughness with weld metal yield strength overmatch.

REFERENCES

(1)   C. Thaulow, "Fracture mechanics testing of weldments", Eng. Frac. Mech., Vol. 31, No. 1, 1988, 181 - 188

(2)   S. Wästberg, "Engineering significance of short cracks located in local brittle zone", Seventh ASME Conference on OFFSHORE MECHANICS AND ARTIC ENGINEERING, 7-12 February 1988, Houston

(3)   C. Thaulow, A. J. Paauw, Å. Gunleiksrud, O. J. Næss, "Heat affected zone toughness of a low carbon micro-alloyed steel", Metal Construction, Vol. 12, No. 2, February 1985, 94-100R

(4)   C. Thaulow, A. J. Paauw, K. Guttormsen, "The heat affected zone toughness of low carbon micro-alloyed steels", Welding Journal, Vol. 9, 1987, 266-s - 279-s

(5)   C. Thaulow, A. J. Paauw, "Materials characterization with respect to HAZ local brittle zones", TWI Conference on WELD FAILURES, 21-24. November 1988, London

(6)   M. Hauge, "A probabilistic failure assessment of defects in welded joints containing brittle zones", International Conference on EVALUATION OF MATERIALS PERFORMANCE IN SEVERE ENVIRONMENTS, 20-23 November 1989, Kobe

(7)   C. Thaulow, A. J. Paauw, "Effect of weld metal overmatch and notch location on the HAZ fracture toughness of high strength steels", Second ISOPE Conference on OFFSHORE AND POLAR ENGINEERING, 14-19 June 1992, San Francisco

(Remark : Table 1 is located after fig. 8)

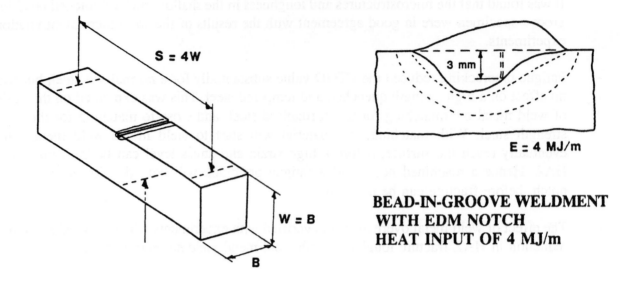

**BEAD-IN-GROOVE WELDMENT
WITH EDM NOTCH
HEAT INPUT OF 4 MJ/m**

CTOD TEST SPECIMEN SURFACE NOTCHED
BXB
a/W = 0.06

Fig. 1　　The surface notched CTOD specimen (BxB and a/W = 0.06) and the location of the electro discharged machined (EDM) notch in the bead-in-groove weldment.

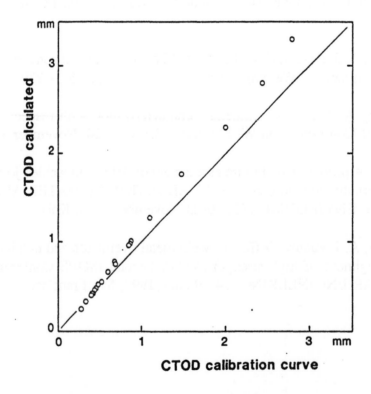

Fig. 2　　Relation between CTOD values from a calibration curve, which was based on measurements with a feeler gauge, and CTOD results calculated with BS 5762:1979.

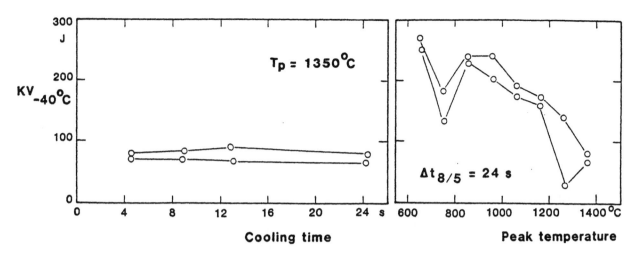

Notch toughness as a function of the cooling time ($\Delta t_{8/5}$) for a peak temperature of 1350°C.

Notch toughness as a function of the peak temperature for a cooling time ($\Delta t_{8/5}$) of 12 s.

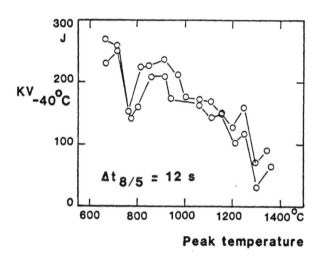

Notch toughness as a function of the peak temperature for a cooling time ($\Delta t_{8/5}$) of 24 s.

Fig. 3   TMCP TITANIUM-OXIDE STEEL - "Steel 18" : The notch toughness results for the weld thermal simulated specimens.

Single cycle :

$\Delta t_{8/5} = 12s$

Steel 11	————
Steel 12	— — —
Steel 13	—·—·—·
Steel 14	—··—··—··
Steel 15	············

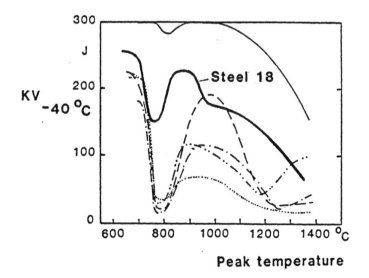

Fig. 4   TMCP TITANIUM-OXIDE STEEL - "Steel 18" :   Comparison of the notch toughness results for weld thermal simulated specimens with previous results.

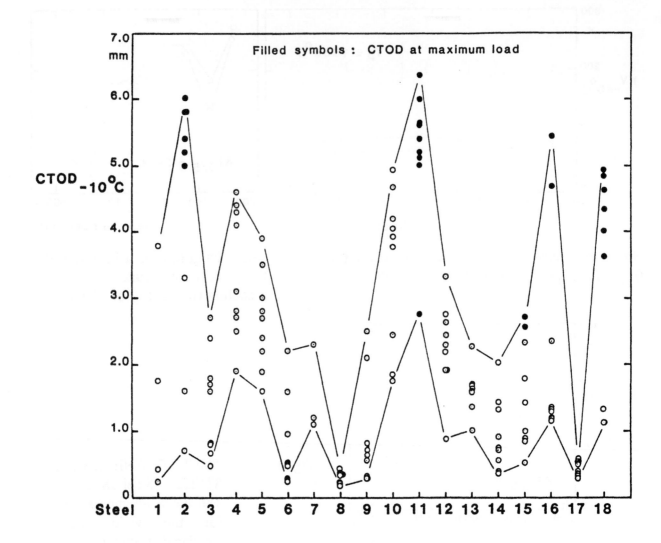

Fig. 5 TMCP TITANIUM-OXIDE STEEL - "Steel 18" : Comparsion of bead-in-groove CTOD results with previous CTOD results.

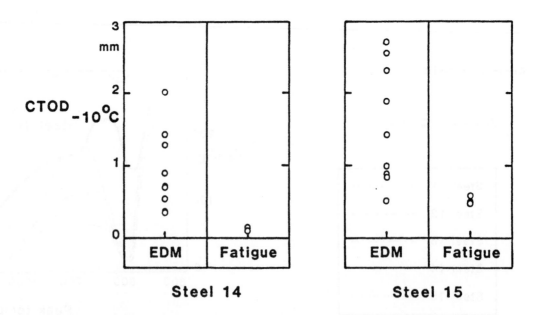

Fig. 6 Comparison between EDM and Fatigue notched bead-in-groove CTOD specimens (BxB and a/W = 0.06) for a normalised steel, "Steel 14", and a QT steel, "Steel 15".

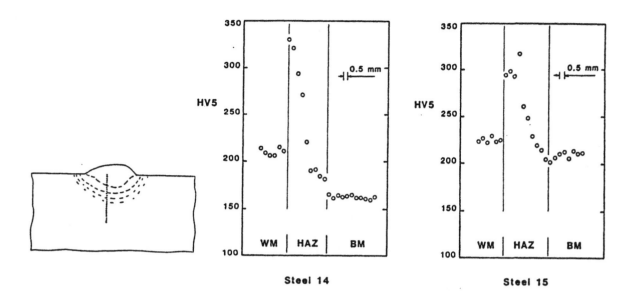

Fig. 7   Hardness distribution in the bead-in-groove weldment at the location of the notch in the bead-in-groove specimens for the normalised and QT steel, see Fig. 5.

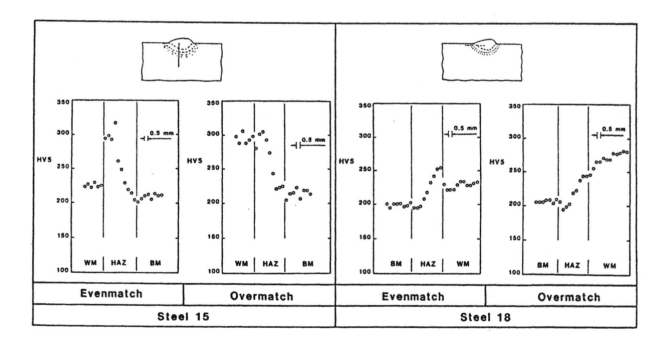

Fig. 8   The hardness distribution in the bead-in-groove weldment for the two welding consumables resulting in an evenmatch or overmatch weld metal compared to the base material for the QT "Steel 15" and TMCP "Steel 18" high strength steels.

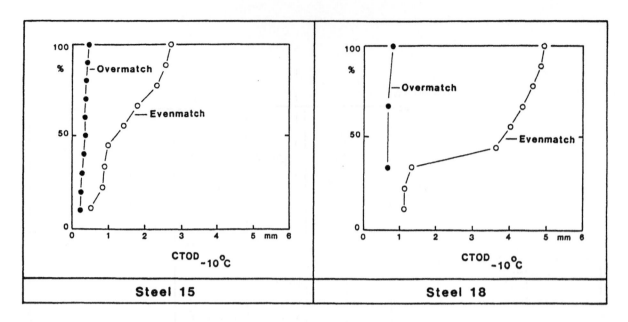

Fig. 9   The influence of the weld metal matching condition on the bead-in-groove CTOD results for the QT "Steel 15" and TMCP "Steel 18" high strength steels, see Fig. 8.

Table 1   The chemical compostion and mechanical properties of the investigated steels.

Chemical compostion.

Steel	Elements in wt%															
	C	Si	Mn	P	S	Cu	Ni	Cr	Mo	V	Nb	Ti	Al	N	O	Ca
Normalised steel																
14	0.15	0.43	1.42	0.009	0.001	0.19	0.19	0.02	0.01	0.01	0.030	0.004	0.024	0.004	0.0037	0.0024
TMCP steels																
11	0.09	0.18	1.49	0.004	0.001	0.12	0.38	0.02	0.01	0.01	0.011	0.008	0.027	0.005	0.0060	0.0027
18	0.09	0.19	1.61	0.005	0.004	0.25	0.37	0.03	0.01	0.01	0.014	0.015	0.003	0.002	0.002	0.0002
QT steel																
15	0.13	0.31	1.37	0.011	0.001	0.12	0.32	0.16	0.04	0.06	0.003	0.014	0.053	0.008	0.0023	0.0013

Mechanical properties in the transverse direction and as-delivered condition.

Steel	R_e MPa	R_m MPa	A_5 %	Z %	aver. $KV_{-40°C}$ J
Normalised steel					
14	356	507	33	72	219
TMCP steels					
11	449	522	27	75	221
18	454	546	30	84	290
QT steel					
15	585	670	22	75	135

PAPER 20

CTOD toughness evaluation of hyperbaric repair welds with shallow and deep notched specimens

I Rak, Prof, Dr, M Koçak, BSc, PhD, MEng, M Golesorkh, Dipl-Ing and
J Heerens, Dr-Ing (GKSS Research Centre Geesthacht, FRG)

SUMMARY

The CTOD toughness evaluation of repaired weld joints (given two weld deposits, complex mechanical heterogeneity and high residual stresses) requires modification of commonly used specimen geometry and testing procedure. This paper presents the results of the SENB specimens with shallow (a/W=0.16) and deep notches (a/W=0.5) extracted from 30 mm thick multipass submerged arc welded (SAW) joints which were repaired with hyperbaric (16 bar) welding process at the toe region of the original SAW weld deposit.

The results of this study show that the CTOD (δ_5) measurements are consistent with the calculated CTOD values according to the BS 5762 standard for both deep and shallow cracked specimens (a/W=0.16) if the plastic rotation factor (r_p) value of 0.25 is used in the latter. The problem of selection of yield stress for calculation of the CTOD according to the BS 5762 standard for highly heterogeneous repaired weld joints can be eliminated using the CTOD (δ_5) technique.

It is important to realise that the through thickness deep and shallow notched specimens can produce lower bound toughness values if the notch accurately samples the local brittle zones. The latter produced its lower bound values under the influence of overmatching hyperbaric repair weld joint despite its lower crack tip constraint. Surface notched specimens (a/W= 0.4) where the very crack tip is controlled by one microstructure did not give lower bound toughness values.

INTRODUCTION

The shallow defects or cracks at the toe or root regions of the weld joints of offshore structures sometimes have to be repaired under harsh service conditions with wet or dry-hyperbaric welding processes. Generally, the volume and depth of the deposited repair welds are smaller compared with the defected original weld. The CTOD toughness evaluation of these complex weld joints (given two weld deposits and higher residual stresses) certainly requires some modification of the usual specimen geometry and testing procedure.

However, the CTOD test standards BS 5762:1979 (1) ASTM E 1290-91 (2) and EGF P1-90 procedure (3) recommend to use deep cracked SENB specimens. Fracture toughness data determined on such specimens are bound to lead to the use of conservative toughness data on material selection, welding qualification and defect assessment procedures. Yet, in welded joints, defects are often found to be in the form of shallow toe or root cracks. Obviously, the significance of such defects may thus be assessed in an unduly conservative manner, especially if very low toughness values are used. Also, having two weld deposits in the repaired joints further complicates the fracture mechanics testing and analysis of the results. A direct application of the fracture toughness testing practise of unrepaired original welds to repaired joints is not a straightforward task, since repair welding can produce additional local brittle zones (LBZ) to be tested. The use of shallow cracked specimens can be essential for testing the LBZ of some repair weld deposits as well as new HAZ/LBZ that appeared.

Various studies have already shown that the elastic-plastic fracture toughness values at crack initiation (δ_i) or J_i are higher for shallow cracks than for deep crack (4-10). It is known that the CTOD formula used in testing standards is based on the plastic hinge model of bend specimens. The value of the plastic rotation factor (r_p) significantly affects the calculated CTOD values. A considerable decrease of the notch depth (a/W) can cause a shift of the plastic hinge point location $(W-a)r_p$ in the unnotched ligament ahead of the crack tip. Hence, the determination of r_p values for shallow cracked specimens plays an important role in CTOD testing, particularly of shallow notched weld specimens where material heterogeneity (mismatching) should particularly be taken into account. Several studies revealed that the plastic rotation factor, can have a higher value than 0.4 for deep notched homogeneous specimens and a much lower value for shallow notched specimens (4-6, 9-12). In fact, the new ASTM E-1290-91 standard (2) uses the r_p values of 0.44 for SENB and 0.46 and 0.47 values for CT specimens (r_p = 0.47 for 0.45 ≤ a/W ≤ 0.50 or r_p= 0.46 for 0.50 ≤ a/W ≤0.55) which is higher than the one used in the BS 5762 standard.

In this sense the specimen size or geometry dependent fracture toughness data have caused increasing interest in elastic-plastic toughness testing procedures. The CTOD or J testing of the weld metal and heat affected zone (HAZ) further complicate the issue due to their micro- and macro-heterogeneities. Therefore, it has become necessary to conduct CTOD or J testing of weldments with various specimen geometries more extensively to avoid misleading results. The present work focuses on CTOD toughness measurements in multipass hyperbaric repair welds to determine the effects of the weld joint micro- and macro heterogeneities and crack depth/location on CTOD testing procedure.

EXPERIMENTAL PROCEDURE

Multipass submerged arc welds (SAW) were prepared on 30 mm thick low carbon HSLA pipeline steel plates under atmospheric condition. The repair welding procedure was simulated for toe and embeded defects and hence half-K repair weld grooves were machined on some of these plates giving a 13 mm depth with a straight fusion line as shown in 'Fig.1'. Repair weld was deposited in flat position by using the GMA welding procedure at 16 bar (160 m water depth) in the welding chamber of the Bundeswehrhochschule Hamburg. The welding consumable and procedure are given in Table 1. A macro-section of the repair weld joint is shown in 'Fig. 2'. For the original weld metal (SAW) and hyperbaric repair weld deposit, consumables were selected to obtain a considerable overmatching strength compared with the base material. The mechanical properties of the base and two weld metals are given in Table 2. Extensive microstructural examination and a microhardness survey were carried out to establish the locations of the LBZs.

Three point bend SENB specimens (BxB, B=28 mm) were extracted from the original SAW and repair welded plates for various notch locations and notch lengths as shown in Table 3. The shallow cracked specimens were prepared with a/W ratio of 0.16 and deep notched specimens were fatigue precracked with a/W=0.4 and 0.5. These notch locations and configurations were selected for two reasons: first, to conduct the screening tests for potentially embrittled zones of this complex weld joint, second, to compare the shallow and deep notched specimen results. All CTOD tests were conducted at -10°C. The CTOD values were calculated in accordance with BS 5762 (δ_{BS}) and also with the GKSS developed δ_5 clip gauges (δ_5) measured directly on the side surface of the specimen at the fatigue crack tip over the 5 mm length of the gauge (13-14). Since there is a very good agreement between the two CTOD measurements (15), it is thought that the δ_5 values can be substituted into the δ_{BS} formula to determine the plastic rotation factor, r_p , experimentally for deep and shallow notched bend specimens:

$$\text{CTOD}_{BS} = \frac{K^2(1-\upsilon^2)}{2\sigma_y E} + \frac{r_p(W-a)V_p}{r_p(W-a)+a+Z} = \delta_5 \qquad [1]$$

From eq. (1), the plastic rotation factor r_p can be obtained as:

$$r_p = \frac{2\sigma_y E V_p(a+Z)}{2\sigma_y E(V_p - \delta_5)(W-a) + K^2(1-\upsilon^2)(W-a)} - \frac{a+Z}{W-a} \qquad [2]$$

The rotation factors obtained in this manner for various shallow (a/W≅0.1) and deep notched (a/W≅0.3-0.5) specimens produced values of about 0.2 and 0.45 respectively (15). It is clear that these r_p values were found under the equality condition of the δ_{BS} and δ_5 CTOD values. The plastic rotation factor determined for shallow and deep cracked specimens in other studies (7,10) also showed the value of about 0.2 for shallow cracked (a/W≈0.1) specimens as a similar value was obtained by Koçak et al (15) using the δ_5 clip gage technique in conjunction with BS 5762 CTOD formula. By following up this approach and substantial test results developed at the GKSS, the r_p value of 0,25 was used in this study to calculate the δ_{BS} for shallow cracked specimens (a/W=0.16).

During the test, the d.c. potential drop technique (16, 17) was applied for monitoring stable crack growth. The load-line-displacement was also measured with reference bar (15) to minimize the effects of indentations that may occur at the rollers. Fatigue precracking was carried out with "step-wise high R-ratio" (SHR) procedure (18) for all specimens. This technique is successfully used at GKSS Research Centre for as welded specimens to obtain a valid uniformly shaped crack front. The SHR technique uses two stress ratios, R=0.1 for crack initiation and growth of about 1 mm then, R=0.7 with allowable maximum load, until the required a/W ratio is obtained. On 'Fig. 3' fracture surface of the specimen fatigue precracked with SHR technique showing improved crack front profile is presented (Note: Light coloured area showing irregular crack front development with R=0.1, but then increasing R-ratio to 0.7 improves the profile). After CTOD tests, post-test sectioning and metallographic examinations were conducted for all HAZ specimens to identify the fatigue crack tip microstructure.

RESULTS AND DISCUSSION

Micro-hardness survey

'Fig. 4' shows the micro-hardness paths (a-f) which were conducted to determine the hardness distribution across the critical zones recognized by metallographic examination. The position of the most brittle zone of the ICCGHAZ region (19-20) of the whole joint (established by microstructural examination of all potentially critical zones) is also shown in this diagram. 'Fig. 5' presents the distributions obtained from path a and b which indicates the increased hardness at the root pass of the repair weld metal (RW b) and root-HAZ compared to its cup passes (RW a). The characteristic maximum hardness values obtained in thickness direction (paths d, e and f) and corresponding calculated yield stresses (21) are are given in Table 4. The calculated yield stresses are found to be lower than the values determined by tensile testing, Table 2. The mismatching factor, M corresponding to the individual zones can also show some difference compared with the global M values of Table 2. Hence, this complex mechanical heterogeneity at the vicinity of the crack tip will certainly influence the characteristics of crack initiation and growth.

CTOD evaluations

The present CTOD testing standards have originally been prepared for the testing of homogeneous materials. Therefore, the selection of a yield strength value (in the elastic part of the CTOD formula) for the testing of welded joints presents some difficulties. It is general practise to use the average of the base and weld metal yield strength for HAZ notched specimens. However, with this practise the effects and contribution of the base and weld metals on the crack tip plastic zone development and consequently on the toughness values are assumed to be equal. This solution is an oversimplification of the problem particularly for mismatched joints. 'Fig. 6' schematically shows the difficulty of the yield stress selection for the HAZ/fusion line notched specimens of mismatched welds because of the presence of a mechanical property gradient.

The BS 4515 standard (22) prescribes the recommendations and acceptance criteria for Charpy and CTOD testing of hyperbaric repair welds. According to this standard, the CTOD tests should be conducted with deep notched specimens extracted from notch positions 1b, 3 and 4a (see Table 3). However, this standard does not give any guidance with respect to the mechanical heterogeneity of the crack tip vicinity and respective selection of yield stress. For the thickness and yield strength of the plate used in this study the minimum permissible CTOD value according to this standard is 0.12 mm.

Various yield strength values were used (in relation to the notch position) in the calculation of CTOD values to BS 5762 formula as given in Tables 5 and 6. The so-called "equivalent yield strength" value was obtained on the basis of the average material combination at the vicinity of the crack tip, for example for notch position 4a and 4b:

$$\sigma y = 1/3[\sigma_y^{RW} + \sigma_y^{OW} + \sigma_y^{BM}]) \tag{3}$$

Tables 5 and 6 give further all fracture toughness results in terms of CTOD (BS), CTOD (δ_5) and J for shallow and deep notched specimens with various notch positions. Analysis of the J results and their relationship with both CTOD values will be a topic of up-coming communications and hence will not be discused in this paper. The CTOD (BS) results from these Tables are plotted in 'Fig. 7' according to the respective notch positions. Because of differences in constraint, large differences in toughness occur between shallow and deep (a/W=0.5) through thickness notched weld metal specimens, according to the position of the notch, see 'Fig. 7' position 1a 1b. However, a similar difference is not observed with specimens notched as in positions 2a and 2b with a/W ratios of 0.4 and 0.16 respectively. Shallow cracked specimens (position 2b) showed slightly higher toughness values than the test pieces of position 2a. Clearly, the lowest toughness data (lower than the permissible CTOD value of 0.12 mm) was obtained for specimens having notches at position 4a. In this notch position the fatigue crack was sectioning the repair weld metal and HAZ of the SAW deposit, Table 3. The shallow cracked specimens at this notch position (4b) also produced low pop-in toughness values comparable with the deep notched results (4a). Obviously, this notch position is sampling the most critical regions of the repaired weld joint. Contrary to the expectations, test results obtained from notch position 3 indicate that the repair HAZ region developed in the original SAW weld metal has higher toughness properties than the regions sampled by position 4a, see Table 6 and 'Fig. 7'. Therefore, extensive examination was carried out to determine the location for cleavage crack initiation at these two notch positions. Typical fracture surfaces of the deep and shallow notched specimens for positions 4a and 4b are shown in 'Fig. 8'. For both cases, brittle fracture occured locally at the ICCGHAZ region of the original SAW weld (schematically shown in 'Fig. 4'). Ductile crack initiation and growth was observed at the repair weld metal side of the specimens. The metallographic examination of the ICCGHAZ on sectioned specimens revealed the embrittled grain boundary microstructure shown in 'Fig. 9a'. 'Fig. 9b' suggests that the presence of a coarse grain boundary structure containing a large amount of cementite particles can cause interfacial decohesions and cleavage crack initiation in accordance to dislocation piling up mechanisms suggested by RKR local fracture criterion (23-24).

It is interesting to note that the ICCGHAZ of the SAW weld metal showed this low toughness behaviour only in specimens with repair weld deposit. The CTOD results of the SAW HAZ specimens without a repair weld showed higher values compared with the position 4a results, 'Fig. 7', notch position 4a. This implies that the thermal cycles of the repair weld may caused further embrittlement of the CGHAZ of the original SAW weld and hence low toughness values were only obtained after repair weld deposition. On the other hand the effect of large strength differences along the crack tip in the thickness direction plays significant role in the localised cleavage crack initiation. The highly overmatched repair weld metal protects the crack tip portion which samples the repair weld metal from applied deformation and hence forces the other part of the crack front to accommodate the applied deformation. This asymmetric strength distribution along the crack front of the position 4a and 4b specimens can enhance the attainment of the critical stress for cleavage crack initiation at the ICCGHAZ region of the original SAW weld metal. In consequence, it is believed that the shallow cracked specimens (notch position 4b) also showed low CTOD results despite to their generally lower crack tip constraint. Therefore it can be concluded that the mechanical heterogeneity along the crack front (in this case the presence of overmatching hyperbaric weld metal) can significantly influence the CTOD values arising from the assymmetric distribution of the applied deformation along the crack front.

In order to test the zone of the intersection of two HAZs, Position 5, Table 3, surface notched specimens with a/W=0.4 were prepared. The results from this notch position are shown in 'Fig. 7' which indicates a high toughness level compared with the through thickness notched 4a and 4b specimens. 'Fig. 7' further indicates a certain similarity of the surface notched specimen results, positions 2a, 2b and 5. It is believed that the higher toughness values of the position 5 were obtained under the influence of the neighbouring soft base metal (which readily provides some crack tip relaxation due to the plastic zone development at the undermatched base metal side). Evidently, these high CTOD results are being influenced by a high base metal toughness level included in 'Fig. 7' at position 4a.

Comparison of the CTOD(BS) and CTOD (δ_5) results

A comparison of two CTOD measurement values for deep and shallow notched specimens is shown in Figure 10. The individual values were also given in Tables 5 and 6. Generally, a good agreement between two measurements can be observed if the respective plastic rotation factor is used in the CTOD(BS) formula as shown in Figure 10.

Furthermore, Table 7 presents the CTOD(BS) results calculated by using different yield stresses for the shallow and deep notched specimens at two notch positions 4a and 4b to determine the effect of yield stress selection on CTOD(BS) values. The directly measured CTOD (δ_5) values are also included for comparison purpose. Five yield stress levels were used to calculate the CTOD(BS) values. For this notch position, a significant contribution from the soft (and high toughness) base metal the plastic zone development around the crack tip can be expected. Apparently, if the yield stress of the base metal is being used, the CTOD(BS) results are nearer to the CTOD (δ_5) values. These results are indicate the potential advantage of the directly measured CTOD (δ_5) values since they do not need to be calculated by use of yield strength values. The direct application of the CTOD (δ_5) technique on shallow and deep notched specimens without any plastic rotation factor correction offers another advantage of this technique.

CONCLUSIONS

Characterization of the multipass SAW weld joint repaired with hyperbaric MAG welding process at 16 bar pressure (160 m water depth) was conducted using shallow and deep notched CTOD specimens and via microstructural investigation. The CTOD was determined according to the BS 5762 standard and the GKSS developed δ_5 clip gauge procedure.

The analyses of the experimental results led to the following conclusions:

1. Prior to the extraction of CTOD test pieces and notching, detailed microstructural examination of the repaired joint should be carried out to establish the most critical zones, since additional weld deposition with different welding process, heat input etc. can create new local brittle zones.

2. The hyperbaric repair weld metal presented in this paper has shown good CTOD toughness level. The most critical zone at the root region of the original SAW weld joint was depicted by through thickness shallow and deep notched specimens (position 4a and 4b) under the influence of the overmatched repair weld deposit.

3. The CTOD (δ_5) measurements are consistent with the calculated CTOD values according to the BS 5762 standard for both deep and shallow cracked specimens (a/W=0.16) if the r_p value of 0.25 is used in the latter. The application of the CTOD (δ_5) technique on the shallow and deep notched specimens extracted from repaired weld joints offers a simple and quick toughness estimation technique.

Acknowledgements

The authors wish to acknowledge the financial support of the German Science Foundation (Deutsche Forschungsgemeinschaft, DFG). This work has been conducted as a part-project (C3) in the DFG funded Research Group on "Schadensforschung und Schadensbeseitigung an Stahlkonstruktionen im Wasser". The authors would like thank Prof. H. Hoffmeister and Dr. Huismann, Bundeswehrhochschule Hamburg, for preparation of the welds and Prof. G. Valtinat, TUHH-Hamburg for his intense efforts in the DFG-Research Group. The technical assistance of H. Mackel is gratefully appreciated.

REFERENCES

(1)   BS 5762:1979, Methods for Crack Opening Displacement (COD) Testing, The British Standards Institution (1979).

(2)   ASTM E 1290 - 91, Standard Test Method for Crack-Tip Opening Displacement (CTOD) Fracture Toughness Measurement, 1991.

(3)   EGF Recommendations for Determining the Fracture Resistance of Ductile Materials, EGF P1-90, European Group on Fracture, December 1989.

(4)   Qing-Fen Li, "A Study about Ji and di in Three-Point Bend Specimens with Deep and Shallow Notches", Eng. Fracture Mechanics, Vol. 22, No.1, pp. 9-15, 1985.

(5)   M. Koçak, et al, "Evaluation of HAZ Toughness by CTOD and Tensile Panels", European Symp. on EPFM : Elements of Defect Assessment, Oct. 1989, Freiburg, F.R.G.

(6)   G. Matsoukas, B. Cotterel and Y.-W. Mai, "The Effect of Shallow Cracks on Crack Tip Opening Displacement", Eng. Fracture Mechanics, Vol. 24, No.6, pp. 837-842, 1986.

(7)   Shang-Xian Wu, "Evaluation of CTOD and J-integral for Three-Point Bend Specimens With Shallow Cracks" Proc. of the 7th Int. Conf. on Fracture , ICF 7, Houston, Texas, March 1989, Vol.1, pp. 517-524.

(8)   W. A. Sorem, et al, "An Analytical Comparison of Short Crack and Deep Crack CTOD Fracture Specimens of an A36 Steel", To be Published.

(9)   Li Q.-F et al, "The Effect of a/w Ratio on Crack Initiation Values of COD and J-integral", Eng. Fracture Mechanics, Vol.23, No.5,1986, pp.925-928.

(10)   Zhang, D. Z. and Wang,H., "On the Effect of the Ratio a/W on the Values of δi and Ji in a Structural Steel", Eng. Fracture Mechanics, Vol. 26, No.2, 1987, pp. 247-250.

(11)   O. Kolednik, "On the Calculation of COD from the Clip-gage Displacement in CT and Bend Specimens", Eng. Fracture Mechanics, Vol. 29, No.2, 1988, pp.173-188

(12)   Wu Shang-Xian, "Plastic Rotational Factor and J-COD Relationship of Three Point Bend Specimen", Eng. Fracture Mechanics, Vol.18, No.1, 1983, pp. 83-95.

(13)   D. Hellmann and K.-H. Schwalbe, "Geometry and Size Effects on J-R and δ-R Curves under Plane Stress Conditions", ASTM STP 833, 1984, pp. 577-605.

(14)   "GKSS-Displacement Gauge Systems for Applications in Fracture Mechanics", GKSS-Forschungszentrum Geesthacht, 1991.

(15)   M. Koçak, S. Yao, K-H. Schwalbe, F. Walter, "Effect of Crack Depth (a/W) on Weld Metal Fracture Toughness", Proc. Int. Conf. Welding 90, Geesthacht, FRG, pp. 255-268, Ed. M. Koçak, 1990

(16)   K.-H. Schwalbe, D. Hellmann, "Application of the Electrical Potential Method to Crack Length Measurements using Jonhson's formula", JTEVA, Vol.9, Nr.3, 1981, pp. 218-221.

(17)   K.-H. Schwalbe, D. Hellmann, J. Hereens, J. Knack, J, Müller-Rose, "MeasuremenStable Crack Growth Including Detection of Initiation of Growth Using the DC Potential Drop and the Partial Unloading Methods", ASTM STP 856, E.T. Wessel and F. Loss, Eds., American Society for Testing and Materials, 1985, pp. 3-22.

(18)   M. Koçak, K. Seifert, S. Yao, H. Lampe, "Comparison of Fatigue Precracking Methods for Fracture Toughness Testing of Weldments: Local Compression and Step-Wise High R-Ratio", Proc. Int. Conf. Welding 90, Geesthacht , FRG, pp. 307-318, Ed. M. Koçak, 1990.

(19)   API Spec.2Z, First Edition 1987, "Preproduction Qualification for Steel Plates for Offshore Structures".

(20)   H.I. McHennry, R.M. Denys, "Measurement of HAZ Toughness in Steel Weldments", 5th International Fracture Mechanics Summer School, Dubrovnik 1989,pp.211-222.

(21)   R.J. Pargerter, "Yield Strength from Hardness-a Reappraisal for Weld Metal", Welding Research Bulletin Nov. 1978.

(22)   BS 4515:1984, "Process of Welding of Steel Pipelines on Land and Offshore".

(23   R.O. Ritchie, J.F. Knott, J.R. Rice, " On the Relationship Between Critical Tensile Stress and Fracture Toughness in Mild Steel", JMPS, Vol. 21, 1973, pp. 395-410.

(24)   J.F. Knott, " Macroscopic/Microscopic Aspects of Crack Initiation", Advances in Elasto-plastic Fracture Mechanics, L.H. Larson, Ed., Commission of the European Communities, Joint Research Center, Ispra Establishment, Italy, 1979, pp.1-41.

Table 1 Welding procedure parameters

Parameters	SAW welding	Hyperbaric repair welding (16 bar)
Number of passes	11	7
Wire/flux or gas	LS3MoNi/LW330	SG 3, φ 1 mm/CO2-ArS3
Heat input	27kJ/cm, Δt8l50=12.5 sec	18kJ/cm, Δt8/5=6 sec
Interpass temperature	80°C	80°C

Table 2 Mechanical properties of the base and weld metals

Material	Yield Stress, (N/mm^2)	Tensile Strength, (N/mm^2)	Elongation (%)	Mis-matching factor M
Base Material	388	550	29.7	-
SAW Weld Metal,(1 bar)	582	701	23.4	1.50
Repair Weld Metal, (16 bars-dry)	630	704	24.7	1.62

M=Weld metal yield stress/Base metal yield stress

Table 3 Various notch position for repair welded joint

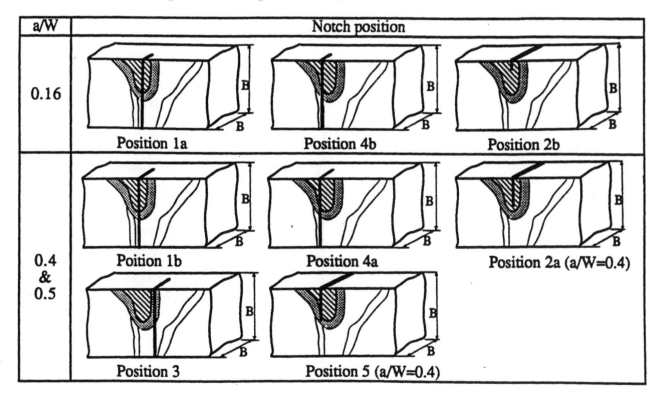

a/W	Notch position
0.16	Position 1a Position 4b Position 2b
0.4 & 0.5	Poition 1b Position 4a Position 2a (a/W=0.4) Position 3 Position 5 (a/W=0.4)

Table 4 Weld metal yield stresses in thickness direction calculated from max.hardness values

Material	HV Micro-hardness	Yield stress-calculated* (MPa)	Mis-matching factor M
RW-cup pass	204	474	1.28
RW-root pass	238	582	1.58
RW-OW HAZ	250	620	1.69
OW	214	506	1.37
BM	170	367.5	-

*σ_{y_w}=3.15HV-168

Table 5. CTOD(BS), CTOD(δ5) and J data of shallow notched specimens (showing notch position and "Equivalent σy "used in CTOD(BS) formula)

Notch position and σy used in CTOD(BS)	CTOD(BS) (mm)	CTOD(δ5) (mm)	J (N/mm)
σy=606 N/mm2 $\sigma_y = 1/2\left[\sigma_y^{RW} + \sigma_y^{OW}\right]$ Position 1a (a/W=0.16)	1.555 (δm) 0.966 (δm) 0.991 (δm)	1.114 δ(m) 1.106 δ(m) 1.107 δ(m)	942 1197 1202
σy=630 N/mm2 $\sigma_y = \sigma_y^{RW}$ Position 2b (a/W=0,16)	0.546 (δm) 0.583 (δm) 0.634 (δm)	0.576 (δm) 0.593 (δm) 0.689 (δm)	748 821 896
σy=503 N/mm2 $\sigma_y = 1/3\left[\sigma_y^{RW} + \sigma_y^{OW} + \sigma_y^{BM}\right]$ Position 4b (a/W=0.16)	0.082 (δc-p) 0.464 (δm) 0.202 (δu) 0.013 (δc-p) 0.309 (δu) 0.577 (δu)	0.092 (δc-p) 0.387 (δm) 0.207 (δu) 0.026 (δc-p) 0.255 (δu) 0.565 (δu)	125 421 310 45 353 775

p - "pop in"

Table 6. CTOD(BS), CTOD(δ5) and J data of deep notched specimens (showing notch position and "Equivalent σy" used in CTOD(BS) formula)

Notch position and σy used in CTOD(BS)	CTOD(BS) (mm)	CTOD(δ5) (mm)	J (N/mm)
σy=606 N/mm2 $\sigma_y = 1/2\left[\sigma_y^{RW} + \sigma_y^{OW}\right]$ Position 1b (a/W=0.5)	0.340 (δm) 0.325 (δm) 0.306 (δm) 0.299 (δm) 0.376 (δm)	0.416 (δm) 0.384 (δm) 0.380 (δm) 0.387 (δm) 0.462 (δm)	384 411 413 393 474
σy=630 N/mm2 $\sigma_y = \sigma_y^{RW}$ Position 2a (a/W=0.4)	0.405 (δm) 0.654 (δm) 0.474 (δm) 0.439 (δm) 0.443 (δm)	0.480 (δm) 0.811 (δm) 0.593 (δm) 0.562 (δm) 0.535 (δm)	479 826 577 581 554
σy=606 N/mm2 $\sigma_y = 1/2\left[\sigma_y^{RW} + \sigma_y^{OW}\right]$ Position 3 (a/W=0.5)	0.270 (δm) 0.368 (δm) 0.370 (δm) 0.366 (δm) 0.332 (δm) 0.371 (δm) 0.372 (δm) 0.349 (δm) 0.342 (δm) 0.391 (δm)	0.263 (δm) 0.466 (δm) 0.443 (δm) 0.388 (δm) 0.420 (δm) 0.418 (δm) 0.440 (δm) 0.452 (δm) 0.425 (δm) 0.466 (δm)	360 525 489 553 445 498 519 484 456 590
σy=503 N/mm2 $\sigma_y = 1/3\left[\sigma_y^{RW} + \sigma_y^{OW} + \sigma_y^{BM}\right]$ Position 4a (a/W=0.5)	0.124 (δu) 0.154 (δu) 0.250 (δu) 0.138 (δu) 0.110 (δu) 0.042 (δc) 0.074 (δc) 0.053 (δc-p) 0.137 (δu) 0.059 (δc) 0.356 (δm)	0.159 (δu) 0.188 (δu) 0.305 (δu) 0.193 (δu) 0.150 (δu) 0.061 (δc) 0.108 (δc) 0.069 (δc-p) 0.164 (δu) 0.074 (δc) 0.441 (δm)	155 192 329 207 158 54 114 74 169 77 443
σy=485 N/mm2 $\sigma_y = 1/2\left[\sigma_y^{BM} + \sigma_y^{OW}\right]$ Position 5 (a/W=0.4)	0.799 (δm) 0.715 (δm) 0.698 (δm) 0.685 (δm) 0.411 (δu) 0.607 (δm) 0.614 (δu) 0.729 (δm)	0.886 (δm) 0.812 (δm) 0.806 (δm) 0.842 (δm) 0.490 (δu) 0.722 (δm) 0.738 (δu) 0.855 (δm)	815 726 742 759 438 637 698 782

Table 7 CTOD(BS)-CTOD(δ5) comparison for a/W=0.16 and 0.5 specimens

	CTOD(BS)-Pos 4b a/W=0.16		CTOD(BS)-Pos 4a a/W=0.5	
	Specimen 10.12	Specimen 11.2	Specimen 10.2	Specimen 10.7
1. σy=388, BM	0.0960	0.0180	0.1390	0.0540
2. σy=582, OW	0.0819	0.0133	0.1228	0.0414
3. σy=630, RW	0.0797	0.0126	0.1204	0.0396
4. σy=606 av.*	0.0807	0.0129	0.1216	0.0405
5. σy=576 av.**	0.0822	0.0134	0.1232	0.0417
CTOD(δ5)	0.0920	0.0250	0.1590	0.0610

*   1/2(RW+OW)

** 1/3(RW+OW+BM)

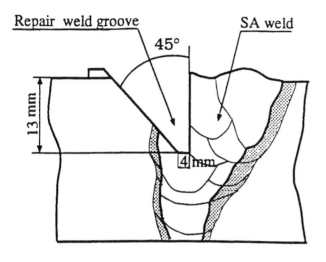

Figure.1 Repair weld groove preparation

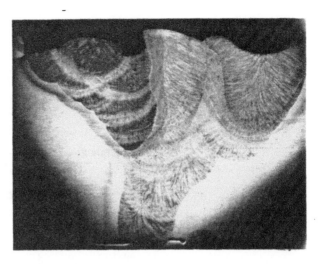

Figure 2 Macro-section of repair weld joint

Figure 3 Fracture surface of the specimen fatigue precracked by SHR technique, position 4a, a/W=0.5

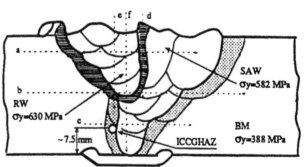

Figure 4 Hardness measurement directions, weld metal mis-matching and ICCG-HAZ region of the original SAW

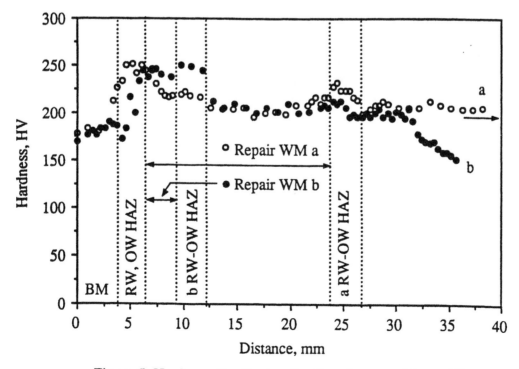

Figure 5 Hardness distribution for directions a and b, see Fig.4

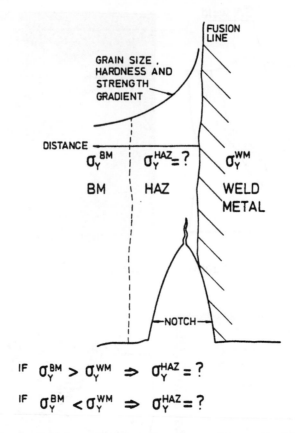

IF $\sigma_Y^{BM} > \sigma_Y^{WM} \Rightarrow \sigma_Y^{HAZ} = ?$

IF $\sigma_Y^{BM} < \sigma_Y^{WM} \Rightarrow \sigma_Y^{HAZ} = ?$

Figure 6 Schematic showing the problem of yield
stress selection for HAZ notched specimens

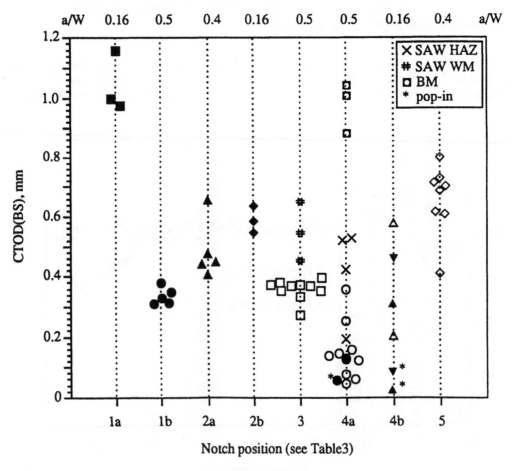

Figure 7 The CTOD(BS) values for all notch positions (T=-10°C)

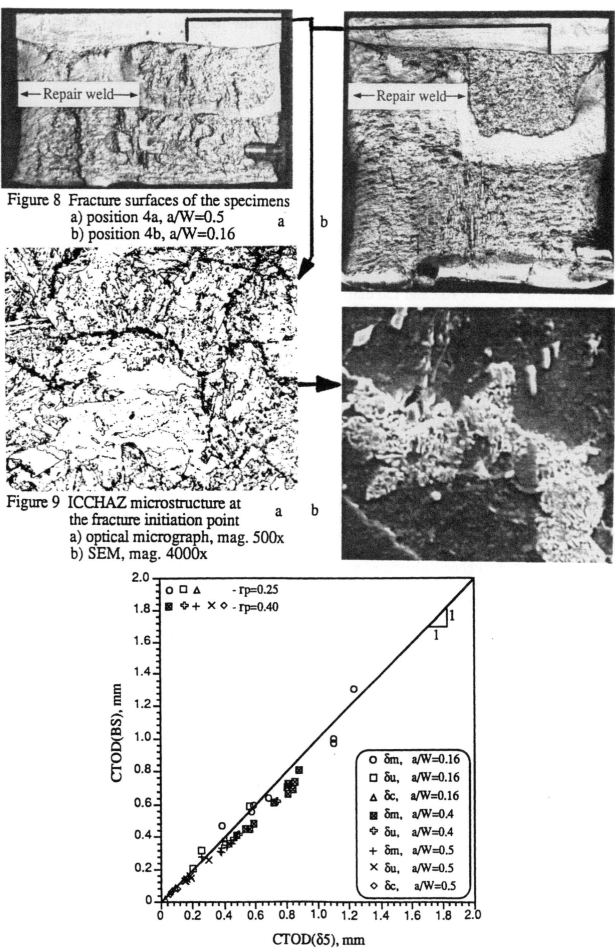

Figure 8 Fracture surfaces of the specimens
   a) position 4a, a/W=0.5
   b) position 4b, a/W=0.16

a   b

Figure 9 ICCHAZ microstructure at
   the fracture initiation point
   a) optical micrograph, mag. 500x
   b) SEM, mag. 4000x

a   b

Figure 10 Relationship between CTOD(BS) and CTOD($\delta5$) for shallow
and deep notched specimens showing a good agreement

TWI

PAPER 37

Shallow-crack toughness results for reactor pressure vessel steel*

T J Theiss, MS (Oak Ridge National Laboratory, USA), S T Rolfe, PhD, PE (University of Kansas) and
D K M Shum, PhD (Oak Ridge National Laboratory, USA)

SUMMARY

The Heavy Section Steel Technology Programme (HSST) is investigating the influence of flaw depth on the fracture toughness of reactor pressure vessel (RPV) steel. To complete this investigation, techniques were developed to determine the fracture toughness from shallow-crack specimens. A total of 38 deep and shallow-crack tests have been performed on beam specimens about 100 mm deep loaded in 3-point bending. Two crack depths (a ≈ 50 and 9 mm) and three beam thicknesses (B ≈ 50, 100, and 150 mm) have been considered. Techniques were developed to estimate the toughness in terms of both the J-integral and crack-tip-opening displacement (CTOD). Analytical J-integral results were consistent with experimental J-integral results, confirming the validity of the J-estimation schemes used and the effect of flaw depth on fracture toughness. Test results indicate a significant increase in the fracture toughness associated with the shallow-flaw specimens in the lower transition region compared to the deep-crack fracture toughness. The increase in shallow-flaw toughness compared with deep-flaw results appears to be well characterised by a temperature shift of 35°C. There is, however, little or no difference in toughness on the lower shelf, where linear-elastic conditions exist for specimens with either deep or shallow flaws.

INTRODUCTION

The HSST programme, sponsored by the U. S. Nuclear Regulatory Commission (NRC), is investigating the influence of crack depth on the fracture toughness of A533B material under conditions prototypic of a pressurised-water reactor (PWR) vessel. Specifically, HSST is quantifying the magnitude of the increase in fracture toughness associated with decreasing flaw depth. The elevated toughness associated with shallow flaws (i.e. shallow-flaw effect) arises from a loss of constraint at the crack-tip because of the proximity of the crack-tip to the specimen surface. This paper presents the final toughness data from the HSST shallow-crack fracture toughness testing programme, post test analysis, and interpretation of the results. More detailed information on the motivation and objectives of the programme, experimental set-up, and verification of the test techniques used can be found in previously published reports by Theiss et al (1-4).

The HSST shallow-flaw programme is a joint experimental/analytical programme which has produced a limited data base of shallow-flaw fracture toughness values and analysis to aid in the transferability of the specimen data to an RPV. The experimental portion of the programme was divided into a development phase and a production phase. The development phase established the techniques appropriate for shallow-crack testing, verified the existence of a shallow-flaw effect in A533B beams, and compared beams of three thicknesses to choose the optimum for the production phase of the programme. Broken ends of the development-phase beams were subsequently remachined and tested, yielding six additional deep-crack beam tests. The production phase of the experimental portion of the programme involved developing a limited

*Research sponsored by the Office of Nuclear Regulatory Research, U.S. Nuclear Regulatory Commission under Interagency Agreement 1886-8011-9B with the U.S. Department of Energy under Contract DE-AC05-84OR21400 with Martin Marietta Energy Systems, Inc.

The submitted manuscript has been authored by a contractor of the U.S. Government under Contract No. DE-AC05-84OR21400. Accordingly, the U.S. Government retains a nonexclusive, royalty-free license to publish or reproduce the published form of this contribution, or allow others to do so, for U.S. Government purposes.

data base of shallow-crack toughness values at various temperatures. The analytical portion of the shallow-flaw programme consisted of both pretest and post test analyses of the test specimens. The pretest analysis was used to size the instrumentation for the tests and to select an appropriate shallow-crack depth. Post test analysis verified the techniques developed to determine the toughness for the shallow-crack specimens and quantified the degree of constraint in the deep- and shallow-crack test specimens.

EXPERIMENTAL PROGRAMME

The specimen configuration chosen for all testing in the shallow-crack programme is the single-edge-notch-bend (SENB) specimen with a through-thickness crack (as opposed to the 3D surface crack). A beam approximately 100 mm deep (W) was selected for use in the HSST shallow-crack project. To maintain consistency with ASTM standards, the beams were tested in three-point bending. All testing was conducted on unirradiated reactor material (A533 Grade B, Class 1 steel) with the cracks oriented in the thickness (S) direction to simulate the material conditions of an axial flaw in an RPV. Specimens were taken from the central, homogeneous region of the source plate to minimise metallurgical differences between the material surrounding a shallow and deep flaw.

The specimen thickness was varied in the development-phase tests to examine the influence of thickness on toughness. Three beam thicknesses were used: $B \approx 50$, 100, and 150 mm. The span (S) for the 50-mm-thick beam was 4W or 406 mm. The spans for the 100- and 150-mm beams were increased to assure failure without exceeding the load capacity of the loading fixture. Figure 1 shows three of the beam sizes used. Both shallow- and deep-crack specimens were tested at each thickness. Beams 100-mm thick were used for the production phase tests.

The development and production phases of the testing programme resulted in 14 and 18 data points, respectively, and an additional 6 deep-crack beams of varying thickness were tested, providing a total of 38 data points. All but one of the development-phase tests were conducted at -60°C, and the 6 additional deep-crack beams were tested at -45°C using beams with different thicknesses. The production-phase tests used one beam geometry (100 x 100 mm) but were conducted at various temperatures. Two crack depths (one shallow and one deep) were used for the shallow-crack fracture toughness testing programme. The nominal shallow-crack depth chosen was $\approx$ 10 mm, which was representative of the flaw depth of interest for RPVs, as discussed by Cheverton et al (5), and yields a normalised crack depth (a/W) of 0.10. All deep-crack specimens were cracked to an a/W value of approximately 0.5.

Instrumentation was attached to the specimens to permit independent determination of both J-integral and crack-tip opening displacement (CTOD). The J-integral was determined from the load-line-displacement (LLD) vs. load diagram. The LLD was measured using a reference bar attached to the beam loading fixture and a micrometer attached to the neutral axis of the beam. CTOD was determined from crack-mouth-opening-displacement (CMOD) gauges mounted directly on the crack-mouth of the specimen. Toughness data are expressed in terms of CTOD according to ASTM E1290-89, Crack-Tip Opening Displacement (CTOD) Fracture Toughness Measurement. ASTM E399, Plane-Strain Fracture Toughness of Metallic Materials, was used to analyse the deep-crack specimens to determine if the test results could be considered "valid" plane-strain (K_{Ic}) data. ASTM E813, J_{Ic}, A Measure of Fracture Toughness, is not applicable to these tests since typically the failures were cleavage events; however, critical J-integral cleavage values (J_c) were determined for each test. The shallow-crack toughness formulations are as similar as possible to the deep-crack ASTM standard toughness formulations.

Material Properties

Two heats of unirradiated A 533 B material were tested in this programme. The development-phase and six additional deep-crack beams were taken from the HSST-CE plate and were tested in the T-S orientation. The production-phase beams were taken from HSST Plate 13B, were given a final heat treatment prior to machining, and were tested in the L-S orientation. Material properties used in the analysis of the shallow-crack test results for both the development and production phases are included in Table 1. Additional information on the shallow-crack production-phase-material characterisation is being published by Iskander[*], and properties of the source plates have been reported by Naus et al (6) & (7).

Table 1 Material properties for A 533 B steels used in HSST shallow-crack programme

Development phase and six deep-crack beams	Production phase
HSST CE–WP	HSST Plate 13B after postweld heat treatment
$E = 202 - 0.0626\,T$, GPa	$E = 202 - 0.0626\,T$, GPa
$\nu = 0.3$	$\nu = 0.3$
$\sigma_0 = 211 + 55{,}000 / (T + 273)$, MPa	$\sigma_0 = 430 - 0.223\,T + 0.014\,T^2$, MPa
$\sigma_u = 371 + 55{,}000 / (T + 273)$, MPa	$\sigma_u = 609 + 0.618\,T + 0.00927\,T^2$, MPa
$\sigma_f = 1/2\,(\sigma_0 + \sigma_u)$	$\sigma_f = 1/2\,(\sigma_0 + \sigma_u)$
$RT_{NDT} = -35°C$	$RT_{NDT} = -15°C$ (centre material)

T = temperature, °C.

Crack Tip Opening Displacement (CTOD) Determination

The plastic component of CTOD is determined experimentally from the plastic component of CMOD and the rotation factor (RF), according to ASTM E1290. The plastic displacement of the crack flanks is assumed to vary linearly with distance from the plastic centre of rotation. In this way, the plastic CMOD can be related to the plastic CTOD. The plastic centre of rotation is located ahead of the crack tip a distance equal to the rotation factor multiplied by the remaining ligament (W-a). The rotation factor in ASTM E1290 is 0.44 but is a function of specimen geometry and material. RF values determined for deep-crack beams are not necessarily appropriate for otherwise identical shallow-crack beams.

An experimental technique was utilised in this programme to locate the neutral axis of the beam ahead of the crack tip, using strain gauges on each face of the beam. Assuming the plastic centre of rotation is located at the neutral axis of the beam, the RF can be determined. Since the rotation factor relates the plastic component of CMOD to the plastic component of CTOD, only plastic strains were used to determine the rotation factor. The rotation factors determined using this technique were relatively insensitive to load once plastic strains became nontrivial and were consistent upon each face of the beam. The RF for a beam was taken as the average calculated RF from each face. Four deep-crack beams were strain gauged, yielding an average RF of 0.44. Eight shallow-crack beams were gauged to yield an average RF of 0.49. The rotation factor used for the CTOD toughness calculation is the average of the values from this technique for the two crack depths.

A parametric evaluation was performed to assess the sensitivity of the RF upon the calculated CTOD toughness. This evaluation indicated that the plastic component of CTOD is not sensitive to the value of the rotation factor. Shallow-crack beams are less sensitive to the rotation factor than deep-crack beams. A 25% increase in rotation factor increases the plastic CTOD by about 5% and 17% for the shallow and deep-crack geometries, respectively. The rotation factor is insensitive to beam thickness and absolute beam dimensions, varying only with a/W ratios for a given material and specimen depth. Based on the comparison of deep and shallow RF and the insensitivity of CTOD to RF, the ASTM E1290 value of the RF of 0.44 would appear to be appropriate for both deep and shallow-cracked A533B specimens.

Critical CTOD (δ_c) values calculated using the RF values just described and ASTM E1290-89 are included in Table 2. The ratios of the shallow-to-deep-crack lower-bound δ_c at T-RT_{NDT} of -25°C and -10°C are 3.3 and 4.9, respectively, which is consistent with the A36 and A517 results

---

[*] S. K. Iskander, "Preliminary Report on the Characterization of HSST Plate 13B in the L-S Orientation," to be issued as a USNRC NUREG report.

published by Sorem et al (8) and Smith & Rolfe (9), respectively. Further examination of these data indicate little variation of δ_c as a function of beam thickness for either the deep-crack or shallow-crack beams.

Table 2 HSST shallow-crack test data

HSST beam No.	Temperature (°C)	S (mm)	B (mm)	W (mm)	a (mm)	Failure load (kN)	CTOD total (mm)	J integral (MPa-mm)	K_c from CTOD (MPa·$\sqrt{m}$)	K_c from J (MPa·$\sqrt{m}$)
Development phase										
3	−36	406	51	100	10.0	600.0	0.586	261	265	243
4	−61	406	51	100	51.8	128.1	0.048	42	96	97
5	−55	406	51	99	51.2	139.7	0.049	48	97	105
6	−59	406	51	100	51.9	184.6	0.117	102	149	152
7	−59	406	51	94	10.2	483.5	0.137	92	132	144
8	−60	406	51	94	9.6	657.4	0.476	284	245	254
9	−62	406	51	94	9.5	552.4	0.352	173	212	198
10	−60	406	51	94	14.0	489.3	0.235	143	180	180
11	−57	864	102	94	8.4	472.4	0.196	101	157	152
12	−57	864	102	95	49.8	116.5	0.061	50	108	106
13	−60	864	102	94	8.8	501.7	0.357	208	213	217
14	−60	864	152	93	8.7	723.2	0.346	225	209	226
15	−59	864	153	94	8.7	684.1	0.146	85	136	139
16	−58	864	153	94	50.0	170.4	0.060	46	107	102
Six Additional Deep-Crack Beams										
12A	−44	406	102	94	51.0	251.8	0.077	60	120	117
13A	−46	406	102	94	50.8	293.1	0.111	86	144	140
14A1	−44	406	51	93	50.2	135.2	0.121	93	150	145
14A2	−44	406	51	93	50.8	102.7	0.043	39	90	94
15A	−47	406	153	94	50.7	435.0	0.096	79	133	134
16A	−43	406	153	94	51.9	348.3	0.062	51	107	108
Production phase										
17	−6	610	102	102	52.6	245.1	0.116	98	144	147
18	−24	610	101	102	10.6	777.1	0.468	238	239	231
20	−4	610	101	101	10.8	823.3	1.733	987	453	469
21	−23	610	101	102	10.7	724.1	0.306	152	194	185
22	−7	610	101	102	10.9	793.5	0.942	566	334	355
24	−7	610	102	102	52.0	269.1	0.367	270	255	245
25	−39	610	102	102	52.0	238.4	0.110	85	145	138
26	−40	610	102	102	11.0	740.1	0.355	175	213	199
27	−22	610	101	102	10.7	787.3	0.559	242	261	233
28	−6	610	101	102	10.3	832.7	1.242	788	384	419
31	−40	610	102	102	51.5	205.5	0.063	51	110	108
32	−103	610	102	102	11.1	417.7	0.016	20	69	68
33	−103	610	102	102	10.7	339.8	0.009	13	53	54
34	−106	610	101	102	10.4	431.0	0.017	21	72	70
35	−7	610	102	102	51.7	244.2	0.121	97	147	147
36	−38	610	102	102	51.6	176.1	0.042	35	89	89
37	−39	610	102	102	10.8	745.9	0.263	135	183	175
38	−39	610	102	102	10.8	755.3	0.206	106	162	155

J-Integral, J_c, Determination

J-integral fracture toughness (J_c) was determined for each beam using LLD data. Little or no crack growth took place in these tests so ASTM E813 was not strictly applicable. The technique used to calculate the J-integral divides the elastic and plastic components of J and uses only the plastic area under the load v. LLD curve and a plastic η factor as presented by Sumpter (10). The equations used to determine the shallow-crack J-integral toughness are as follows :

$$J_c = J_{el} + J_{pl},$$

where [1]

$$J_{el} = K_c^2(1-v^2)/E \quad \text{and} \quad [2]$$

$$J_{pl} = \eta_{pl}U_{pl}/(B(W-a)), \quad [3]$$

where U_{pl} is plastic area under load v. LLD curve.

The J-integral toughness values for each beam are given in Table 2. J-integral results are consistent with the CTOD results. The ratios of the shallow-to-deep-crack lower-bound J_c at T-RT_{NDT} of -25°C and -10°C are 2.4 and 2.9 respectively, which is consistent with the δ_c results.

Comparison of δ_c and J_c Values

CTOD toughness values can be converted into J-integral values as discussed by Barsom and Rolfe (11) according to $J_c = m \cdot \sigma_f \cdot \delta_c$, where σ_f (flow stress) is the average of the yield and tensile strengths, and m is the constraint parameter. Since J_c and δ_c are known for each specimen, comparison of J_c and δ_c allows m to be determined as a function of crack depth. Plots of J v. CTOD show a linear relationship between the two toughness expressions. The constraint parameter, m, for each test was determined using the critical toughness (J_c and δ_c) values. The constraint parameter as a function of crack depth yields repeatable results as shown in Fig. 2. The average deep-crack constraint parameter is 1.5. The average shallow-crack constraint parameter is 1.0 except for three beams which resulted in a significantly elevated m value. These three shallow-crack beams were tested on the lower shelf, where CTOD calculations are based on a value of twice the yield stress. An average constraint parameter value of 1.9 was found for these beams. These constraint parameter values are consistent with results of Kirk & Dodds (12). Critical CTOD was converted into J-integral expressions using the average values of m shown in Fig. 2. J-integral values converted from CTOD will be referred to as J_c (CTOD); J_c (LLD) refers to J-integral values determined directly from LLD records.

Stress-Intensity Factor, K_c, Determination

Typically RPV fracture toughness values are expressed in terms of the critical stress-intensity factor, K_{Jc}. The two J-integral toughness expressions were converted into elastic-plastic K_{Jc} values according to $K_{Jc} = \sqrt{(J E')}$. The plane-strain value of E', $E/(1-v^2)$, is justified because thickness had little influence on the resulting toughness values. Figure 3 and Table 2 contain comparisons of K_c from the two J estimation techniques used (CTOD and LLD). As shown in Fig. 3, the two J estimation techniques give similar values of K_{Jc}. The maximum difference between the two techniques is about 10%. The averaged difference is less than 1%.

The toughness data expressed in terms of K_{Jc} (CTOD) vs. normalised temperature (T-RT_{NDT}) are presented in Table 2 and Fig. 4. The data show a significant increase in the fracture toughness for shallow-crack specimens in the transition region of the A533B toughness curve. All but one (data point in Fig. 4 with the arrow) of the specimens failed in cleavage. As expected, the shallow-crack specimens tested on the lower shelf, where linear-elastic behaviour occurs, showed little or no toughness increase. The specimens had crack depths that were deep ($a \approx 50$ mm) or shallow ($a \approx 10$ mm), except for one beam with a crack depth of 14 mm. This intermediate-crack-depth specimen also appears to show the shallow-crack-toughness elevation.

The shallow-crack toughness increase can be quantified in terms of a ratio of toughness values at one temperature or as a temperature shift. In terms of K_{Jc}, the shallow-crack toughness increase is approximately 60% at T-$RT_{NDT} \approx$ -25°C. Figure 4 shows the shallow-crack and deep-crack test data with approximate lower bound curves. The shallow-crack lower-bound curve was formed using the deep-crack lower bound curve shifted by 35°C. The shifted deep-crack lower-bound curve fits the shallow-crack data well at all test temperatures.

Toughness data in terms of $K_{Jc}(CTOD)$ are plotted as a function of beam thickness for all of the tests conducted at T - $RT_{NDT} \approx$ -25°C and -11°C in Fig. 5. As indicated in Figs. 4 and 5, the toughness values for the shallow- and deep-crack specimens from the 100- and 150-mm-thick beams generally are consistent with the 50-mm-thick data. However, there appears to be slightly more data scatter associated with the 50-mm-thick beams than with the 100- and 150-mm-thick beams. None of the deep-crack tests strictly meet the requirements of ASTM E399 for a valid plane-strain K_{IC} result because of insufficient crack depth. The beams which had otherwise linear-elastic test records and were sufficiently thick for valid results are marked in Fig. 5.

POST TEST SPECIMEN ANALYSIS

Detailed finite-strain, finite element analyses were performed for six specimens that were tested at -40°C. Three of the specimens (Beams # 36, 31, & 25) are deep-flaw specimens with $a/W \approx$ 0.5, while the remaining three are shallow-flaw specimens (Beams # 38, 37, & 21) with nominal $a/W \approx$ 0.1. One of the primary objectives of these analyses was to evaluate the J estimation techniques developed to determine shallow-crack toughness with analytical J-integral values. Additional details of the post test analysis are discussed by Theiss et al (4).

Material Models

Two material models have been adopted in the analysis of the test specimens. The first material model, known as the unadjusted model, simulates the unirradiated tensile properties of A533B (HSST Plate 13B) at -40°C as determined from limited material characterisation. The unadjusted model assumed a value of E = 207.5 GPa and a uniaxial yield stress of 454 MPa. The unadjusted material model underestimated the experimentally measured displacements of the specimens by approximately 20%. The second model, known as the adjusted model, was used to reduce the stiffness of the unadjusted material model based on comparisons with measured values by reducing both the Young's modulus and the uniaxial yield stress from their pretest characterisation values. The magnitudes of the reductions were consistent with scatter in tensile material properties. Within the linear-elastic region the Young's modulus was reduced by 5% such that E = 196.5 GPa. The yield stress was reduced by 9% such that $\sigma_0 = 413$ MPa.

Finite Element Models and Analyses Assumptions

The finite-strain, elastic-plastic post test analyses are performed using the finite element code ABAQUS (13). The analyses assume a rate-independent, J_2 (isotropic-hardening) incremental plasticity theory as implemented in ABAQUS. The planform for both the shallow- and deep-flaw specimen is 102 mm x 610 mm. The initial flaw-depth is 10.2 mm for the shallow-flaw specimen and 50.8 mm for the deep-flaw specimen. The shallow-flaw specimen geometry is modeled with the finite element mesh which is made up of 914 10-node generalised-plane-strain isoparametric elements with a total of 2883 nodes. The deep-flaw specimen geometry is modeled with the finite element similarly refined mesh made up of 922 10-node generalised-plane-strain isoparametric elements with a total of 2903 nodes. These 10-node elements behave as conventional 8-node isoparametric elements except for an extra degree-of-freedom (DOF) that allows for uniform straining in a direction perpendicular to the plane of the mesh (13). In a plane strain analysis the out-of-plane DOF is not active. The integration order of the elements is 2x2.

J-integral values were determined from up to 29 (shallow-flaw) or 31(deep-flaw) paths surrounding the crack tip to verify path independence. A measure of the mesh refinement is that the elastically determined K value using these meshes was within 99.5% of the value reported by Tada (14). Convergence requirements of the elastic-plastic finite element results to be presented were specified by means of limiting the maximum value of the residual nodal force per unit thickness at any node.

Comparison of Calculated and Measured Mechanical Responses

Experimental measurements for the load (P), LLD and CMOD are available for the six specimens considered in these analyses. Comparison of the calculated values and experimental data provides an additional basis for establishing confidence in the experimental estimates of the fracture toughness. Results of the comparison are detailed by Theiss et al (4). As an example, Fig. 6 indicates the extent of the agreement between the calculated and measured P-LLD response for the shallow-flaw specimens.

Figure 6 presents two sets of calculated responses along with the measured responses for the three shallow-flaw specimens (Beams # 38, 37, 26). The measured responses of these specimens appear to indicate the presence of general-yielding conditions at the onset of crack initiation. The two sets of calculated curves correspond to two cases of analysis conditions labeled as Case A and B. The calculated P-LLD curve corresponding to Case A was determined based on a = 10.2 mm and the unadjusted material model. The finite element analysis was carried out under "load-control" in that reaction forces were specified along the back-side of the specimen ahead of the crack tip. Analysis results for Case B were determined based on the actual flaw depth of a = 10.8 mm and the adjusted material model described previously. The finite element analysis was carried out under "displacement-control" as displacements were specified along the back-side of the specimen ahead of the crack tip. As evident from Fig. 6, analysis conditions for Case B appear to result in better agreement between the calculated and measured mechanical responses both in the elastic and plastic regimes.

Comparison for the a/W = 0.5 geometry have been carried out in a similar fashion with details presented by Theiss et al (4). Discrepancies are observed between results based on the unadjusted material model, a/W = 0.5 and the measured responses. Post-test examination of the fracture surfaces for the three deep-flaw specimens indicate that the actual flaw depth is 51.6 mm (a/W = 0.502) rather than the assumed value of 51 mm (a/W = 0.50) or an increase of only 1%. Analysis results determined based on the nominal flaw depth of a/W = 0.50 and the adjusted material model appear to result in better agreement between the calculated and measured mechanical responses both in the elastic and plastic regimes. In subsequent discussions these are referred to as Case D conditions.

Comparison of Analytical J-Integral Values and J-Estimation Schemes

The J-integral values have been determined as a part of the post test analysis of the specimens. The magnitude of critical values of P and LLD (P_c, LLD_c) for the three shallow-flaw specimens at crack initiation are indicated in Table 3. The magnitude of the analytical J-integral based on attaining LLD_c are denoted as J_{LLDc}. Since the calculated P-LLD curve for the shallow-flaw specimen underestimates the measured value of LLD at a given value of P, J_{LLDc} can be regarded as an upper bound to the actual value of the J-integral at the onset of crack initiation. On the other hand, the magnitude of the J-integral based on attaining P_c can be regarded as a lower bound to the actual value of the J-integral. These J-integral values are denoted as J_{Pc}. Analogous results for the deep-flaw geometry based on Case D conditions are listed in Table 4.

J-estimation schemes based on the magnitude of the experimentally determined LLD and converted from CTOD for both the shallow- and deep-flaw geometry have been presented. The J-integral values based on these estimation schemes, denoted here as J_{EXP} (LLD) and J_{EXP} (CTOD) are listed in Tables 3 and 4 for the shallow and deep-crack beams.

Results in Tables 3 and 4 indicate that both values of J_{EXP} calculated from measured values of LLD compare favourably with the finite element results. The general accuracy of the LLD-based J-estimation scheme for the deep-flaw geometry is verified by the observation that all of the deep-flaw J_{EXP} (LLD) values are between J_{LLDc} and J_{Pc}. A similar degree of accuracy is observed for the case of the shallow-flaw geometry, although one of the J_{EXP} (LLD) values is slightly higher than the upper-bound J_{LLDc} value. The J-integral estimation scheme based on CTOD appears to overestimate the fracture toughness for these shallow-flaw specimens since all three values of J_{EXP} (CTOD) were above the upper-bound value of J_{LLDc}.

Table 3 Experimental and analytical results of fracture toughness for the shallow-flaw (a/W = 0.1) specimen based on Case B conditions

Beam No.	P_C (kN)	LLD_C (mm)	$JLLD_C$ (kN/m)	JP_C (kN/m)	J_{EXP} (LLD) (kN/m)	J_{EXP} (CTOD) (kN/m)
38	756	2.71	115	112	106	116
37	746	3.08	142	108	135	148
26	740	3.45	169	105	175	201

Table 4 Experimental and analytical results for the fracture toughness for deep-flaw (a/W = 0.5) specimens based on Case D conditions

Beam No.	P_C (kN)	LLD_C (mm)	$JLLD_C$ (kN/m)	JP_C (kN/m)	J_{EXP} (LLD) (kN/m)	J_{EXP} (CTOD) (kN/m)
36	176	1.24	44	35	35	34
31	206	1.41	57	49	51	53
25	238	1.82	91	71	85	93

CONCLUSIONS

Several conclusions can be drawn from the results of the HSST shallow-crack fracture toughness programme. First, the estimation techniques developed to determine the shallow-crack CTOD and J-integral toughness are valid for these tests. Second, the results showed conclusively that A 533 B shallow-flaw beam specimens have a significant increase over deep-crack specimens in CTOD or J_C toughness and K_C toughness in the transition region. Shallow-crack beams had crack depths ranging from 9 to 14 mm (a/W $\approx$ 0.1 to 0.14), while deep-crack beams had 50-mm-deep cracks (a/W $\approx$ 0.5). Third, little or no difference in toughness on the lower shelf takes place where linear-elastic conditions exist for specimens with either deep or shallow flaws. Next, variations in the beam thickness from 50 to 150 mm had little or no influence on the toughness in both the shallow- and deep-crack specimens. Finally, in the transition region, the increase in shallow-flaw toughness compared with deep-flaw results appears to be well characterised by a temperature shift of 35°C.

REFERENCES

(1)   T. J. Theiss, "Recommendations for the Shallow-Crack Fracture Toughness Testing Task Within the HSST Program," USNRC Report NUREG/ CR-5554 (ORNL/TM-11509), August 1990.

(2)   T. J. Theiss, G. C. Robinson, and S. T. Rolfe, "Preliminary Test Results from the Heavy-Section Steel Technology Shallow-Crack Toughness Program," *Proceedings of the ASME Pressure Vessel & Piping Conference,* Pressure Vessel Integrity, PVP Vol. 213/MPC-Vol. 32, pp. 125–129, ASME 1991.

(3)   T. J. Theiss and J. W. Bryson, "Influence of Crack Depth on Fracture Toughness of Reactor Pressure Vessel Steel,"*Constraint Effects in Fracture ASTM STP 1171*, E. M Hackett, Ed., American Society for Testing and Materials, Philadelphia, (to be published).

(4)   T. J. Theiss. D. K. M. Shum, and S. T. Rolfe, "Experimental and Analytical Investigation of the Shallow-Flaw Effect in Reactor Pressure Vessels" USNRC Report NUREG/ CR-5886 (ORNL/TM-12115), July 1992.

(5)   R. D. Cheverton, S. K. Iskander, and D. G. Ball, "Review of Pressurized-Water-Reactor-Related Thermal Shock Studies," *Fracture Mechanics: Nineteenth Symposium, ASTM STP 969,* T. A. Cruse, Ed., American Society for Testing and Materials, Philadelphia, 1988, pp. 752-766.

(6)   D. J. Naus et al., "SEN Wide-Plate Crack-Arrest Tests Using A 533 Grade B Class 1 Material: WP-CE Series," USNRC Report NUREG/CR-5408 (ORNL/TM-11269), November 1989.

(7)   D. J. Naus et al., "Crack Arrest Behavior in SEN Wide Plates of Quenched and Tempered A 533 Grade B Steel Tested Under Nonisothermal Conditions," USNRC Report NUREG/CR-4930 (ORNL-6388), August 1987.

(8)   W. A. Sorem, R. H. Dodds, Jr., and S. T. Rolfe, "An Analytical Comparison of Short Crack and Deep Crack CTOD Fracture Specimens of an A36 Steel," *WRC Bulletin 351*, Welding Research Council, New York, February 1990.

(9)   J. A. Smith, and S. T. Rolfe, "The Effect of Crack Depth to Width Ratio on the Elastic-Plastic Fracture Toughness of a High-Strength Low-Strain Hardening Steel," *WRC Bulletin 358,* Welding Research Council, New York, November 1990.

(10)   J. D. G. Sumpter and J. W. Hancock, "Shallow Crack Toughness of HY80 Welds: An Analysis Based on T Stresses," *Int. J. Pres. Ves. & Piping* 45, 207–221 (1991).

(11)   J. M. Barsom and S. T. Rolfe, *Fracture and Fatigue Control in Structures*, Prentice-Hall, Englewood Cliffs, N.J., 1987.

(12)   M. T. Kirk, and R. H. Dodds, Jr., "J and CTOD Estimation Equations for Shallow Cracks in Single Edge Notch Bend Specimens," Civil Engineering Studies, Structural Research Series No. 565, Dept. of Civil Engineering, Univ. of Illinois, UILI-ENG-91-2013, Jan. 1991.

(13)   ABAQUS *Theory Manual*, Version 4-8, Hibbitt, Karlson and Sorensen, Inc., Providence, R.I., 1989.

(14)   H. Tada, P. C. Paris and G. R. Irwin, *The Stress Analysis of Cracks Handbook*, Del Research Corporation, Hellertown, Pa., 1973.

Fig. 1.   Three specimen thicknesses used in the shallow-crack programme.

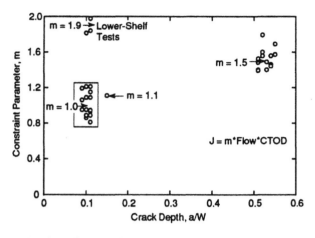

Fig. 2. Constraint parameter (m) values as a function of crack depth (a/W).

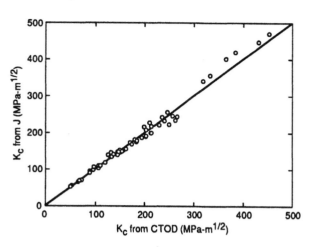

Fig. 3. Comparison of K_C (LLD) and K_C (CTOD) toughness values.

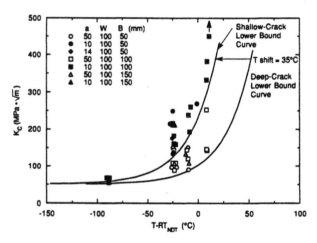

Fig. 4. Toughness (K_C) data vs normalised temperature with lower bound curves.

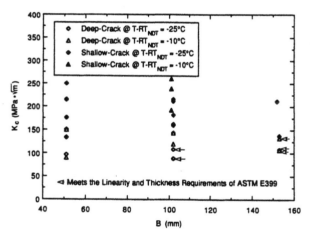

Fig. 5. Toughness (K_C) data vs beam thickness at $T\text{-}RT_{NDT} = -25°C$, and $-10°C$.

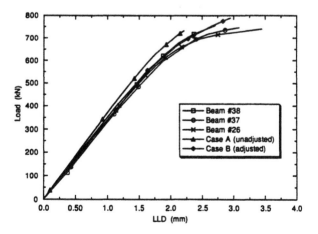

Fig. 6. P-LLD comparison for a/W = 0.1.

Fig. 3. Comparison of K_Q (LLD) and K_Q (CTOD) roughness values.

Fig. 2. Constraint parameter (m) values as a function of crack depth (a/W)

Fig. 5. Toughness (K_Q) data vs beam thickness at RTNDT = 25°C and -10°C.

Fig. 4. Toughness (K_Q) data vs normalised temperature with lower bound curves.

Fig. 6. R-LLD comparison for a/W = 0.1.

SESSION C: Applied conditions

3D elastic-plastic finite element analysis for the determination of the J-integral in single edge notched bend specimens (Paper 36)
A SPROCK and W DAHL

J and CTOD estimation equations for shallow cracks in single edge notch bend specimens (Paper 2)
M T KIRK and R H DODDS Jr

3D elastic-plastic finite element analysis for CTOD and J in SENB, SENAB and SENT specimen geometries (Paper 14)
R H LEGGATT and J R GORDON

The limits of applicability of J and CTOD estimation procedures for shallow-cracked SENB specimens (Paper 28)
Y-Y WANG and J R GORDON

The influence of weld strength mismatch on crack tip constraint in single edge notch bend specimens (Paper 29)
M T KIRK and R H DODDS Jr

Emission and reflections of surface waves from shallow surface notches (Paper 35)
C THAULOW, M HAUGE and J SPECHT

Shallow surface cracks in welded T butt plates — a 3D linear elastic finite element study (Paper 22)
B FU, J V HASWELL and P BETTESS

3D elastic-plastic finite element analysis for the determination of the J-integral in single edge notched bend specimens

A Sprock, Dipl-Ing and W Dahl, Prof Dr.rer.nat. Dr-Ing (Institute for Ferrous Metallurgy, University of Aachen, FRG)

SUMMARY

The determination of J from global parameters load, load-line displacement and CMOD is possible if the existing BSI and ASTM-
standards are used. These standards are valid for deep notched
specimens (a/W>0.45) (1,2). For shallow crack specimens however
there are not any sufficient standards which are suitable to determine J from the global parameters. Finite element analyses are
therefore required to determine accurate values for J at shallow
cracks. It is possible by taking the results of the finite element
analyses to develop a function η dependent on the a/W-ratio, which
gives the connection between J and the global parameters.

This work tries to confirm the value of J obtained from the analyses with a finite element program. First of all the necessity
of exact modeling of the investigated specimen is emphasised. Then
a comparison between two different stress analysis procedures is
made. After this the path independance of the J-integral will be
proved and the J-distribution versus the specimen thickness will
be shown.

INTRODUCTION

As part of the project "An International Research Project To
Develop Shallow Crack Fracture Mechanics Tests" the Technical
University of Aachen as an in-kind contributor has carried out
some finite element analyses of SENB fracture mechanics specimens. The object of this project was to develop shallow crack
fracture mechanics toughness test methods that would be complementary to the existing BSI and ASTM standard deep notch test
methods.

One relevant aspect was the quantitative determination of the
J-integral. Therefore 3D finite element analyses were performed
with SENB specimens with different a/W-ratios.

MATERIAL PROPERTIES AND SPECIMEN GEOMETRIES

The material input data used for the finite element analyses based on measured stress strain curve for BS 4360 Grade 50 EE steel plate, with post yield work hardening gradient factored by R=3. Table 1 shows the material data and Fig.1 the stress strain curve used (3).

Table 1

Material Data

Steel	Yield strength N/mm²	Tensile strength N/mm²	Tensile/Yield	Young's modulus N/mm²	Poisson's ratio
M1/HI	388	814	2.1	203000	0.3

The specimen configuration to be analysed by finite element method is single edge notched bend. The specimen dimensions are given in Fig.2 (3).

NUMERICAL INVESTIGATIONS

The FE-results were used to develop methods of deriving J from the finite element analyses. The methods should consider a form of the relationship between J and the measureable specimen parameters and describe the local behaviour at shallow cracks. The determination of the 'right' J values can be achieved with finite element analyses.

1. Firstly it is necessary to develop a good model that describes the investigated specimen exactly. This requires the development of a 3D model. 2D meshes are not able to predict the global behaviour correctly. Plane strain or plane stress would oversimplify the analysis. Fig.3 shows the finite element model of SENB with an a/W=0.05 ratio. It contains 485 elements in 5 layers and 2650 nodes. The model has 7950 degrees of freedom.

The model allows a good prediction of the global behaviour and also of the local behaviour near the crack tip. This was shown in the first year of the project.

2. By using the FE-program ABAQUS (4) it is possible to consider nonlinear material behaviour, when the NLGEOM-option (Nonlinearity-GEOMetrie) is chosen or to define a problem as a 'small displacement' analysis (in this case the MNO-option (Material-Nonlinearity-Only) is chosen). This means that the geometric nonlinearity is ignored - the kinematic relationship are linearised. The errors in such an approximation are of the order of the strains and rotations compared to unity. For the determination of J the MNO-option seems to be more convenient, since large strains are not expected in the finite element analyses.

Figure 4 shows the global behaviour of the investigated 3 point single edge notched bend specimen. It compares an analysis using tho MNO- and the NLGEOM-option. There are only small differences at large displacements. The load values of the load displacement curve for the MNO-option are slightly higher because the MNO-option does not consider the geometric nonlinearity.

Local behaviour near the cracktip is characterized by the J-integral. As figure 5 shows, there is a good agreement between the MNO- and NLGEOM-analysis. The differences of the results gained by the use of both options are very small; so it is possible to determine the J value with each of the options. The MNO-analysis as well as the NLGEOM-analysis supplies convenient J-values.

3. The determination of a convenient J from the finite element analysis should be checked carefully. This is done by proving the path independence of J and by showing the J distribution versus the specimen thickness. Because of the path independence of the J-integral several contours should be defined around the crack tip. Each contour provides an evaluation of the J-integral. Since the integral should be path independent, these values should be the same. In fact they will differ because of

the numerical approximation of the geometric model used. Figure 6 shows the path independence of J. After just a few contours the specimen with the a/W=0.5 ratio reaches a constant value. Also the specimen with the a/W=0.05 ratio shows a constant value, so that the path independence is sufficiently guaranteed.

The J-distribution versus the specimen thickness for deep notched specimens shows the maximum values in the middle of the specimen, see Fig.7, whereas the shallow crack specimen has a distinct maximum near the specimen surface, see Fig.8. Figure 7 and 8 also show the integrated average values J_m. In the deep notched specimen, see Fig.7 the integrated value J_m in the middle of the specimen is smaller than the value of the J distribution, whereas J_m in the shallow crack specimen lies above the J distribution, see Fig.8. Although J_m deviates in an insignificant way from the J value in the middle of the specimen, it is questionable which value should be used to develop the eta-function. The J value inside of the specimen characterizes the local behaviour in the middle of the specimen.

Because of the specimen geometries a load behaviour similar to the plane strain can be assumed. The constraint in the middle of the specimen leads to an increase of the stress inside the specimen which is the reason for the rise of the maximum stress triaxiality. The exact load behaviour in the middle of the specimen will be underestimated if only J_m is considerd, see Fig.7, because J_m represents all J values from the middle to the surface. Because of a decrease of the stresses at the surface of the specimen a load behaviour similar to the plane stress can be assumed. The J values are considerably lower at the surface, so the integrated value J_m considers the whole load behaviour, whereas the J value in the middle of the specimen describes the load behaviour similar to a plane strain assumption.

In the case of the shallow crack specimens shown in Fig.8, the maximum J values are near to the surface. Because of this maximum at the surface the integrated value J_m now lies above the J distribution in the middle of the specimen.

The question of the more convenient J which is necessary to give a formula that describes the eta function could not be clearly determined. The difference between both J values is 9% at a/W = 0.5 and 5% at a/W = 0.05, but it should be possible with both values to find out an appropriate eta function which is also convenient for shallow crack specimens.

CONCLUSIONS

The performed analyses have shown that it is possible to give a convenient J value for shallow crack specimens by the finite element analyses. The results of the finite element analyses are used to develop a function for η, which gives the connection to the global measurable parameters. Further investigations have to be done to complete the existing ASTM and BSI standards for shallow crack test methods.

REFERENCES

1. ASTM E 813-89 "Standard Test Method for J_{IC}, a Measure of Fracture Toughness"

2. BS 7448, Part 1, 1991: "Fracture Mechanics Toughness Tests, Part 1. Method for Determination of K_{Ic}, Critical CTOD and Critical J Values of Metallic Materials"

3. R.H. Leggatt: "An International Research Project to Develop Shallow Crack Fracture Mechanics Tests - European Contribution". Report No. 5574/22/91, The Welding Institute, February 1991

4. Hibbitt, H.D., B.I. Karlsson and P. Sorensen (1989) ABAQUS User's Manual, Theory Manual Version 4.8.5., Providence, R.I., U.S.A., 1989

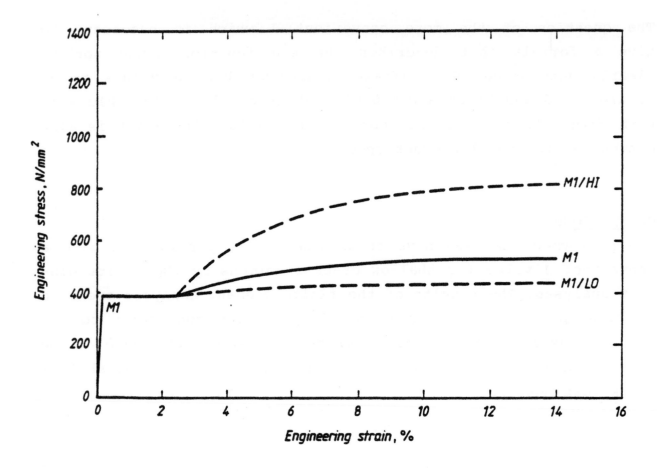

Fig.1: Stress strain curve M1/H1 used for the finite element analyses

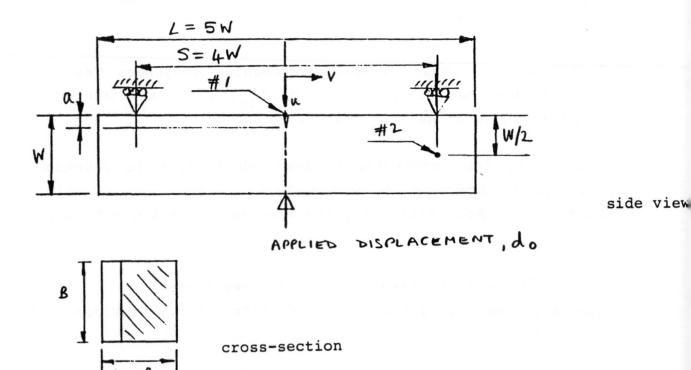

Fig.2: Specimen geometry SENB W = B = 25 mm

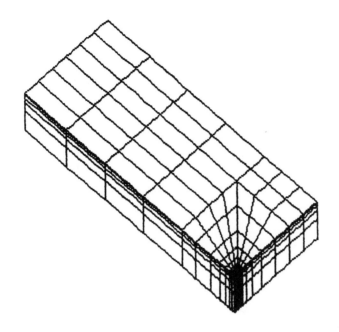

Fig.3: 3D finite element model SENB B = W = 25 mm, a/W=0.05

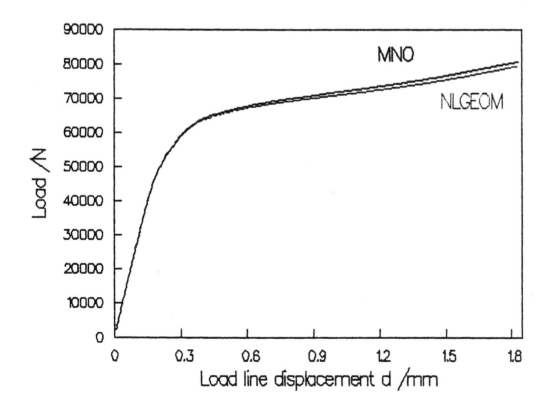

Fig.4: Comparison load-displacement MNO- and NLGEOM-analysis SENB
       a/W=0.05

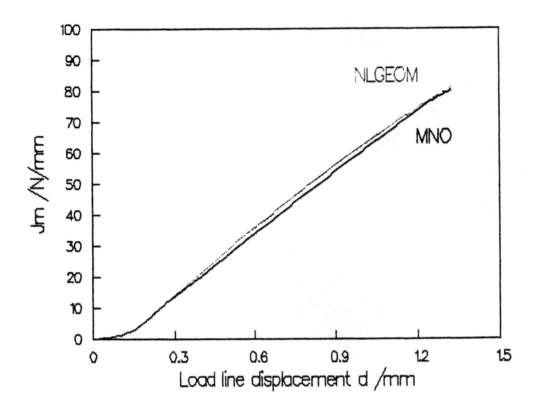

Fig.5: Comparison J-Integral MNO- and NLGEOM-analysis SENB
       a/W=0.05

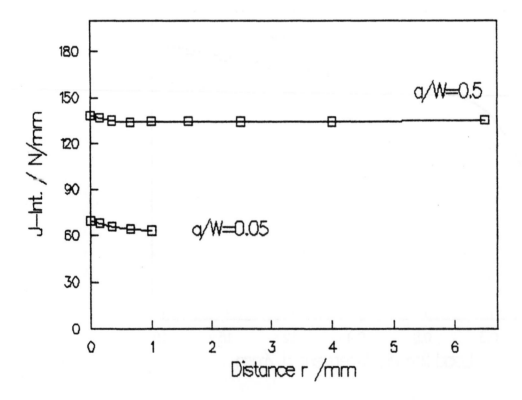

Fig.6: J-Integral dependent on the distance r from the crack tip
       SENB

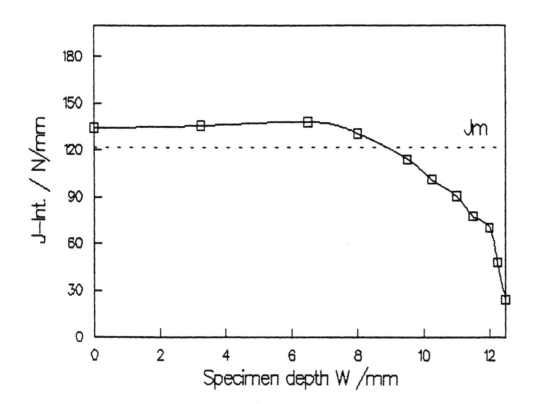

Fig.7: J-distribution versus specimen thickness SENB a/W=0.5

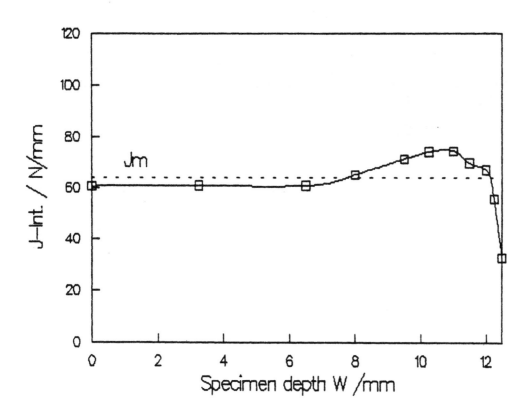

Fig.8: J-distribution versus specimen thickness SENB a/W=0.05

TWI

PAPER 2

J and CTOD estimation equations for shallow cracks in single edge notch bend specimens

M T Kirk, BS, MS (US Naval Surface Warfare Center) and R H Dodds Jr, BS, MS, PhD (University of Illinois, USA)

SUMMARY

Fracture toughness values determined using shallow cracked single edge notch bend, SE(B), specimens of structural thickness are useful for structural integrity assessments. However, testing standards have not yet incorporated formulas that permit evaluation of J and CTOD for shallow cracks from experimentally measured quantities (i.e. load, crack mouth opening displacement (CMOD), and load line displacement (LLD)). Results from two dimensional plane strain finite−element analyses are used to develop J and CTOD estimation strategies appropriate for application to both shallow and deep crack SE(B) specimens. Crack depth to specimen width (a/W) ratios between 0.05 and 0.70 are modelled using Ramberg Osgood strain hardening exponents (n) between 4 and 50. The estimation formulas divide J and CTOD into small scale yielding (SSY) and large scale yielding (LSY) components. For each case, the SSY component is determined by the linear elastic stress intensity factor, K_I. The formulas differ in evaluation of the LSY component. The techniques considered include: estimating J or CTOD from plastic work based on load line displacement $(A_{pl}|_{LLD})$, from plastic work based on crack mouth opening displacement $(A_{pl}|_{CMOD})$, and from the plastic component of crack mouth opening displacement (CMOD_{pl}). $A_{pl}|_{CMOD}$ provides the most accurate J estimation possible. The finite−element results for all conditions investigated fall within 9% of the following formula:

$$J = \frac{K^2(1 - \nu^2)}{E} + \frac{\eta_{J-C}}{Bb}A_{pl}|_{CMOD}; \text{ where } \eta_{J-C} = 3.785 - 3.101\frac{a}{W} + 2.018\left(\frac{a}{W}\right)^2$$

The insensitivity of η_{J-C} to strain hardening permits J estimation for any material with equal accuracy. Further, estimating J from CMOD rather than LLD eliminates the need to measure LLD, thus simplifying the test procedure. Alternate, work based estimates for J and CTOD have equivalent accuracy to this formula; however the η coefficients in these equations depend on the strain hardening coefficient. CTOD estimates based on scalar proportionality of CTOD_{lsy} and CMOD_{pl} are highly inaccurate, especially for materials with considerable strain hardening, where errors up to 38% occur.

INTRODUCTION AND OBJECTIVE

Standardized procedures for fracture toughness testing require both sufficient specimen thickness to insure predominantly plane strain conditions at the crack tip and a crack depth of at least half the specimen width (1−3). Within certain limits on load level and crack growth, these restrictions insure the existence of very severe conditions for fracture as described by the Hutchinson Rice Rosengren (HRR) crack tip fields (4,5). These conditions make the applied driving force needed to initiate fracture in a laboratory specimen lower than the value needed to initiate fracture in common civil and marine structures where such severe geometric conditions are not present. As a consequence, structures often carry greater loads without failure than predicted from fracture toughness values measured using standardized procedures.

Both Sumpter (6) and Kirk and Dodds (7) achieved good agreement between the initiation fracture toughness of single edge notched bend, SE(B), specimens and structures containing part−through semi−elliptical surface cracks by matching thickness and crack depth between specimen and structure. These results demonstrate that toughness values determined from shallow cracked SE(B) specimens are appropriate for assessing the fracture integrity of structures. However, testing standards have not yet incorporated formulas permitting evaluation of J and CTOD for shallow cracks from experimental measurements (i.e. load, crack mouth opening displacement (CMOD), and load line displacement (LLD)). This investigation develops J and CTOD estimation procedures applicable for both shallow and deep crack fracture toughness testing for materials with a wide range of strain hardening characteristics.

APPROACH

Two dimensional, plane strain finite element analyses of SE(B) specimens are performed for crack depths from 0.05 to 0.70 a/W with Ramberg–Osgood strain hardening coefficients (n) between 4 and 50. Table 1 details the conditions considered. The analyses provide load, CMOD, and LLD records to permit evaluation of coefficients relating J and CTOD to measurable quantities. The range of parameters considered in these analyses allows evaluation of the dependence of these coefficients on a/W and n. The estimation formulas divide J and CTOD into small scale yielding (*SSY*) and large scale yielding (*LSY*) components. In each formula, the *SSY* component is defined by the linear elastic stress intensity factor, K_I. The formulas differ only in the *LSY* component. Procedures to estimate the *LSY* component include:

1. J_{lsy} from plastic work (area under the load vs. LLD_{pl} curve, or $A_{pl}|_{LLD}$)
2. $CTOD_{lsy}$ as a fraction of $CMOD_{pl}$ using a rotation factor
3. $CTOD_{lsy}$ from plastic work (area under the load vs. LLD_{pl} curve, or $A_{pl}|_{LLD}$)
4. J_{lsy} and $CTOD_{lsy}$ from plastic work (area under the load vs. $CMOD_{pl}$ curve, or $A_{pl}|_{CMOD}$)
5. $CTOD_{lsy}$ as a fraction of $CMOD_{pl}$ without the notion of a rotation factor

Existing standards employ the first two techniques (1–3); the remainder are new proposals.

Table 1: SE(B) specimens modelled.				
	Ramberg–Osgood Strain Hardening Coefficient (n)			
a/W	4	5	10	50
0.05	✔	✔	✔	✔
0.15	✔	✔	✔	✔
0.25	✔	✔	✔	✔
0.50	✔	✔	✔	✔
0.70	✔	✔	✔	✔

J AND CTOD ESTIMATION PROCEDURES

Current standards

Existing test standards for J and CTOD (1–3) employ the following estimation formulas:

$$J = \frac{K^2(1 - v^2)}{E} + \frac{\eta_{pl}}{Bb}A_{pl}|_{LLD} \qquad [1]$$

$$CTOD = \frac{K^2(1 - v^2)}{m\sigma_{flow}E} + \frac{r_{pl}b\,CMOD_{pl}}{r_{pl}b + a} \qquad [2]$$

where

K	linear elastic stress intensity factor	v	Poisson's ratio
η_{pl}	plastic eta factor	B	specimen thickness
$CMOD_{pl}$	plastic component of CMOD	b	remaining ligament, $W - a$
$A_{pl}|_{LLD}$	area under the load vs. LLD_{pl} curve	m	constraint factor
σ_{flow}	flow stress, average of yield and ultimate*	r_{pl}	plastic rotation factor

Values of $\eta_{pl}, m,$ and r_{pl} are well established for perfectly plastic materials based on closed form solutions. For deeply cracked specimens ($a/W \geq 0.5$), current test standards use $\eta_{pl} = 2$, $m = 2$, and $r_{pl} = 0.44$ Sumpter (8) and Wu., et al. (9) have proposed the following relations to account for crack depth less than 0.5 a/W:

$$\eta_{pl} = 0.32 + 12\frac{a}{W} - 49.5\left(\frac{a}{W}\right)^2 + 99.8\left(\frac{a}{W}\right)^3 \quad \text{for } a/W < 0.282 \qquad [3]$$

$$\eta_{pl} = 2.0 \qquad\qquad\qquad\qquad\qquad\qquad\quad \text{for } a/W \geq 0.282$$

* ASTM E1290 and BS 5762 both use yield stress in eqn. [2]. In this investigation, flow stress is used instead.

$$r_{pl} = 0.5 + 0.42\frac{a}{W} - 4\left(\frac{a}{W}\right)^2 \qquad \text{for } a/W < 0.172 \qquad [4]$$

$$r_{pl} = 0.463 - 0.04\frac{a}{W} \qquad \text{for } a/W \geq 0.172$$

Sumpter derived the η_{pl} equation from limit analyses of the SE(B), while Wu, Cotterell, and Mai used a slip line field analysis to determine the variation of r_{pl} with a/W. Material strain hardening alters the deformation characteristics of the specimen, thereby altering η_{pl}, m, and r_{pl}. Existing procedures neglect any influence of strain hardening.

New proposals

Estimation formulas used in existing test standards have received the greatest attention as the coefficients relating J and CTOD to experimental measurements are amenable to closed form solution, at least in the non−hardening limit. For hardening materials, closed form solution is not possible, therefore either experimental techniques (10) or finite element analyses (11) are used to provide data from which η_{pl}, m, and r_{pl} are calculated. Quantities other than $CMOD_{pl}$ and $A_{pl}|_{LLD}$ measured during a test can also be related to J or CTOD, if the proper proportionality coefficient is known. The following are some alternatives:

1. Estimate $CTOD_{lsy}$ from plastic work ($A_{pl}|_{LLD}$):

    $$CTOD = \frac{K^2(1 - v^2)}{m\sigma_{flow}E} + \frac{\eta_{C-L}}{Bb\sigma_{flow}}A_{pl}|_{LLD} \qquad [5]$$

    This formula is analogous to eqn. [1] for J testing

2. Use plastic work defined by the area under the load vs. $CMOD_{pl}$ curve ($A_{pl}|_{CMOD}$) to estimate either J_{lsy} or $CTOD_{lsy}$:

    $$J = \frac{K^2(1 - v^2)}{E} + \frac{\eta_{J-C}}{Bb}A_{pl}|_{CMOD} \qquad [6]$$

    $$CTOD = \frac{K^2(1 - v^2)}{m\sigma_{flow}E} + \frac{\eta_{C-C}}{Bb\sigma_{flow}}A_{pl}|_{CMOD} \qquad [7]$$

    This technique eliminates the need for LLD measurement, which simplifies J testing.

3. Express $CTOD_{lsy}$ as a fraction of $CMOD_{pl}$:

    $$CTOD = \frac{K^2(1 - v^2)}{m\sigma_{flow}E} + \eta_\delta CMOD_{pl} \qquad [8]$$

    Equations [8] and [2] are functionally the same, thus η_δ and r_{pl} are related:

    $$\eta_\delta = \frac{r_{pl}b}{r_{pl}b + a} \qquad [9]$$

    Sorem (11) found r_{pl} to be extremely sensitive to the CTOD−CMOD relationship for shallow cracks. This estimation procedure was proposed to circumvent this sensitivity. The validity of this approach is based on the observed, nearly linear dependence of $CTOD_{lsy}$ on $CMOD_{pl}$ in finite−element solutions.

In this investigation, finite element analyses provide data from which η_{pl}, m, r_{pl}, η_{C-L}, η_{J-C}, η_{C-C}, and η_δ are calculated.

FINITE ELEMENT MODELLING

Two dimensional, plane strain finite element analyses of SE(B) specimens are performed using conventional small strain theory. The analyses are conducted using the POLO−FINITE analysis software (12) on an engineering workstation. Dodds, et al. (13) previously detailed the finite−element modelling procedure.

Finite element models are constructed for a/W ratios of 0.05, 0.15, 0.25, 0.50, and 0.70. The SE(B) specimens have standard proportions; the unsupported span is four times the specimen width. Ramberg Osgood strain hardening coefficients of 4, 5, 10, and 50 model materials ranging from highly strain hardening to nearly elastic − perfectly plastic. Figure 1 illustrates these stress − strain curves. Symmetry of both geometry and loading permit use of a half−symmetric model. Each model contains approximately 400 elements and 1300 nodes; the $a/W = 0.25$ model is shown in Figure 2. The same half−circular core of elements surrounds the crack tip in all models. Crack tip element sizes range from 0.2% to 0.02% of the crack length depending on the a/W modelled. The J−integral is computed at each load step using a domain integral method (14−15). J values calculated over domains adjacent to and remote from the crack tip are within 0.003% of each other, as expected for deformation plasticity. CTOD is computed from the blunted shape of the crack flanks using the $\pm 45°$ intercept procedure. LLD is taken as the relative displacement in the loading direction of a node on the symmetry plane located approximately $0.4b$ ahead of the crack tip and of a node located above the support. This procedure eliminates the effect of spuriously high displacements in the vicinity of both the load and support points. The η, m, and r_{pl} coefficients are determined from these results by calculating the slope of the quantities indicated in Table 2 at each load step. Slope calculation is initiated with data from the final three load steps. Data from earlier load steps are included in this calculation until the linear correlation coefficient (r) falls below 0.999. This procedure eliminates data from the first few load steps, which are predominantly elastic, and therefore not expected to provide reliable relationships between plastic quantities.

Table 2:	Calculation of coefficients in J and CTOD estimation formulas.		
Eqn.	Coefficient	X	Y
[1]	η_{pl}	$\dfrac{A_{pl}\vert_{LLD}}{Bb}$	J_{pl}
[6]	η_{J-C}	$\dfrac{A_{pl}\vert_{CMOD}}{Bb}$	J_{pl}
[2] [5] [7] [8]	m	$\delta\sigma_{flow}$ †	J
[5]	η_{C-L}	$\dfrac{A_{pl}\vert_{LLD}}{Bb\sigma_{flow}}$	δ_{pl}
[7]	η_{C-C}	$\dfrac{A_{pl}\vert_{CMOD}}{Bb\sigma_{flow}}$	δ_{pl}
[2]	r_{pl}	$CMOD_{pl}$ *	δ_{pl}
[8]	η_δ	$CMOD_{pl}$	δ_{pl}

*: μ is the slope of this line, $r_{pl} = \dfrac{\mu a}{b(1-\mu)}$

†: δ = CTOD

RESULTS AND DISCUSSION

The variation of the η, m, and r_{pl} coefficients with a/W and n determined from the finite−element results is summarized in Figures 3−4, and in the Appendix. Solutions for non−hardening materials, where available, are indicated on the figures. Each coefficient shows considerable variation with crack depth. The variation with material strain hardening is also a common feature of all coefficients except η_{J-C}, which relates J_{lsy} to $A_{pl}\vert_{CMOD}$. η_{J-C} is essentially independent of n for $a/W \geq 0.15$. The remainder of this section examines the differences between perfectly plastic and finite element solutions, and the errors associated with each estimation procedure. Finally, recommendations of J and CTOD estimation formulas for use in fracture testing of SE(B) specimens are made.

Perfectly plastic and finite element proportionality coefficients

The variation of both r_{pl} and η_δ with a/W for a low strain hardening material, Figure 3 e−f, agrees well with the slip line field solution of Wu, et al. (9) above $a/W=0.15$. However, at smaller a/W the elastically dominated response, ignored in the slip line field solution, causes a deviation between the slip line field and finite−element r_{pl} and η_δ values.

The variation of η_{pl} with a/W determined by finite element analysis has a different functional form than determined by Sumpter (8) using a limit load solution, Figure 3a. The limit load derivation employs the following approximation for plastic work:

$$U_{PL} = P_{LIM} \cdot LLD_{PL} \qquad [10]$$

where

$$P_{LIM} = \left(\zeta BW^2\sigma_{flow}\right)/S$$

$$\zeta = 1 - 0.33a/W - 6(a/W)^2 + 15.5(a/W)^3 - 19.8(a/W)^4$$

S = unsupported bend span

Thus, the accuracy of η_{pl} values determined by limit analysis depends on the equivalence of plastic work calculated by eqn. [10] and the actual plastic work (area under a load vs. LLD_{pl} diagram) for a strain hardening material. This equivalence is not achieved even for the low strain hardening $n=50$ material, as illustrated in Figure 5.

J and CTOD estimation errors

Figure 6 illustrates the variation of J and CTOD with LLD and CMOD for an $a/W=0.15$, $n=5$ SE(B) determined by finite element analysis. This dependence of fracture parameters on measurable quantities is contrasted with that predicted by the J and CTOD estimation procedures using η and m coefficients calculated from the finite element results. Work−based J and CTOD estimates (eqns. [1], [6], [5], and [7]) match the finite element results much more closely than do formulas that calculate $CTOD_{lsy}$ as a fraction of $CMOD_{pl}$ (eqns. [2] and [8]). Figure 7 shows J and CTOD estimation errors, more clearly illustrating the differences between the estimation procedures. To evaluate the effects of both a/W and n on estimation accuracy, the following error measure is defined:

$$\overline{ERR} = \sum_{i=1}^{N} |E_i FP^{fe}{}_i| \bigg/ \sum_{i=1}^{N} FP^{fe}{}_i \qquad [11]$$

where

$E_i = \dfrac{FP^{est}{}_i - FP^{fe}{}_i}{FP^{fe}{}_i} 100$, error at load step i

N total number of load steps

$FP^{est}{}_i$ estimated J or CTOD at load step i

$FP^{fe}{}_i$ J or CTOD at load step i from finite−element analysis

For an $a/W=0.15$ / $n=5$ SE(B), the $\overline{ERR}$ value for CTOD estimation using r_{pl}, eqn. [2], is 21%. Comparison of this value with the data in Figure 7 demonstrates that $\overline{ERR}$ is a root mean square error measure.

The variation of $\overline{ERR}$ with a/W and n for the six estimation procedures is shown in Figure 8. Errors associated with work−based J and CTOD estimates (work calculated from CMOD) are below 5% for all a/W and n. If work is instead calculated from LLD, J and CTOD estimation errors are also generally below 5%, with the exception of shallow cracks in a very low strain hardening material ($a/W=0.05$, $n=50$). However, equations that express $CTOD_{lsy}$ as a fraction of $CMOD_{pl}$ are inaccurate for all a/W ($\overline{ERR}>17\%$) in highly strain hardening materials ($n \leq 5$). As the maximum estimation error can exceed $\overline{ERR}$ by up to a factor of 2 (Figure 7), $\overline{ERR}$ values above 17% are clearly excessive. Accuracy improves ($\overline{ERR}<12\%$) for materials with less strain hardening ($n \geq 10$). However, these estimates have accuracy comparable to work−based CTOD estimates only for deep cracks in essentially non−hardening materials. Thus, the validity of assumptions made in deriving the various estimation procedures directly affects their accuracy. J and CTOD estimation from plastic work is achieved by partitioning total work into SSY and LSY components. Additive separation is exact because, for a linear elastic body, $K^2(1 - v)/E$ is the elastic strain energy. Conversely, the linear relation between $CTOD_{lsy}$ and $CMOD_{pl}$ assumed in eqns. [2] and [8] cannot exist (exactly) for any body with an elastic component that varies with load (i.e. for any amount of strain hardening). Strain hardening strongly influences the linearity of the $CTOD_{lsy}$ − $CMOD_{pl}$ relationship, as illustrated in Figure 9. Thus, eqns. [2] and [8] work best for minimally strain hardening materials.

Recommended J and CTOD estimation procedures

Requirements for accurate estimation

The formulas used to evaluate fracture parameters from experimental data should not introduce substantial errors into the J and CTOD estimates. This need for accuracy favors estimating J_{lsy} and $CTOD_{lsy}$ from plastic work. Even though estimation of the LSY component from plastic work requires numerical integration of experimental data, this seems warranted to reduce errors by up to five fold; compare Figure 8d to Figure 8f. In addition to using inherently accurate formulas, selecting η, m, and r_{pl} coefficients corresponding to a specific a/W and material should not be a potential error source. In view of the ambiguity attendant to fitting experimental stress−strain data with a power law curve, insensitivity of η, m, and r_{pl} to material strain hardening would be extremely advantageous.

J Estimation

The only procedure that meets both of the aforementioned requirements is J estimation from plastic work based on CMOD. By fitting the data in Figure 3b, the variation of η_{J-C} with a/W is expressed as follows:

$$\eta_{J-C} = 3.785 - 3.101\frac{a}{W} + 2.018\left(\frac{a}{W}\right)^2 \text{ for all } n, \ 0.05 \le \frac{a}{W} \le 0.70 \quad [12]$$

Figure 10 shows this fit together with the η_{J-C} data. The use of η_{J-C} values from eqn. [12] produces estimation errors of at most 9%, and generally much less, as illustrated in Figure 11. In situations where fracture toughness in terms of a critical J value is desired, estimation using eqns. [6] and [12] is clearly superior to estimating J from plastic work based on LLD, where η_{pl} depends on material strain hardening coefficient. Further, estimating J from CMOD rather than LLD eliminates the need to measure LLD, which simplifies the test procedure.

Despite the clear advantages of estimating J from plastic work based on CMOD, estimation based on LLD may be necessary for very shallow cracks due to experimental complexities associated with clip gage attachment (16). If J estimation using LLD is unavoidable, η_{pl} can be indexed less ambiguously to the ratio of the ultimate strength to the yield strength than to the strain hardening coefficient. The ultimate tensile strength for a Ramberg Osgood material is obtained by solving for the tensile instability point, converting true stress to engineering stress, and taking the ratio of this value with 0.2% offset yield stress. This calculation gives:

$$\mathbf{R} = \frac{\sigma_u}{\sigma_o} = \left(\frac{1}{0.002n}\right)^{\frac{1}{n}} \bigg/ \exp\left(\frac{1}{n}\right) \quad [13]$$

This equation, along with the information in the Appendix, is used to determine the appropriate η_{pl} value for the experimental conditions of interest based on data from a simple tensile test.

CTOD estimation

As noted previously, CTOD estimation from plastic work is considerably more accurate than CTOD estimation directly from $CMOD_{pl}$. Use of eqn. [5] or [7] is therefore preferred to eqn. [2] or [8]. However, the η, m, and r_{pl} coefficients in all of these equations depend strongly on n. The strain hardening coefficient is estimated from $\mathbf{R}$ as described previously. Appropriate m and η_{C-L} or η_{C-C} values for the experimental conditions of interest are then determined from information in the Appendix.

SUMMARY AND CONCLUSIONS

Results from two−dimensional, plane strain finite element analyses are used to develop J and CTOD estimation strategies appropriate for application in both shallow and deep crack SE(B) specimens. Crack depth to specimen width (a/W) ratios between 0.05 and 0.70 are modelled using Ramberg Osgood strain hardening exponents (n) between 4 and 50. The estimation formulas divide J and CTOD into small scale yielding (SSY) and large scale yielding (LSY) components. For each case, the SSY component is determined by the linear elastic stress intensity factor, K_I. The formulas differ in evaluation of the LSY component. The techniques considered include: estimating J or CTOD from plastic work based on load line displacement ($A_{pl}|_{LLD}$), from plastic work based on crack mouth opening displacement ($A_{pl}|_{CMOD}$), and from the plastic component of crack mouth opening displacement ($CMOD_{pl}$). $A_{pl}|_{CMOD}$ provides the most accurate J estimation possible. The finite element results for all conditions investigated fall within 9% of the following formula:

$$J = \frac{K^2(1-v^2)}{E} + \frac{\eta_{J-C}}{Bb}A_{pl}|_{CMOD}; \text{ where } \eta_{J-C} = 3.785 - 3.101\frac{a}{W} + 2.018\left(\frac{a}{W}\right)^2$$

The insensitivity of η_{J-C} to strain hardening permits J estimation for any material with equal accuracy. Further, estimating J from CMOD rather than LLD eliminates the need to measure LLD, thus simplifying the test procedure. Alternate, work based estimates for J and CTOD have equivalent accuracy to this formula; however the η coefficients in these equations depend on the strain hardening coefficient. CTOD estimates based on scalar proportionality of $CTOD_{lsy}$ and $CMOD_{pl}$ are highly inaccurate, especially for materials with considerable strain hardening, where errors up to 38% occur.

ACKNOWLEDGEMENTS

The authors are pleased to acknowledge the suggestion of J R Gordon and M G Dawes to index η, r, and m coefficients to a strength ratio rather than to the strain hardening coefficient. This research was conducted as part of the Surface Ship and Submarine Materials Block under the sponsorship of I. L. Caplan (Carderock Division, Naval Surface Warfare Center (CDNSWC), Code 011.5). The work supports CDNSWC Program Element 62234N, Task Area RS345S50. The work of the first author was conducted at the University of Illinois as part of a training program administered by CDNSWC. Support for R.H. Dodds was provided by CDNSWC under contract number N61533−90−K−0059. Computational support was made possible by CDNSWC Contract No. N61533−90−K−0059.

REFERENCES

(1) ASTM Standard Test Method for J_{Ic}, A Measure of Fracture Toughness, E813−89.

(2) ASTM Standard Test Method for Crack−Tip Opening Displacement (CTOD) Fracture Toughness Measurement, E1290−89.

(3) BS 5762: 1979, "Methods for Crack Tip Opening Displacement (*COD*) Testing," British Standards Institution, London, 1979.

(4) Hutchinson, J.W., *Journal of Mechanics and Physics of Solids*, Vol. 16, pp. 13−31, 1968.

(5) Rice, J.R., and Rosengren, G.F., *Journal of Mechanics and Physics of Solids*, Vol. 16, pp. 1−12, 1968.

(6) Sumpter, J.D.G., *ASTM STP 995*, pp. 415−432, 1989.

(7) Kirk, M.T., and Dodds, R.H., *Engineering Fracture Mechanics*, Vol. 39, No. 3, pp. 535−551, 1991.

(8) Sumpter, J.D.G., *Fatigue and Fracture of Engineering Materials and Structures*, Vol. 10, No. 6, pp. 479−493, 1987.

(9) Wu, S.X., Cotterell, B., and Mai, Y.W., *International Journal of Fracture*, Vol. 37, pp. 13−29, 1988.

(10) Wu, S.X., *Engineering Fracture Mechanics*, Vol. 18, No. 1, pp. 83−95, 1983.

(11) Sorem, W.A., Dodds, R.H., and Rolfe, S.T., *International Journal of Fracture*, Vol. 47, pp. 105−126, 1991.

(12) Dodds, R.H., and Lopez, L.A., *International Journal for Engineering with Computers*, Vol. 13, pp. 18−26, 1985.

(13) Dodds, R.H., Anderson, T.L., and Kirk, M.T., *International Journal of Fracture*, Vol. 48, pp. 1−22, 1991.

(14) Li, F.Z., Shih, C.F., and Needleman, A., *Engineering Fracture Mechanics*, Vol. 21, pp. 405−421, 1985.

(15) Shih, C.F., Moran, B., and Nakamura, T., *International Journal of Fracture*, Vol. 30, pp. 79−102, 1986.

(16) Theiss, T.J., and Bryson, J.R., to appear in the ASTM STP resulting from the *Symposium on Constraint Effects in Fracture*, held May 8−9 1991, Indianapolis, Indiana.

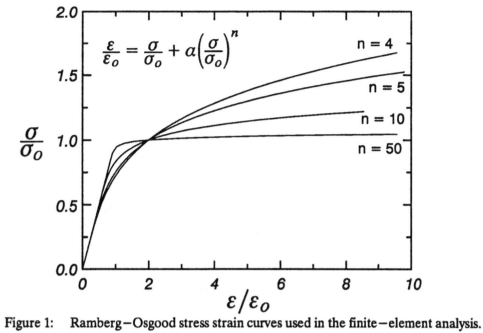

Figure 1:   Ramberg–Osgood stress strain curves used in the finite–element analysis.

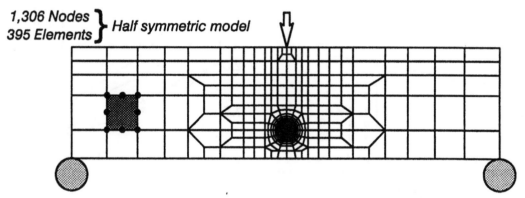

Figure 2:   Finite–element model of the a/W=0.25 SE(B) specimen.

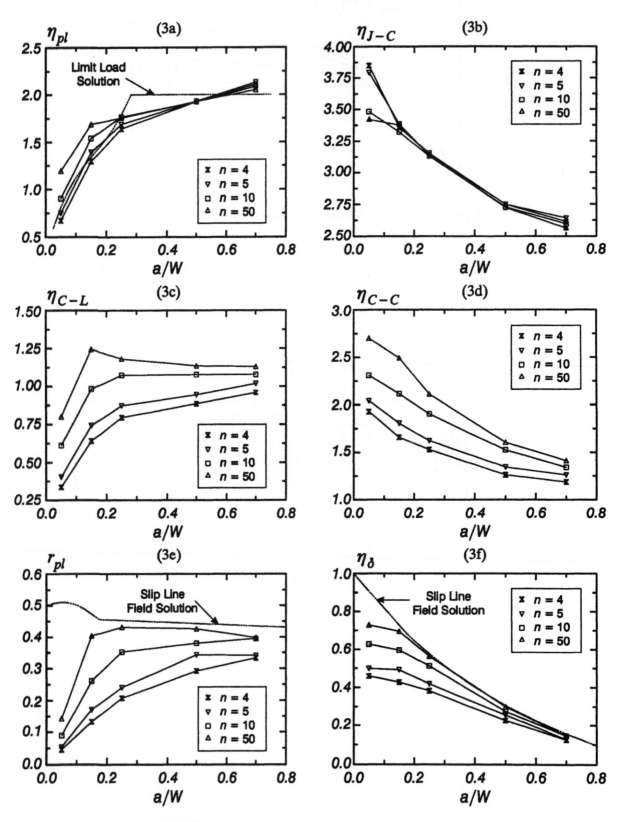

Figure 3: Variation of coefficients in *J* and CTOD estimation equations with *a/W* and *n*. (a) eqn. [1], (b) eqn. [6], (c) eqn. [5], (d) eqn. [7], (e) eqn. [2], (f) eqn. [8].

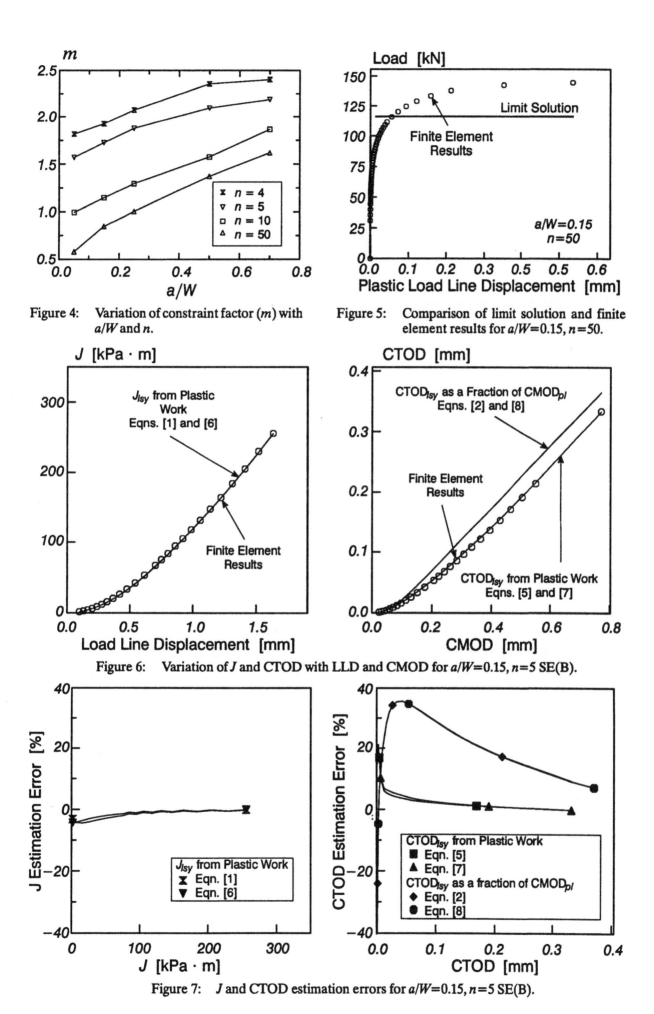

Figure 4: Variation of constraint factor (m) with a/W and n.

Figure 5: Comparison of limit solution and finite element results for a/W=0.15, n=50.

Figure 6: Variation of J and CTOD with LLD and CMOD for a/W=0.15, n=5 SE(B).

Figure 7: J and CTOD estimation errors for a/W=0.15, n=5 SE(B).

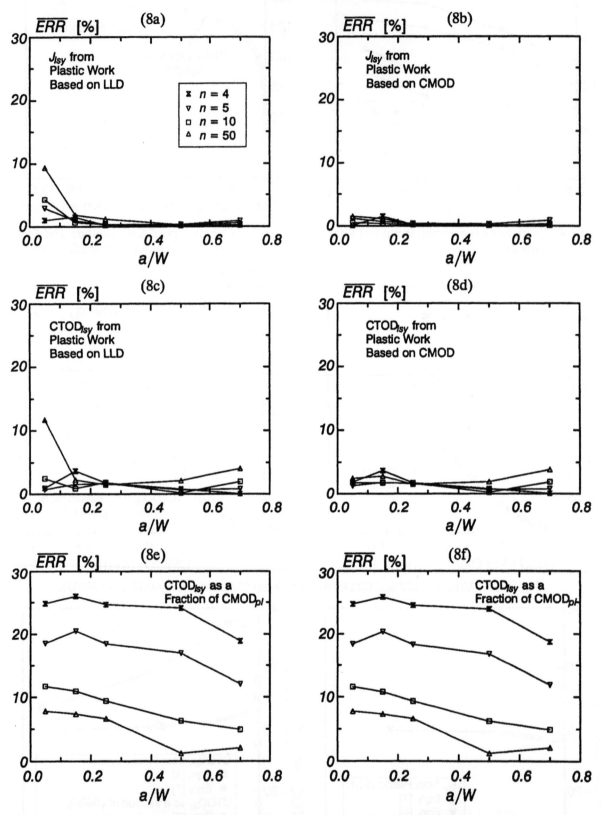

Figure 8: Variation of *J* and CTOD estimation errors with *a/W* and *n*. Symbols represent the same conditions in each figure. (a) eqn. [1], (b) eqn. [6], (c) eqn. [5], (d) eqn. [7], (e) eqn. [2], (f) eqn. [8].

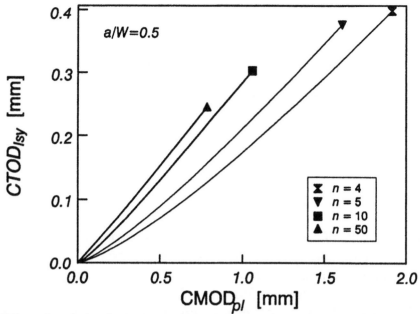

Figure 9: Effect of strain hardening on the linearity of the $CTOD_{lsy}$ − $CMOD_{pl}$ relation for a/W=0.50.

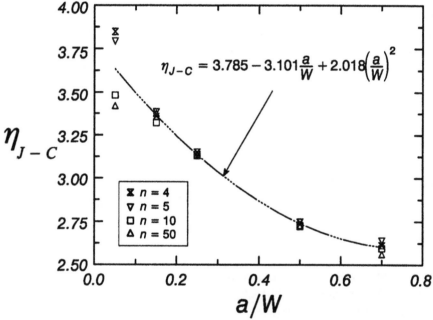

$$\eta_{J-C} = 3.785 - 3.101\frac{a}{W} + 2.018\left(\frac{a}{W}\right)^2$$

Figure 10: Comparison of eqn. [12] to finite−element data.

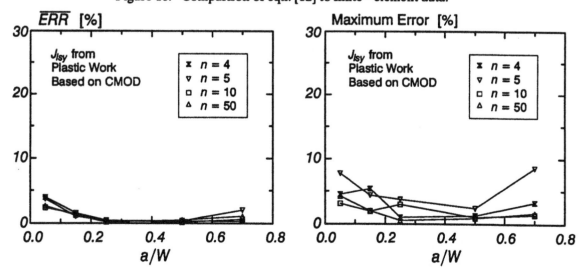

Figure 11: Error associated with using η_{J-C} values from eqn. [12].

PAPER 14

3D elastic-plastic finite element analysis for CTOD and J in SENB, SENAB and SENT specimen geometries

R H Leggatt, MA, PhD, EurIng, CEng, MIMechE, SenMWeldI (TWI, UK) and
J R Gordon, BSc, PhD, SenMWeldI, CEng (EWI, USA)

INTRODUCTION

This paper describes the results of a series of over 90 three–dimensional elastic–plastic finite element analysis of fracture mechanics specimens with crack depths between 5 and 50% of the specimen depth. The analyses were carried out by seven research centres in Europe and North America as contributions to the International Research Project to Develop Shallow Crack Fracture Mechanics Tests managed by TWI and EWI on behalf of a group of sponsors. The International project also included the development of experimental techniques for the testing of shallow–crack fracture specimens (1) and an extensive series of fracture tests on specimens with a range of crack depths. The objective of the numerical analyses was to investigate the relationship between the local parameters CTOD and J which are used to characterise the behaviour at the crack tip, and the global parameters load, load–line–displacement and CMOD which can be measured during fracture tests. This paper describes the range of cases analysed, the experimental validation of the numerical analyses and the investigation of the relationships between local and global parameters.

CONTRIBUTING ORGANISATIONS

Ir H Braam, ECN, The Netherlands
Dr M Twickler, Dr R Twickler and Dipl –Ing. A Sprock, Technical University, Aachen, Germany
Ir R W J Koers, Shell, The Netherlands
Mr O P Sovik and Mr M Hauge, SINTEF, Norway

Professor R H Dodds and Mr M Kirk, University of Illinois, USA
Dr J C Newman and Dr K N Shivakumar, NASA, USA
Dr G W Wellman, Sandia National Laboratories, USA

MATERIALS

The stress–strain curves used in the Phase 1 finite element analyses were the measured stress–strain curves for test materials M1 to M4 (Table 1).

The stress–strain curves used in the Phase 2 finite element analyses were based on the stress–strain curves for materials M1 and M3, but with modified behaviour to give increased or decreased gradients in the work hardening range, and modified ultimate tensile strengths. The modified stress–strain curves were identified as M1/LO, M1/HI, M3/LO and M3/HI, the suffix 'LO' indicating a reduced work hardening gradient and tensile strength and the suffix 'HI' indicating increased strength and work hardening capacity.

The tensile properties of all the materials studied in the project are summarised in Table 1. The true stress – true strain curves are compared in Fig.1–3.

GEOMETRIES AND LOAD CASES

The specimen geometries analysed include three specimen configurations (SENB, SENT and SENAB, i.e. single edge notch bend, tension and arc bend), three cross sections (2B × B, B × B and B × 3B), seven crack depths (a/W = 0.5, 0.1, 0.15, 0.2, 0.3, 0.4 and 0.5), four real stress–strain curves (M1 to M4) and four modified stress–strain curves (M1/LO, M1/HI, M3/LO and M3/HI).

All the finite element analyses were performed using three–dimensional elastic-plastic models.

COMPARISON BETWEEN COMPUTED AND EXPERIMENTAL RESULTS

Load Versus Load–Line Displacement

The accuracy of the computed load versus load–line displacement behaviour was assessed in terms of the normalised yield moment for the SENB and SENAB specimens and the normalised yield load for the SENT specimens. The agreement between experimental and computed results was found to be good in most cases, i.e., discrepancies less than 10%. However, the predictions of load for the deeply–notched (a/W = 0.5) SENT specimens were in serious error (up to 37%) after the specimens started to yield. This may have been due to secondary bending stresses not included in small displacement finite element models.

CMOD Versus Load–Line Displacement

The prediction of the CMOD versus load–line displacement behaviour was found to be fairly poor for the SENB and SENT specimens with a/W = 0.1 to a/W = 0.5, with discrepancies up to about 20%. The agreement was very poor for the SENB and SENT specimens with a/W = 0.05, and for all the SENAB specimens, with discrepancies up to 50%. It appears that CMOD may be very sensitive to small variations in yield properties. This was confirmed, to some extent, as duplicate blunt–notched tests performed at TWI exhibited significant differences in CMOD versus load–line displacement behaviour.

Comparison of Blunt and Sharp Notched Specimens

Analyses were performed for blunt- and sharp–notched SENB specimens with a/W ratios of 0.1 and 0.5. It was found that although the finite element predictions for the blunt- and sharp–notched specimens exhibited minor differences beyond general yield, they were nevertheless in excellent agreement (i.e. generally within 4–5%) for both a/W ratios studied.

CMOD versus CTOD

A limited study was made of computed and experimental values of the ratio $CTOD_p/CMOD_p$, the plastic components of the crack tip and crack–mouth opening displacements. The experimental values of the ratio were found by sectioning four specimens after testing and making measurements of the crack opening. The computed values of the ratio $CTOD_p/CMOD_p$ agreed with the experimental values to within 3% for SENB specimens with a/W = 0.05 and 0.5, to within 7% for a SENT specimen with a/W = 0.05 and within 15% for a SENT specimen with a/W = 0.5. Hence, it was concluded that, despite the relatively poor agreement between computed and experimental values of CMOD as a function of load–line displacement, the finite element analyses did give an adequate representation of the relationship between CMOD and CTOD, apart from the case of deeply–notched SENT specimens, the analyses of which have already been noted to be problematic.

CTOD ESTIMATION PROCEDURES

Initially two definitions of CTOD were investigated, the 90 degree intercept opening, $CTOD_{90}$, and the extrapolated crack flank opening, $CTOD_{CF}$, see Fig.4. The crack flank CTOD was introduced because it was thought that it would be possible to obtain it directly from experimental results using a double–clip gauge arrangement. However, it was found that the double–clip gauge knife edge fixtures tended to rotate about their points of attachment at the crack mouth. Hence, the measured displacements at the knife edges could be extrapolated to the specimen surface to give the CMOD but not below the surface to give the CTOD. It was, therefore, decided to discontinue investigation of $CTOD_{CF}$ and use the well–established $CTOD_{90}$ to characterise the crack–tip opening behaviour.

The following CTOD estimation procedures were evaluated in this program to determine how accurately they could predict CTOD for both shallow and deeply–cracked bend and tension specimens.

1. Standard CTOD estimation procedure (e.g. ASTM E1290)

    The CTOD estimation formula presented in the BSI (2) and ASTM (3) CTOD Standards has the following format:

$$\delta = \frac{K^2(1-v^2)}{2E\sigma_{YS}} + \frac{r_p(W-a)V_p}{r_p(W-a)+a+z} \qquad [1]$$

    where

K	=	stress intensity factor
E	=	Young's modulus
σ_{YS}	=	yield strength
v	=	Poisson's ratio
W	=	specimen width
a	=	crack length
V_p	=	plastic component of CMOD measured at knife edge height, z
r_p	=	plastic rotational factor
z	=	knife edge height.

    ASTM E1290 specifies a plastic rotational factor of 0.44 to calculate the fully plastic component of CTOD, whereas BS 5762 specifies an r_p value of 0.4.

2. Eta–delta (η_δ) CTOD estimation procedure

    An alternative CTOD estimation formula has been proposed by Dodds (4). This approach, which is referred to as the η_δ method, is identical in concept to the general estimation formula, i.e., the total CTOD is separated into small scale yielding and fully plastic components. The plastic component, however, is reformulated to avoid the use of a plastic rotational factor. Instead, the η_δ method calculates the plastic component of CTOD directly from the plastic component of CMOD as shown in Eq.[2]. If the crack opens with no crack front rotation η_δ would assume the value 1.0.

$$\delta = \frac{K^2(1-v^2)}{2E\sigma_{YS}} + \eta_\delta \, CMOD_{pl} \qquad [2]$$

J ESTIMATION PROCEDURES

The following J-estimation procedures were studied in the current programme

1. Standard J-estimation procedure (i.e. ASTM E813 or E1152)

The generally accepted procedure for calculating J (5,6) is given by the following expression:
where U_{pl} = plastic component of area under load versus load-point displacement record.

$$ J = \frac{K^2(1-v^2)}{E} + \frac{\eta_p U_{pl}}{B(W-a)} \qquad [3] $$

2. J calculated from area under load versus CMOD record.

An alternative J estimation approach, based on CMOD, was also studied in this project to determine if reliable J results can be obtained for a range of specimen geometries by measuring a local displacement at the notch mouth. This approach has the advantage of measuring a displacement, which reflects crack tip deformation, rather than a far field displacement which frequently needs to be corrected for extraneous effects. The J estimation formula, based on CMOD, has the following form:

$$ J = \frac{K^2(1-v^2)}{E} + \frac{\eta_c A_{pl}}{B(W-a)} \qquad [4] $$

where A_{pl} = plastic component of area under the load versus CMOD record.

RESULTS OF FINITE ELEMENT ANALYSES

General

The results presented in the following sections were obtained from the finite element analysis of the SENB and SENAB specimens.

Results of CTOD Analyses: Eta-delta (η_δ) Method

Values of η_δ were obtained from the finite element results using the following equation:

$$ \eta_\delta = \frac{\delta - \delta_{ssy}}{CMOD_{pl}} \qquad [5] $$

where $\quad \delta_{ssy} = \frac{K^2(1-v^2)}{2E\sigma_{YS}} \qquad [6]$

A typical set of η_δ results are presented in Fig.5 as plots of η_δ versus $CMOD_{pl}$. If the η_δ approach is a valid CTOD calculation procedure, then η_δ should be independent of $CMOD_{pl}$. It is evident from Fig.5 that once plasticity effects dominate, η_δ is relatively independent of $CMOD_{pl}$. Nevertheless, the transition regime before η_δ achieves a constant value is dependent on both the a/W ratio and the work hardening capacity of the material. In general, as the a/W ratio decreases, the degree of plasticity before η_δ achieves a constant value increases. For a given a/W ratio, the degree of plasticity before η_δ achieves a constant value increases as the work hardening capacity of the material increases. This indicates that the development of a plastic hinge in SENB and SENAB specimens is achieved at lower values of CMOD in deeply–notched specimens than shallow notched specimens. Furthermore, for a given a/W ratio, plastic hinge development is achieved at lower values of CMOD in low work hardening materials than high work hardening materials. As a result the η_δ CTOD estimation scheme is best suited to low and medium work hardening materials. As the work level increases the η_δ method becomes questionable as the level of plasticity before η_δ achieves a constant value becomes significant.

Wu et al. (7) have proposed a slip–line solution for shallow–cracked bend specimens. The slip–line field solution permits the calculation of plastic rotational factor for different a/W ratios. The closed form solution proposed by Wu et al. for the plastic rotational factor is given by:

$$r_{sl} = 0.5 + 0.42 \, (a/W) - 4.0 \, (a/W)^2 \quad \textit{for } a/W \leq 0.172 \tag{7}$$

$$r_{sl} = 0.463 - 0.04 \, (a/W) \quad \textit{for } a/W > 0.172 \tag{8}$$

The slip–line field solution proposed by Wu et al. (7) can also be used to calculate values of η_δ using the following expression:

$$\eta_\delta = \frac{r_{sl} \, (W-a)}{r_{sl} \, (W-a) + a + z} \tag{9}$$

The predicted r_p and η_δ results are plotted as a function of a/W ratio in Fig.6. It is evident that at a/W ratios of 0.5 the plastic rotational factor is approximately 0.44 which is consistent with the ASTM CTOD testing standard ASTM E1290.

The stabilised SENB and SENAB η_δ results (i.e. the plateau η_δ values) obtained from the North American and European finite element analyses are compared with the slip–line field solution in Fig.7. It is apparent from Fig.7 that at large a/W ratios (e.g. a/W = 0.5) the η_δ results show little scatter and are in good agreement with the slip–line field solution. However, as the a/W ratio decreases the degree of scatter increases. Moreover, the computed η_δ results for the higher work hardening materials show a significant departure from the slip–line field solution.

Closer examination of the results in Fig.7 indicates that increasing the level of work hardening produces two effects:

1. At a/W ratios greater than typically 0.3 the computed η_δ results obtained from all the materials show the same general trend as the slip–line field solution although as the work hardening level increases the computed η_δ values become progressively smaller than the slip–line field value.

2. At smaller a/W ratios the predicted η_δ results appear to reach a limiting value which is dependent on the level of work hardening but relatively independent of a/W.

Although several closed form η_δ CTOD estimation procedures were proposed, based on the results presented in Fig.7, it was found that they could result in errors in the predicted CTOD of more than 50%.

Results of J Analyses: Eta-Plastic (LLD) Method

Values of η_p were obtained from the finite element results by subtracting the calculated small-scale yielding component of J from the total J value to leave the plastic component of J (J_{pl}). Values of η_p were then obtained using the following expression:

$$\eta_p = \frac{J_{pl}B(W-a)}{U_{pl}}$$

[10]

A typical set of η_p results are presented in Fig.8 as plots of η_p versus J_{pl}. It is evident from Fig.8 that once plasticity effects dominate, η_p is relatively independent of J_{pl} although, the trend becomes less pronounced at small a/W ratios.

Sumpter (8) has proposed a J estimation procedure for SENB specimens with a/W ratios down to almost zero. Closed form solutions are presented for η_p as a function of a/W ratio. The J estimation formula is given by:

$$J = \frac{K^2(1-\upsilon^2)}{E} + \frac{\eta_p U_{pl}}{B(W-a)}$$

[11]

where $\eta_p = 0.32 + 12.0\ (a/W) - 49.5\ (a/W)^2 + 99.8\ (a/W)^3\ \textit{for } a/W <0.282$ [12]

$$\eta_p = 2.0\ \textit{for } a/W \geq 0.282$$

[13]

The stabilised η_p results obtained from the North American finite element results are compared with the solutions proposed by Sumpter in Fig.9. It is evident that the Sumpter solutions provide a reasonable fit to the finite element results.

Results of J Analyses: Eta-Plastic (CMOD) Method

Values of η_c were calculated by subtracting the small-scale yielding component of J from the total J to give the plastic component of J (J_{pl}). Values of η_c were then obtained from the following expression:

$$\eta_c = \frac{J_{pl}B(W-a)}{A_{pl}}$$

[14]

where A_{pl} = plastic component of area under the load versus CMOD record.

A typical set of η_c results is presented in Fig.10 as plots of η_c versus J_{pl}, where it is evident that once plasticity effects dominate, η_c is relatively independent of J_{pl}.

Sumpter (8) has also proposed a J estimation formula based on the area under the load versus CMOD record. The estimation formula is given by:

$$J = \frac{K^2(1-\upsilon^2)}{E} + \frac{\eta_p A_{pl}}{B(W-a)} \frac{0.25S}{a+r(W-a)} \quad\quad\quad [15]$$

where r $=$ 0.45 for a/W $\geq$0.3
           r $=$ 0.3 +0.5 a/W for a/W <0.3
           η_p $=$ 2.0 for a/W $\geq$0.282
           η_p $=$ 0.32 +12.0 (a/W) – 49.5 (a/W)^2 + 99.8 (a/W)^3 for a/W <0.282.

The above expression can be reformulated to give the following equation:

$$J = \frac{K^2(1-\upsilon^2)}{E} + \frac{\eta_c A_{pl}}{B(W-a)} \qu\quad\quad\quad\quad [16]$$

where $\eta_c = \eta_p \dfrac{0.25S}{a+r(W-a)}$ $\quad\quad\quad\quad\quad\quad [17]$

The stabilised η_c results obtained from the North American and European SENB and SENAB and European SENB finite element results are compared with the η_c expression derived from the Sumpter analysis in Fig.11. It can be seen that the Sumpter solution underpredicts η_c over the entire range of a/W ratio although it does predict the trend of the computed results.

The η_c results obtained from the North American and European finite element analyses were used to develop an η_c–based J estimation procedure. This was achieved by fitting a fourth order polynomial to the η_c results. The fit was forced to give a smooth curve beyond a/W ratios of 0.5. This was not achievable with a third order polynomial fit. The fourth order polynomial expression for η_c is as follows:

$$\eta_c = 2.632 + 13.687 \, (a/W) - 48.847 \, (a/W)^2 + 57.979 \, (a/W)^3 - 23.343 \, (a/W)^4 \quad [18]$$

The above expression is compared with the computed η_c results in Fig.12.

COMPARISON OF COMPUTED AND PREDICTED RESULTS

CTOD Calculation Procedures

The accuracy of the ASTM and BSI CTOD estimation procedures was evaluated at three a/W ratios (0.05, 0.2 and 0.5) and three levels of work hardening behaviour (M3L, M3 and M3H) by comparing the predicted CTOD behaviour with the computed CTOD behaviour.

It was found that at for a/W = 0.05 the BSI and ASTM CTOD estimation procedures overpredicted the computed CTOD behaviour by approximately 20% for the low work hardening material, M3L. In the case of the a/W = 0.05 results for materials M3 and M3H the ASTM and BSI procedures produced CTOD estimates which overpredicted the computed results by more than 100%.

For a/W = 0.5 the ASTM and BSI CTOD estimation procedures predicted the computed CTOD results to within typically 30% for materials M3L and M3H. For material M3 the BSI and ASTM procedures gave CTOD estimates which at times overestimated the computed CTOD by 50%.

A number of candidate η_δ CTOD estimation procedures were also evaluated. It was found that although they provided much more accurate CTOD predictions than the standard ASTM and BSI procedures (particularly at small a/W ratios) the errors in predicted CTOD could still exceed 50%. For this reason it is the opinion of the authors that a CTOD estimation approach based on the area under the load versus CMOD test record or a procedure which estimates CTOD directly from J is potentially a better overall CTOD estimation approach.

J CALCULATION PROCEDURES

The J estimation procedure developed from the results of the North American finite element analyses was evaluated to determine how accurately it could predict J for both shallow- and deeply-cracked bend specimens. Also included in the evaluation were the ASTM J calculation procedure for deeply-cracked bend specimens (ASTM E1152) and the η_p and η_c procedures proposed by Sumpter. The J estimation equations studied were as follows:

(i)   ASTM E1152   (iii) Sumpter η_c procedure
(ii) Sumpter η_p procedure   (iv) EWI/TWI η_c procedure.

The accuracy of the different estimation procedures was evaluated at three a/W ratios (0.05, 0.2 and 0.5) and three levels of work hardening behaviour (M3L, M3 and M3H) by comparing the predicted J behaviour with the computed J behaviour. The results of the evaluation are summarized in Fig.13–15 as plots showing the error in the predicted J as a function of the computed J.

The results for all three a/W ratios indicate that the best J estimates are given by the EWI/TWI η_c J estimation procedure. This procedure produces J estimates which are within 10% (and in many cases within 5%) of the computed J. This demonstrates that the J estimation procedures are capable of predicting J more accurately then the CTOD procedures can estimate CTOD.

It is evident from Fig.13–15 that the standard ASTM J calculation procedure provided accurate J estimates for a/W = 0.5 (maximum difference between predicted and computed J less than 10%). At a/W = 0.2 the standard ASTM procedure overpredicted the computed behaviour by up to 40% although the difference was more typically less than 20%. At a/W = 0.05 the standard ASTM J estimation procedure produced dramatic overestimates (e.g. more than 100%) of the computed.

RECOMMENDATIONS

Based on the results of the numerical analyses undertaken in this project the following J estimation procedure is recommended for SENB specimens with a/W ratios in the range of 0.05–0.6:

$$J = \frac{K^2(1-\upsilon^2)}{E} + \frac{\eta_c A_{pl}}{B(W-a)}$$ [19]

where A_{pl} = plastic component of area under the load versus CMOD record.

$$\eta_c = 2.632 + 13.687(a/W) - 48.847(a/W)^2 + 57.979(a/W)^3 - 23.343(a/W)^4$$ [20]

The results from the finite element analyses indicate that the above estimation scheme should predict J to within 5–10% for the a/W range 0.05–0.6 for a range of work hardening levels.

Although a number of candidate CTOD estimation procedures were evaluated it was found that they could result in significant errors in the predicted CTOD. In the opinion of the authors a CTOD estimation approach based on the area under the load versus CMOD test record or a procedure which estimates CTOD directly from J is potentially a better overall CTOD estimation approach.

REFERENCES

(1) Dawes M G, Gordon J R, McGaughy T H and Slater G: 'Shallow Crack Test Methods for the Determination of K_{IC}, CTOD and J Fracture Toughness'. Proceedings of the International Conference on Shallow Crack Fracture Mechanics : Tests and Applications, Published by Abington Publishing, 1992.

(2) BS7448:Part 1:1991: 'Fracture Mechanics Toughness Tests, Part 1. Method for the Determination of K_{Ic}, Critical CTOD and Critical J Values of Metallic Materials'. (This standard replaced BS5447:1977:K_{Ic} tests, and BS5762:1979:CTOD tests.)

(3) ASTM E1290-89: 'Standard Test Method for Crack-Tip Opening Displacement (CTOD) Fracture Toughness Measurement'.

(4) Dodds R H: 'Plastic Rotation Factors for Shallow Crack CTOD Testing'. Private communication.

(5) ASTM E813-89: 'Standard Test for J_{Ic}, a Measure of Fracture Toughness'.

(6) ASTM E1152-87: 'Standard Test Method for Determining J-R Curves'.

(7) Wu S X, Cotterell B, and Mai T W: 'Slip Line Field Solutions for Three-Point Notch Bend Specimens'. Int. Journal of Fracture, Vol. 37, No. 1, May 1988, pp. 13-29.

(8) Sumpter J D G: 'J_c Determination for Shallow Notched Welded Bend Specimens'. Fatigue, Fract. Eng. mater. Struct. 10 (6) pp. 579-493, 1987.

Table 1 Summary of mechanical properties for materials used in finite element analyses

Material	M1	M2	M3	M4	M1/LO	M1/HI	M2/HI
Specification	BS4360	AISI	ASTM	API	–	–	–
Grade	50EE	4145H	A36-84a	5CT L80			
Yield strength, σ_{ys}, N/mm²	388	850	283	590	388	388	850
Tensile strength, σ_{TS},N/mm²	530	980	491	705	435	814	1240
Tensile/Yield, σ_{TS}/σ_{YS}	1.37	1.15	1.73	1.19	1.12	2.10	1.46
Young's modulus, E,N/mm²	203,000	209.000	207,000	202,500	203.000	203,000	209.000
Poisson's Ratio, v	0.3	0.3	0.3	0.3	0.3	0.3	0.3

The stress-strain curves for materials M3/HI and M3/LO are specified by the Ramberg-Osgood formula $\varepsilon/\varepsilon_0 = \sigma/\sigma_0 + \alpha(\sigma/\sigma_0)^n$, with $\sigma_0 = 283$ N/mm², $\alpha = 1.0$, ε_0/E, $E = 207,000$ N/mm² and n = 5.0 for M3/HI and n = 50.0 for M3/LO.

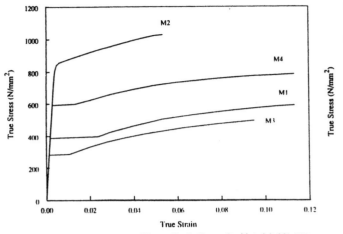

Figure 1: Comparison of Stress-Strain Curves for Materials M1-M4

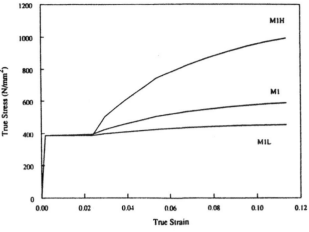

Figure 2: Comparison of Stress-Strain Curves for Materials M1L, M1 and M1H

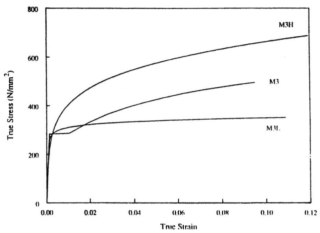

Figure 3: Comparison of Stress-Strain Curves for Materials M3L, M3 and M3H

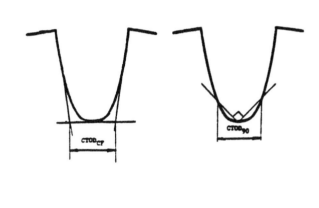

Figure 4: Definition of extrapolated crack flank CTOD and 90°-intercept CTOD

SENB SPECIMENS

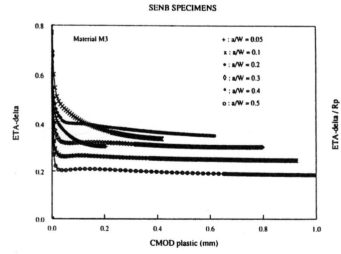

Material M3

+ : a/W = 0.05
x : a/W = 0.1
• : a/W = 0.2
◊ : a/W = 0.3
* : a/W = 0.4
o : a/W = 0.5

Figure 5: Variation of η_δ with $(Mo1)_{pl}$ (Material M3)

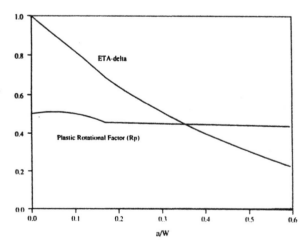

Figure 6: Variation of r_p and η_δ with a/w (Slip line field solutions)

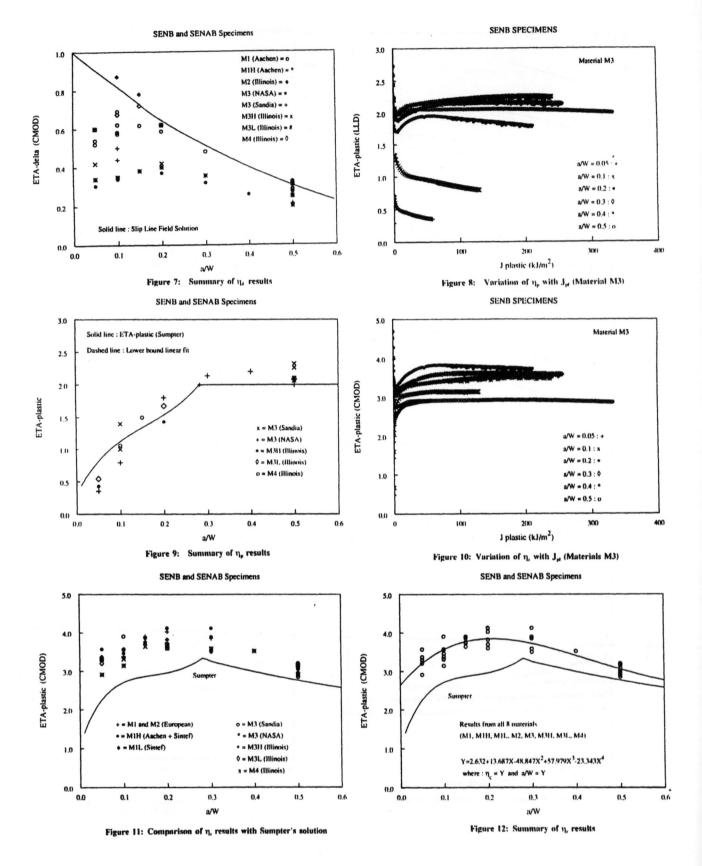

Figure 7: Summary of η_δ results

Figure 8: Variation of η_p with J_{pl} (Material M3)

Figure 9: Summary of η_p results

Figure 10: Variation of η_c with J_{pl} (Materials M3)

Figure 11: Comparison of η_c results with Sumpter's solution

Figure 12: Summary of η_c results

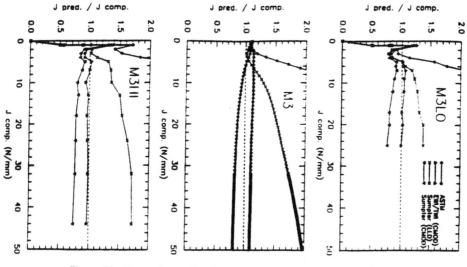

Figure 13: Comparison of J estimation procedures for a/w = 0.05 SENB specimens

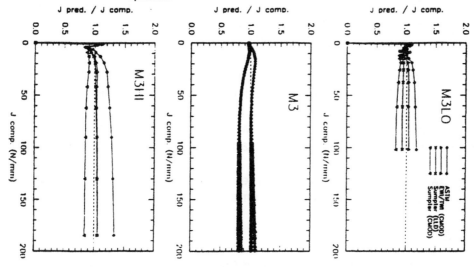

Figure 14: Comparison of J estimation procedures for a/w = 0.2 SENB specimens

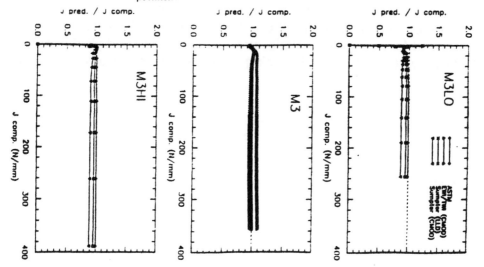

Figure 15: Comparison of J estimation procedures for a/w = 0.5 SENB specimens

PAPER 28

The limits of applicability of J and CTOD estimation procedures for shallow-cracked SENB specimens

Y-Y Wang, BS, MS, PhD and J R Gordon, BSc, PhD, CEng, SenMWeldI (Edison Welding Institute, USA)

INTRODUCTION

Over the last ten years standard fracture toughness test procedures have been developed which cover the measurement of toughness in terms of K, J and CTOD (1-5). These test procedures include minimum specimen size requirements, including limits on crack size and crack depth to specimen width ratio (a/W) to ensure that the measured toughness is independent of specimen size. Although the minimum specimen size requirements ensure lower-bound toughness estimates for homogeneous materials these requirements can lead to excessively conservative assessments for structures which contain shallow cracks and/or are loaded in tension.

Ideally, when assessing the integrity of engineering structures containing shallow cracks the toughness used in the assessment should be obtained by testing shallow-cracked samples in which the loading mode and constraint are representative of the structure. Furthermore, in situations where there is a gradient of material properties the testing of shallow-cracked specimens will ensure that the measured toughness is representative of the microstructure sampled by the cracks in the structure.

This paper summarises the results of an ongoing research project at EWI to develop J and CTOD estimation procedures for SENB specimens with small a/W ratios.

ANALYSIS MATRIX

Plane-strain finite element analyses were performed for SENB specimens with a/W ratios of 0.1, 0.2, 0.3 and 0.5. In addition analyses were performed for a range of specimen widths to determine the effect of "W" (and "a") and identify the limits of applicability of the J and CTOD estimation procedures developed in the project. To enable the effect of work hardening to be assessed, finite element analyses were performed for a range of material stress-strain curves and yield strength levels. In all cases the stress-strain curves were modelled using a Ramberg-Osgood power law hardening relationship.

$$\frac{e}{e_O} = \frac{\sigma}{\sigma_O} + \alpha\left(\frac{\sigma}{\sigma_O}\right)^n \qquad [1]$$

where σ = true stress
σ_O = yield strength
ε = true strain
ε_O = yield strain ($\varepsilon_O = \sigma_O/E$)
α = dimensionless parameter
n = strain hardening exponent.

The stress-strain curves were selected to provide a range of work hardening levels ranging from essentially elastic perfectly plastic to high work hardening. Details of the finite-element analysis matrix are summarized in Table 1.

Table 1. Finite-element analysis matrix

a/W	W (mm)	σ_{YS} (N/mm^2)	α	n
0.1, 0.2, 0.3 and 0.5	50	500	1	5, 10, 20 and 50
0.1, 0.2, 0.3 and 0.5	10, 25, 50 and 100	500	1	10
0.1, 0.2, 0.3 and 0.5	50	400, 500 and 600	1	10

The major objective of this study was to develop J and CTOD estimation procedures for shallow-cracked SENB specimens (i.e., a/W ~0.1) for a range of work hardening behaviours. However, rather than include the Ramberg-Osgood strain hardening exponent in the estimation procedures (which requires a detailed knowledge of the material's stress-strain curve) it was decided to express the work hardening capacity as the ratio of tensile strength to yield strength. In the case of a Ramberg-Osgood true stress-true strain curve the ratio of tensile strength to yield strength can be determined using the following expression (6):

$$\frac{\sigma_{TS}}{\sigma_{YS}} = \frac{\left(\frac{1}{0.002n}\right)^{1/n}}{\exp(1/n)} \qquad [2]$$

A typical mesh for an SENB specimen with a/W = 0.3 is shown in Fig. 1. All of the finite element analyses were performed using ABAQAS. Values of J were obtained using the virtual crack extension method. At each load step J values were computed for 10 contours. The variation of J among the different contours was within 0.1%. Values of CTOD were calculated using the 90 degree intersection definition.

J-ESTIMATION PROCEDURES

General

The following J-estimation procedures were studied in the current programme:

- Standard J-estimation procedure (i.e., ASTM E813 or E1152).
- J calculated from area under load versus CMOD record.

Standard J-Estimation Procedure

The general formula for calculating J is given by:

$$J = \frac{K^2(1-v^2)}{E} + \frac{\eta_p U_{pl}}{B(W-a)} \qquad [3]$$

where U_{pl} = plastic component of area under load versus load-point displacement record.

J Calculation Procedure Based on CMOD

An alternative J estimation approach, based on CMOD, was also studied in this project to determine if reliable J results can be obtained by measuring a local displacement at the notch mouth. The J estimation formula, based on CMOD, has the following form (6):

$$J = \frac{K^2(1-v^2)}{E} + \frac{\eta_c A_{pl}}{B(W-a)} \qquad [4]$$

where A_{pl} = plastic component of area under the load versus CMOD record.

CMOD ESTIMATION PROCEDURES

General

The following three CTOD estimation procedures were evaluated in this programme to determine how accurately they could predict CTOD for both shallow- and deeply-cracked bend and tension specimens:

1. Standard CTOD estimation procedure (i.e., ASTM E1290).
2. Eta-delta (η_δ) CTOD estimation procedure.
3. "J-equivalent" CTOD estimation procedure.

Standard CTOD Estimation Procedure

The CTOD estimation formula presented in the BSI (1) and ASTM (2) CTOD Standards for SENB specimens has the following format:

$$\delta = \frac{K^2(1-v^2)}{2E\sigma_{YS}} + \frac{r_p(W-a)V_p}{r_p(W-a)+a+z} \qquad [5]$$

where K = stress intensity factor
E = Young's modulus
σ_{YS} = yield strength
v = Poisson's ratio
W = specimen width
a = crack length
V_p = plastic component of CMOD measured at knife edge height, z
r_p = plastic rotational factor
z = knife edge height.

The first term in [5] represents the small-scale yielding component of CTOD. The second term is the plastic component of CTOD.

ASTM E1290 specifies a plastic rotational factor of 0.44, whereas BS 7448 specifies an r_p value of 0.4.

The η_δ CTOD Estimation Procedure

In the η_δ method the plastic component of CTOD is calculated directly from the plastic component of CMOD as shown in [6]. If the crack opens with no crack from rotation η_δ would assume the value 1.0.

$$\delta = \frac{K^2(1-v^2)}{2E\sigma_{YS}} + \eta_\delta \, CMOD_{pl} \qquad [6]$$

J-Equivalent CTOD Estimation Procedure

The final CTOD estimation procedure studied in this project is referred to as the J-equivalent CTOD estimation procedure. The J-equivalent CTOD estimation equation is as follows:

$$\delta = \frac{K^2(1-v^2)}{2E\sigma_{YS}} + \frac{\eta_c^\delta A_{pl}}{\sigma_{YS}B(W-a)} \qquad [7]$$

where A_{pl} = plastic component of area under load versus CMOD record
       B = specimen thickness.

RESULTS OF FINITE ELEMENT ANALYSES

Results of J Analyses

Standard J estimation procedure

The η_p results obtained from the numerical analyses are presented in Fig. 2 as a plot of η_p versus a/W ratio. Also included in Fig. 2 are results from similar analyses performed by Kirk and Dodds (6). Although the finite element analyses confirm that once plasticity effects dominate η_p is not a function of yield stress or specimen size, the results presented in Fig. 2 indicate that η_p is dependent on both the a/W ratio and the work hardening capacity of the material (particularly at small a/W ratio, i.e., less than 0.3).

J calculation procedure based on CMOD

The η_c results obtained from the numerical analyses are presented in Fig. 3 as a plot of η_c versus a/W ratio. Also included in Fig. 3 are results from similar analyses performed by Kirk and Dodds (6). The finite element results presented in Fig. 3 indicate that although η_c is dependent on a/W ratio it is almost independent of work hardening level. Furthermore, the results form the analyses with different yield strengths and specimen sizes indicate that once plasticity effects dominate η_c is independent of yield strength and specimen size.

The computed η_c and η_p results demonstrate that in the case of shallow-cracked SENB specimens J estimates should be obtained from a CMOD-based procedure rather than a load line-based method.

Based on the results presented in Fig. 3 the following J estimation procedure is proposed for SENB specimens with a/W ratios in the range 0.1-0.6:

$$J - \frac{K^2(1-v^2)}{E} + \eta_c \frac{A_{pl}}{B(W-a)} \qquad [8]$$

where $\eta_c = 3.5 - 1.4167 \ (a/W)$.

Results of CTOD Analyses

Standard CTOD estimation procedure

The results from the finite element analyses indicate that although the standard ASTM and BSI CTOD estimation procedures provide accurate estimates of CTOD for deeply-notched SENB specimens they can overpredict the actual CTOD for a/W = 0.1 SENB specimens from 20% to more than 100% depending on the work hardening capacity of the material.

Eta-delta (η_δ) CTOD calculation procedure

The η_δ results obtained from the finite element analysis of a/W = 0.1 and 0.5 SENB specimens are presented in Figs. 4 and 5. If the η_δ approach is a valid CTOD calculation procedure, then η_δ should be independent of $CMOD_{pl}$. It is evident from Figs. 4 and 5 that once plasticity effects dominate η_δ is relatively independent of $CMOD_{pl}$ for low work hardening materials (i.e., n > 15-10). However, as the work hardening level increases η_δ becomes increasingly dependent on $CMOD_{pl}$. This indicates that although an η_δ CTOD calculation procedure could be developed for low work hardening materials it may not be suitable for medium and high work hardening materials. The results from the analyses of different sized specimens also demonstrates that as expected η_δ is specimen size-dependent although this only becomes evident for the medium to high work hardening materials where η_δ is dependent on $CMOD_{pl}$. Figure 6 shows the variation of η_δ with $CMOD_{pl}$ for medium work hardening SENB specimens of different sizes (a/W = 0.1, n = 10). The same results are presented in Fig. 7 as a function of $CMOD_{pl}$ normalised by specimen width together with the η_δ results obtained for the other a/W ratios studied. It is clear from Fig. 7 that normalising $CMOD_{pl}$ by W removes the specimen size dependence.

The normalised η_δ results for the high and low work hardening materials (n = 5 and 50) are summarised in Figs. 8-9. These results demonstrate that although normalising $CMOD_{pl}$ by W removes specimen size effects the η_δ method of calculating CTOD is not suited to medium-high work hardening materials.

J equivalent CTOD estimation procedure

The η_c^δ results obtained from the numerical analyses are presented in Fig. 10 as a plot of η_c^δ versus a/W ratio. Also included in Fig. 10 are similar results obtained by Kirk and Dodds. The finite element results presented in Fig. 10 indicate that η_c^δ is dependent on both a/W ratio and work hardening level. Nevertheless, it is evident from Fig. 10 that the η_c^δ results for each work hardening level display the same trend and have approximately the same slope. Furthermore, the results from the analyses with different yield strength and specimen sizes indicate that once plasticity effects dominate η_c^δ is independent of specimen size and almost independent of yield strength, Fig. 11.

Based on the results presented in Fig. 10 the following CTOD estimation procedure is proposed for SENB specimens with a/W ratios in the range 0.1-0.6.

$$\delta = \frac{K^2(1-v^2)}{2E\sigma_{YS}} + \eta_c^\delta \frac{A_{pl}}{\sigma_{YS}B(W-a)}$$ [9]

$$\text{where} \quad \eta_c^\delta = 4.292 - 1.667(a/W) + 0.304 \left(\frac{\sigma_{TS}}{\sigma_{YS}}\right)^2 - 2.00 \left(\frac{\sigma_{TS}}{\sigma_{YS}}\right)$$

COMPARISON OF COMPUTED AND PREDICTED RESULTS

J Calculation Procedures

The proposed CMOD-based J estimation procedure [8] was evaluated to determine how accurately it could predict J for both shallow- and deeply-cracked bend specimens. The proposed J estimation procedure was evaluated at two a/W ratios (0.1 and 0.5) and three work hardening levels (n = 5, 10 and 50) by comparing the predicted J behavior with the computed J behavior. The results of the evaluation are summarised in Fig. 12 as plots showing the error in the predicted J as a function of the computed J.

It is evident from Fig. 12 that for the a/W ratios and work hardening levels studied, the proposed J estimation procedure provides J estimates which are within 10%, and in many cases with within 2-3%, of the computed values.

CTOD Estimation Procedures

The proposed CTOD estimation procedure was evaluated in [9] at two a/W ratios (0.1 and 0.5) and three work hardening levels (n = 5, 10 and 50) by comparing the predicted CTOD behaviour with the computed CTOD behaviour. The results of the evaluation are summarised in Fig. 13 as plots showing the error in the predicted CTOD as a function of the computed CTOD.

It is evident from Fig. 13 that for the a/W ratios and work hardening levels studied, the proposed CTOD estimation procedure provides CTOD estimates which are within 20% for CTOD levels greater than 0.1 mm. For CTOD levels greater than 0.2 mm, the predicted CTOD values are within 10% of the computed values.

Comparison of J and CTOD Results

The J and CTOD results obtained from the a/W = 0.1 and 0.5 SENB finite element analyses are presented as plots of J divided by yield strength versus CTOD in Figs. 14-15. It is evident that in all cases the J-CTOD plots are nearly linear. The slopes of the J-CTOD plots are, however, dependent on both the a/W ratio and material work hardening capacity.

It is generally accepted that J and CTOD can be related by an equation of the form:

$$J = m \cdot \sigma_y \cdot CTOD$$ [9]

where m = dimensionless coefficient
σ_y = yield strength or flow stress.

Values of the "m" factor (based on yield strength) are presented in Fig. 16 as a function of a/W. Also included in Fig. 16 are results from similar analyses performed by Kirk and Dodds. It is evident that for each work hardening level the plot of m versus a/W is approximately linear. Moreover, the slopes of the m versus a/W plots are almost identical. Based on the results presented in Fig. 16 the following J-CTOD correlation is proposed:

$$J = m\sigma_{YS}\delta \qquad\qquad [10]$$

where $m = 0.8016 \ (a/W) + 1.3165 \ (\sigma_{TS}/\sigma_{YS}) - 0.07573$.

The accuracy of the above correlation was assessed by comparing the predicted CTOD using [8] and [10] and the CTOD computed directly from finite element analyses. The results of this evaluation are summarised in Fig. 17 for a/W = 0.1 and 0.5 SENB specimens and three work hardening levels. It is evident from Fig. 17 that the correlation equation predicts CTOD values which are within 20% of the computed values. Indeed, if the results from the deeply-notched high work hardening SEND specimen analysis (a/W = 0.5, n = 5) are ignored, the predicted CTOD values are within 5% of the computed values for CTOD values greater than 0.05 mm.

CONCLUSIONS

This paper presents the results of a project to develop J and CTOD estimation procedures for shallow-cracked SENB specimens. The main conclusions of this study can be summarised as follows:

J estimation procedures

Although the standard ASTM J estimation procedure (η_p = 2.0) provides accurate estimates of J for deeply-notched SENB specimens, the procedure progressively overestimates J as the a/W ratio decreases.

J estimation procedures for shallow-cracked SENB specimens should be based on the area under the load versus CMOD record.

CTOD estimation procedures

Although the standard ASTM and BSI CTOD estimation procedures provide accurate estimates of CTOD for deeply-notched SENB specimens they can overpredict the actual CTOD for a/W = 0.1 SENB specimens from 20% to more than 100% depending on the work hardening capacity of the material.

The eta-delta method of calculating CTOD is not suited to medium and high work hardening materials.

CTOD estimation procedures for shallow-cracked SENB specimens should be based on the area under the load versus CMOD record.

RECOMMENDATIONS

Based on the results of the numerical analyses the following J and CTOD estimation procedures are recommended for SENB specimens with a/W ratios in the range of 0.1-0.6:

J estimation procedure

The recommended *J* estimation procedure is based on the area under the load versus CMOD record (i.e., the η_c method). The resulting estimation equation is:

$$J = \frac{K^2(1-\nu^2)}{E} + \eta_c \frac{A_{pl}}{B(W-a)}$$

[11]

where $\eta_c = 3.5 - 1.4167\ (a/W)$.

CTOD estimation procedure

The recommended CTOD estimation procedure is based on the area under the load versus CMOD record. The resulting estimation equation is:

$$\delta = \frac{K^2(1-\nu^2)}{2E\sigma_{YS}} + \eta_c^\delta \frac{A_{pl}}{\sigma_{YS}B(W-a)}$$

[12]

$$\text{where}\quad \eta_c^\delta = 4.292 - 1.667\ (a/W) + 0.304\left(\frac{\sigma_{TS}}{\sigma_{YS}}\right)^2 - 2.0\left(\frac{\sigma_{TS}}{\sigma_{YS}}\right)$$

REFERENCES

(1) BS7448:Part 1:1991, "Fracture Mechanics Toughness Tests, Part 1. Method for the Determination of K_{Ic}, Critical CTOD and Critical *J* Values of Metallic Materials" (this standard replaced BS5447:1977:K_{Ic} tests, and BS5762:1979:CTOD tests).

(2) ASTM E399-90, "Standard Test Method for Plane-Strain Fracture Toughness of Metallic Materials."

(3) ASTM E813-89, "Standard Test for J_{Ic}, a Measure of Fracture Toughness."

(4) ASTM E1152-87, "Standard Test Method for Determining J-R Curves."

(5) ASTM E1290-89, "Standard Test Method for Crack-Tip Opening Displacement (CTOD) Fracture Toughness Measurement."

(6) Kirk, M. T. and Dodds, R. H., "*J* and CTOD Estimation Equations for Shallow Cracks in Single Edge Notch Bend Specimens," University of Illinois Research Report UILV-ENG-91-2013.

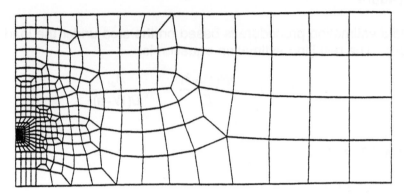

Figure 1 Finite Element Mesh for SENB Specimen (a/W = 0.3)

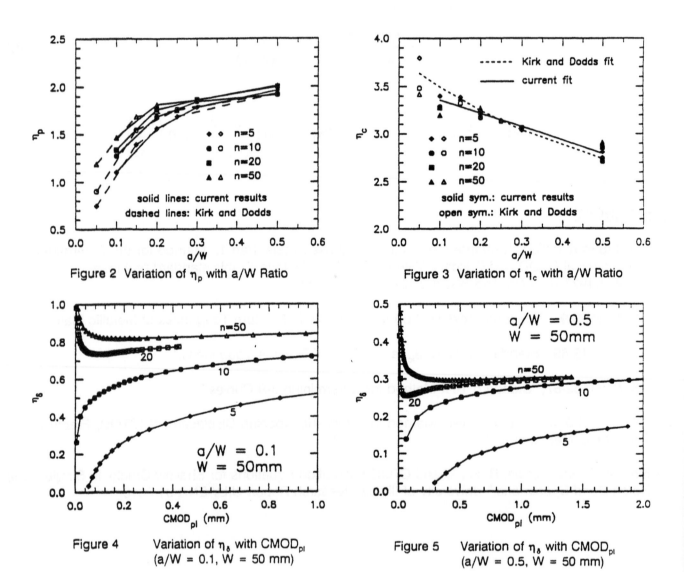

Figure 2 Variation of η_p with a/W Ratio

Figure 3 Variation of η_c with a/W Ratio

Figure 4   Variation of η_δ with $CMOD_{pl}$
(a/W = 0.1, W = 50 mm)

Figure 5   Variation of η_δ with $CMOD_{pl}$
(a/W = 0.5, W = 50 mm)

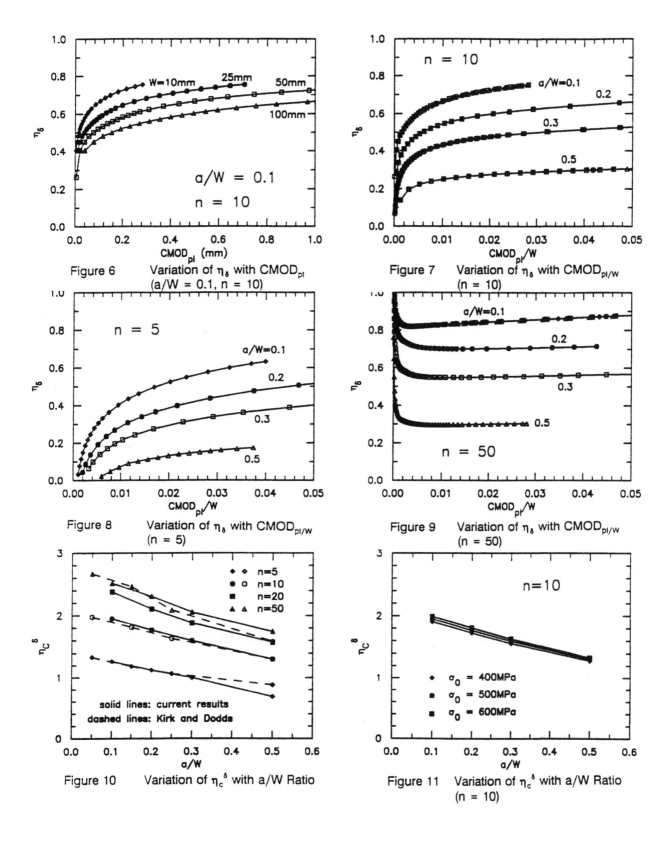

Figure 6   Variation of η_δ with $CMOD_{pl}$
(a/W = 0.1, n = 10)

Figure 7   Variation of η_δ with $CMOD_{pl/w}$
(n = 10)

Figure 8   Variation of η_δ with $CMOD_{pl/w}$
(n = 5)

Figure 9   Variation of η_δ with $CMOD_{pl/w}$
(n = 50)

Figure 10   Variation of η_c^δ with a/W Ratio

Figure 11   Variation of η_c^δ with a/W Ratio
(n = 10)

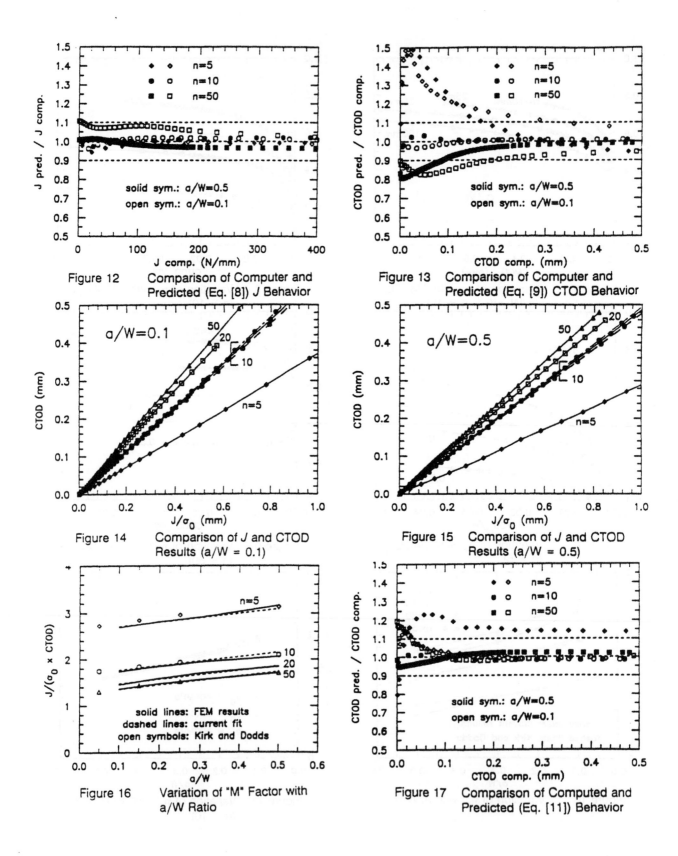

Figure 12   Comparison of Computer and Predicted (Eq. [8]) J Behavior

Figure 13   Comparison of Computer and Predicted (Eq. [9]) CTOD Behavior

Figure 14   Comparison of J and CTOD Results (a/W = 0.1)

Figure 15   Comparison of J and CTOD Results (a/W = 0.5)

Figure 16   Variation of "M" Factor with a/W Ratio

Figure 17   Comparison of Computed and Predicted (Eq. [11]) Behavior

The influence of weld strength mismatch on crack tip constraint in single edge notch bend specimens

M T Kirk, BS, MS (US Naval Surface Warfare Center) and R H Dodds Jr, BS, MS, PhD (University of Illinois)

SUMMARY

Dodds and Anderson provide a framework to quantify finite size and crack depth effects on fracture toughness when failure occurs at deformation levels where J no longer uniquely describes the state of stresses and strains in the vicinity of the crack tip. Size effects on cleavage fracture are quantified by defining a value termed J_{SSY}: the J to which an infinite body must be loaded to achieve the same stressed volume, and thereby the same likelihood of cleavage fracture, as in a finite body. In weld metal fracture toughness testing, mismatch between weld metal and baseplate strength can alter deformation patterns, which complicates size and crack depth effects on cleavage fracture toughness. However, the virtually limitless number of weld joint geometry / crack depth combinations preclude calculation of J_{SSY} for each individual case. This study addresses the accuracy with which J_{SSY} for a welded single edge notch bend, SE(B), specimen can be approximated by previously published results for homogeneous specimens. The case of a crack located on the weld joint centerline is treated. The combined effects of weld groove type, degree of mismatch, and crack depth to specimen width (a/W) ratio are considered by performing plane strain elastic–plastic finite element analyses of SE(B) specimens containing a variety of common weld groove details. These results demonstrate virtually no effect of $\pm20\%$ mismatch on J_{SSY} if the distance from the crack tip to the weld/plate interface (L_{min}) exceeds 5 mm. If L_{min} falls below 5 mm, there exists a deformation (applied–J) dependent value of L_{min} below which reasonably accurate J_{SSY} estimation is possible. At higher levels of overmatch (50% to 100%), it is no longer possible to parameterize departure of J_{SSY} for a weldment from that for a homogeneous SE(B) based on L_{min} alone. Weld geometry significantly influences the accuracy with which J_{SSY} for a welded SE(B) can be approximated by J_{SSY} for a homogeneous specimen at these extreme overmatch levels.

INTRODUCTION

The application of conventional fracture mechanics to assess the integrity of a cracked structure relies on the notion that a single parameter uniquely characterizes material resistance to fracture. Material resistance to catastrophic brittle fracture is characterized by a critical value of the stress intensity factor, K_{Ic}, while resistance to the onset of ductile, or upper–shelf, fracture is characterized by a critical value of the J–integral, J_{Ic}. Testing standards which govern the measurement of K_{Ic} and J_{Ic}, ASTM E399 and ASTM E813 respectively, require sufficient specimen thickness to insure predominantly plane strain conditions at the crack tip and sufficient crack depth to position the crack tip in a highly constrained bending field. These restrictions are designed to insure the existence of severe conditions for fracture as described by the Hutchinson–Rice–Rosengren (HRR) asymptotic fields (1,2). The testing standards thereby guarantee that K_{Ic} and J_{Ic} are lower bound, geometry independent measures of fracture toughness. However, cracks in civil and marine structures are seldom this highly constrained, which makes predictions of the fracture resistance of a structure based on laboratory fracture toughness values overly pessimistic.

Researchers at The Welding Institute (TWI) have long advocated a more pragmatic, engineering approach to assess the fracture integrity of cracked structures (3). This approach requires that constraint in the test specimen approximate that of the structure to provide an "appropriate" toughness for use in a structural analysis. The appropriate constraint is achieved by matching thickness and crack depth between specimen and structure. Experimental studies by Sumpter (4) and by Kirk and Dodds (5), comparing cleavage fracture toughness (J_c) values of shallow crack bend specimens to J_c values for part–through semi–elliptical surface cracks, demonstrate the validity of this approach. These studies show that use of geometry dependent fracture toughness values allows more accurate prediction of the fracture performance of structures than is possible using more traditional approaches. However, the task of characterizing fracture toughness becomes more complex as non−standard specimens are required, and different fracture toughness data are needed for each geometry of interest. Further, this approach cannot be economically applied to thick section structures (e.g. nuclear pressure vessels).

While the TWI approach is useful and frequently applied in engineering practice, it fosters the erroneous impression that, even for homogeneous materials, absolute crack depth (a), relative crack depth (a/W), and thickness (B) all influence fracture toughness. Hereafter the apparent effect of these geometric parameters on fracture toughness are referred to collectively as *size effects*. The observed size effects on J_c (6,7) indicate that a single parameter no longer uniquely describes conditions near the crack tip. The increase of J_c values in shallow crack specimens develops when the in−plane plastic deformation produced by gross bending of the specimen impinges on the local crack tip fields. This relaxes the kinematic constraint against further plastic flow. Once the global and local plastic fields interact, the crack tip stresses and strains no longer increase in proportion to one another with amplitude governed by J alone. In this situation, equivalence of J between specimens of different crack depths does not insure the same crack tip stress and strain fields. As fracture by any micro−mechanism ultimately requires attainment of some critical condition described in terms of stress and/or strain, different values of applied J and CTOD may be required to achieve these critical conditions in different structures. Thus, the observed crack depth dependence of J_c is actually an effect on the relation between macroscopic fracture parameters and micro−scale crack driving force.

The distinction between the fictitious finite size effect on fracture toughness and the actual finite size effect on micro−scale crack driving force is unimportant to the practitioner so long as the effect is consistently treated and fracture toughness experiments are available at the "appropriate" constraint level. However, recognizing that finite size influences driving force rather than toughness suggests that the effect can be quantified analytically. Dodds and Anderson (8,10) demonstrated that combining detailed knowledge of stresses near the crack tip (determined by finite element analysis) with a micro−mechanics model appropriate for cleavage fracture permits quantification of finite size effects on micro−scale / macro−scale crack driving force relations. Application of these techniques has resolved the apparent size dependencies exhibited in J_c data from both bending (7,8) and tension (9) experiments. These techniques are discussed in the following section.

In welded construction, fracture is most likely to occur either in or near the weld as this location contains the most severe residual stresses, potential welding defects, and the lowest fracture toughness. Structural integrity assessments are therefore often based on toughness data derived by testing fracture specimens containing welds. Mismatch between weld metal and baseplate strength can alter the deformation patterns in such specimens, which complicates finite size effects on cleavage fracture toughness. This study addresses the effect of weld strength mismatch on size effects in SE(B) specimens quantified using the techniques developed by Dodds and Anderson. A crack located in the weld metal on the weldment centerline is treated in detail. The combined effects of weld groove type, degree of mismatch, and crack depth to specimen width (a/W) ratio are considered.

MICROMECHANICAL CONSTRAINT CORRECTIONS

Dodds and Anderson (8,10) show that, by quantifying the effects of finite size on micro−scale / macro−scale crack driving force relations, the apparent size effect on fracture toughness can be rigorously predicted without resort to empirical arguments. These size effects become steadily more pronounced as load increases due to the deviation of crack tip region deformations from the small scale yielding conditions essential for single parameter fracture mechanics (SPFM) to apply. Once SPFM becomes invalid, a micro−mechanics failure criteria is required to establish the the geometry invariant conditions at fracture. Finite element analysis provides a means to quantify the geometry dependent relations between these conditions and macro−scale crack driving force. This permits (in principle) prediction of fracture in any body from toughness values measured using standard specimens.

For steels operating at temperatures where cleavage occurs after significant plastic deformation but before the initiation of ductile growth (lower to mid−transition), attainment of a critical stress over a microstructurally relevant volume is an appropriate micro−mechanical failure criteria (11). A number of important engineering structures can fail by this mechanism, including high strength rails, offshore oil platforms, ships, storage tanks, and nuclear pressure vessels after years of neutron irradiation embrittlement. Techniques for predicting the apparent size effects on cleavage fracture toughness developed by Dodds and Anderson are now described.

Crack tip stresses in infinite bodies

An infinite body solution provides the idealized reference needed to quantify the effect of finite size on the crack tip stress state. In classical nonlinear fracture mechanics, the HRR field equations (1–2) serve as this reference solution. However, analysis of the small scale yielding (SSY) problem demonstrates that the HRR solution does not accurately describe the stress state around a crack tip over the length scale needed to characterize cleavage fracture initiation (8). The SSY model (Figure 1), originally proposed by Rice and Tracy (12) and McMeeking (13), consists of a circular region containing an edge crack. Boundary displacements are applied to this region consistent with the linear elastic solution for a Mode I crack in an infinite body. Finite element modelling of SSY permits definition of the *full field* solution for a crack in an infinite body over distances of 2 to 10 times the CTOD. Steady state conditions are achieved wherein stresses and strains at all angles scale with $r/(J/\sigma_o)$ and $r/$CTOD, as do the HRR fields. This steady state condition persists until the plastic zone size becomes a significant portion of the modelled domain radius, $\approx 5\%$, at which point the small scale yield conditions are violated. The difference between the HRR and SSY solutions for a power law hardening material with a Ramberg Osgood strain hardening coefficient (n) of 10 is shown in Figure 2. While both solutions converge as $r/$CTOD $\rightarrow 0$, demonstrating the asymptotic nature of the HRR solution, the HRR solution becomes inaccurate at distances $r/$CTOD ≥ 1.

Both the HRR and SSY solutions are based on conventional small strain theory. Therefore, neither solution models accurately stresses very close to the crack tip where finite blunting deformations, not accounted for within small strain theory, reduce stress. SSY computations employing a finite strain formulation show this reduction at distances $r/$CTOD ≤ 2 (14). As illustrated in Figure 2, the finite strain and small strain calculations are in very close agreement beyond $r/$CTOD $= 2$. These observations indicate that cleavage fracture initiation is unlikely within the zone where finite strain effects dominate due to loss of stress triaxiality, a conclusion supported by experimental observations (15,16). Thus, the stress distribution defined by the SSY model using conventional small strain theory provides an appropriate reference solution to quantify size effects on crack tip stress fields. In practice, the SSY stress distribution is defined for the material of interest by finite element analysis for comparison with the stress distribution of the finite body.

Crack tip stresses in finite bodies

The distribution of stresses in the crack–tip region of a finite body depends on the geometry, on the mode and magnitude of loading, and on the material strain hardening exponent. Finite element analyses for the situation of interest provide stresses in the crack tip region for comparison with the SSY reference solution. Stresses are calculated and stored at each load step as the body is deformed toward a limit state. Sufficient mesh refinement is needed to fully define the stresses over distances of 2–10 times the CTOD at all load levels. Such analyses require a considerably more detailed mesh than does a routine analysis to determine J.

Calculation of constraint corrections

Maximum principal stress contours for the SSY problem are compared to those of an $a/W = 0.15$, $n = 10$ single edge notch bend, SE(B), specimen in Figure 3. The finite body effect reduces the peak stress amplitude with increasing deformation. The normalized size of the stressed contour shrinks relative to the SSY limit, which reduces the likelihood of cleavage fracture. However, the spatial distribution around the crack is nearly identical in the infinite body and in the finite body, as evidenced by the similarity of the contour shapes. This self–similarity indicates that stressed volumes in infinite and finite bodies differ only by a deformation dependent scalar multiple. The information depicted graphically in Figure 3 can be used to determine this multiple and, thereby, the effect of finite size and load intensity on the stress distribution near the crack tip. Although cleavage is driven by stress and stressed volume, the difficulty of measuring critical values of these parameters dictates that fracture driving force, and thereby critical fracture conditions, be expressed in terms of more easily measured macroscopic parameters (e.g. J and CTOD). Thus, an effective macroscopic driving force for cleavage fracture (J_{SSY}) can be defined as follows:

J_{SSY} is the J to which the SSY model (infinite body) must be loaded to achieve the same stressed volume, and thereby the same likelihood of cleavage fracture, as in a finite body.

The variation of J_{SSY} with J is depicted schematically for two finite bodies in Figure 4. Upon initial loading of any finite body, crack tip plasticity is well contained within a surrounding elastic field. Crack tip conditions are well approximated by SSY so, initially, $J_{SSY} \approx J_{FiniteBody}$. Subsequent interaction of plasticity at the crack tip with plasticity resulting from overall deformation of the structure relaxes the kinematic constraint against plastic flow at the crack tip, thus reducing the stresses in the crack tip region below what they are in SSY at the same J. This reduces the micro−scale driving force for cleavage. Consequently, the finite body requires more applied−J to achieve the same conditions for cleavage (same stressed volume) as in the infinite body. This finite size effect on crack−tip stress fields differs for different geometries constructed from the same material; it is indicated by deviation from the 1:1 slope in Figure 4. Information of this type is useful for both analysis of fracture test data and for assessing the defect integrity of structures. Path A−B−C on Figure 4 illustrates the procedure to remove geometric dependencies from experimental cleavage fracture toughness (J_c) data by determining the geometry independent cleavage fracture toughness (J_{SSY}) corresponding to a measured J_c value. Alternatively, Figure 4 permits determination of the effective driving force for cleavage fracture produced by structural loading to a certain $J_{Applied}$ value (path E−D−C). Two different methods have been used to calculate J_{SSY} from finite element results. These are reviewed briefly below; Kirk and Dodds present a more detailed discussion elsewhere (17).

Dodds et al. (8) show that the variation of maximum principal stress on the crack plane with distance from the crack tip in finite bodies is self−similar to that occurring under SSY conditions. J_{SSY} is calculated as the J value required in the SSY model to obtain the same opening mode stress as in the finite body at a distance of 4·CTOD ahead of the crack−tip. This forced equivalence between SSY and finite body stresses at 4·CTOD corresponds to selecting the critical microstructural distance (l_o^*) in the Richie−Knott−Rice (RKR) model (11). However, self−similarity between the SSY and finite body stress distributions makes the specific l_o^* value used unimportant over a wide range of deformation. Predictions of the RKR model depend upon l_o^* because RKR uses the HRR fields as a reference solution. HRR is not self−similar to finite body stress distributions while SSY is. Thus, the ability to determine J_{SSY} irrespective of the actual l_o^* value relies on self−similar stress distributions around the crack in finite and infinite bodies. This technique for calculating J_{SSY} is employed in this study.

The self−similarity of principal stress contours between finite bodies and the SSY condition was illustrated in Figure 3. Anderson and Dodds (10) use the entire principal stress contour, not just the opening stresses on the crack plane, to determine J_{SSY}. J_{SSY} is calculated as the J value required in the SSY model to achieve the same area as in the finite body within the principal stress contour $\sigma_1 = 3\sigma_o$. This forced equivalence between SSY and finite body areas at $\sigma_1 = 3\sigma_0$ corresponds to selecting the critical stress for cleavage fracture initiation (σ_f^*) in the RKR model (11). However, self−similarity between the SSY and finite body stress distributions makes the specific σ_f^* value used unimportant over a wide range of deformation. Predictions of the RKR model depend upon σ_f^* because RKR uses the HRR fields as a reference solution. HRR is not self−similar to finite body stress distributions while SSY is. Thus, the ability to determine J_{SSY} irrespective of the actual σ_f^* value relies on self−similar stress distributions around the crack in finite and infinite bodies.

APPROACH

Currently, J_{SSY} for welded SE(B) specimens can be estimated from published results for homogeneous specimens (7,8). However, mismatch between weld metal and baseplate strength alters the deformation patterns in such specimens, causing potential errors in J_{SSY} determined by assuming specimen homogeneity. In this study, plane−strain elastic−plastic finite element analyses of SE(B) specimens containing a variety of common weld groove details are performed. The crack tip stress fields quantified by these analyses permit evaluation of the variation of J_{SSY} with J for welded SE(B) specimens. This information establishes a baseline to judge the applicability of estimating J_{SSY} for a welded specimen from an analysis that takes no account of weld mismatch. The cases illustrated in Figure 5 are each modelled as 20% overmatched, homogeneous (no weld), and 20% undermatched. Unless indicated otherwise, the constitutive properties detailed in Table 1 are used. The weldment is modelled as a bi−material with no transition zone (heat affected zone, or HAZ) placed between the weld and the plate. The strain hardening exponents in Table 1 are calculated from yield stress based on an experimental correlation applicable to construction steels developed by Barsom and Rolfe[18]:

$$n = \left[\frac{\sigma_o}{103.4}\right]^{\frac{4}{3}} \quad (\sigma_o \text{ is in MPa})$$

[1]

Certain aspects of this approach, adopted for expediency, require justification to ensure the applicability of these results to real weldments. For example, modelling a weldment as a bi−material calls into question the applicability of these results to weldments having a constitutive property gradient across the HAZ. Further issues include

1. The calculation of strain hardening capacity from yield strength using eqn. [1], while physically realistic, provides different absolute hardening capacities dependent upon the plate yield strength selected for analysis. It is not apparent, for example, that the results of an analysis of a 20% undermatched weld joining 414 MPa yield strength steel (plate $n = 6.3$, weld $n = 4.7$) apply to a 20% undermatched weld joining 690 MPa yield strength steel (plate $n = 12.5$, weld $n = 9.3$).

2. Quite large amounts of overmatch can occur in practice (e.g. welding A36 steel with an E7018 electrode producing approximately 50% overmatch) which lie outside the primary focus of this study.

This investigation addresses each of these issues. Related work concerning J and CTOD estimation for SE(B) specimens containing mismatched welds is reported elsewhere [19].

Table 1: Constitutive properties for weldment analyses				
	Weld		Plate	
% Mismatch	σ_o (MPa)	n	σ_o (MPa)	n
20% Over	717	13	593	10
No Weld	---	---	717	13
20% Under	717	13	896	18

Yield strength (σ_o) and strain hardening exponent (n) are coefficients in the Ramberg Osgood constitutive relation:

$$\frac{\varepsilon}{\varepsilon_o} = \frac{\sigma}{\sigma_o} + \alpha\left(\frac{\sigma}{\sigma_o}\right)^n$$

where $\alpha = 1$ and $\varepsilon_o = E/\sigma_o$

FINITE ELEMENT MODELLING

Two dimensional, plane strain finite element analyses of SE(B) specimens and the SSY model are performed using conventional small strain theory. These analyses are conducted using the POLO−FINITE analysis software [20] on an engineering workstation. Dodds, et al., [8] previously described the finite element modelling procedure.

Finite element models are constructed for each combination of a/W ratio and weld joint geometry. These computations apply to SE(B) specimens of standard proportions; the unsupported span is four times the specimen width. Models of symmetric joints contain approximately 900 elements and 2850 nodes, while the non−symmetric mesh of the single bevel weld contains 1414 elements and 4431 nodes. Figure 6 illustrates this model. A semi−circular core of elements surrounds the crack tip in all models. Crack tip element size ranges from 0.2% to 0.02% of the crack depth depending on the crack depth modelled. The J−integral is computed at each load step using a domain integral method [21,22]. J values calculated over domains adjacent to and remote from the crack tip, but not crossing a bi−material interface, are within 0.003% of each other, as expected for deformation plasticity combined with these detailed meshes. All J values reported for weldments are calculated over domains that lie completely within the weld metal [19]. CTOD is estimated from the blunted shape of the crack flanks using the $\pm 45°$ intercept procedure.

RESULTS AND DISCUSSION

Finite element results for ±20% mismatch

The constraint correction curve illustrated schematically in Figure 4 is shown quantitatively in Figure 7 for two welds containing a/W=0.15 cracks: 60°Double−V with a 5.1 mm root gap and 70°Double−V with a 12.7 mm root gap. J_{SSY} for some welds (e.g. 70°Double−V) is independent of ±20% mismatch, indicating that while mismatch alters plasticity distribution throughout the specimen, the near−tip stresses are unaffected. Other welds, e.g. 60°Double−V, have some effect on J_{SSY}. However, the virtually limitless number of weld joint geometry / crack depth combinations preclude calculation of J_{SSY} for each individual case. In view of the small effect of ±20% mismatch on J_{SSY}, even for the 60°Double−V weld, the following therefore seems a useful and reasonable approximation:

$$J_{SSY}|_{WELD} \approx J_{SSY}|_{\substack{HOMOGENEOUS \\ (ALL \; WELD \; METAL)}} \qquad [2]$$

This 'all weld metal' approximation, if sufficiently accurate, enables J_{SSY} estimation from previously published solutions for homogeneous SE(B)s (7,8). As finite element models which fully resolve near−tip fields and account for the weld / plate interface are quite detailed, considerable effort can be saved if the all weld metal approximation is sufficiently accurate.

The following error measure is defined to evaluate the accuracy of the all weld metal approximation:

$$Error = \frac{J_{SSY}|_{WELD} - J_{SSY}|_{\substack{HOMOGENEOUS \\ (ALL \; WELD \; METAL)}}}{J_{SSY}|_{WELD}} \cdot 100 \qquad [3]$$

The variation of J_{SSY} estimation error with normalized $J_{SE(B)}$ is illustrated in Figure 8 for all of the welds analyzed. Even though all of the errors are reasonably small, certain weld joint / crack depth combinations have much greater effects than others. While a considerable number of cases have been analyzed, situations producing a greater effect on J_{SSY} may remain undiscovered. It therefore seems useful to determine what factors promote deviation of J_{SSY} from the homogeneous solution. Certainly the minimum distance from the crack tip to the weld/plate interface (L_{min}) should play an important role. If L_{min} is large, it seems unlikely that weld mismatch could significantly alter the crack tip stress fields used to calculate J_{SSY}. Conversely, narrow weld joints (L_{min} small) may have some influence, particularly as L_{min} approaches the length scale over which J_{SSY} is calculated. The variation of J_{SSY} estimation error with L_{min} at two fixed applied−J levels is shown in Figure 9 for all weldments analyzed. These data demonstrate that, even at very high applied−J, there is virtually no effect of ±20% mismatch on J_{SSY} if L_{min} exceeds 5 mm. For welds having L_{min} smaller than 5 mm, reasonably accurate J_{SSY} estimation is possible by eqn. [2] until the applied deformation becomes so large that significant interaction occurs between the stress field at the weld / plate interface and the stress field at the crack tip. The information in Figure 9 can be used to determine the relationship between L_{min} and the deformation level above which accurate J_{SSY} estimation by eqn. [2] is no longer possible, as illustrated in Figure 10. The variation of weld size to produce less than 10% J_{SSY} estimation error with applied−J for ±20% mismatched weldments is given in Figure 11. This curve can be used as a basis to judge whether accurate determination of J_{SSY} requires finite element analysis of the actual weldment ($J_c - L_{min}$ combinations above the curve) or if published solutions for homogeneous SE(B)s provide sufficient accuracy ($J_c - L_{min}$ combinations below the curve).

Justification of assumptions in approach

HAZ modelling

The results presented in the preceding section are determined using finite element models which do not account for the transition in constitutive properties between the weld and the plate. The HAZ is so remote from a crack on the weld centerline and is so thin that it should have little effect on the stresses near the crack tip used to calculate J_{SSY}. To demonstrate the validity of ignoring the HAZ, analyses of the two shallow cracked square groove weldments including a highly refined HAZ are performed. Groove widths are 0.16W and 0.32W, producing L_{min} values of 2.5 mm and 5 mm, respectively. A detail of the 0.16W square groove model near the crack is shown along with the constitutive properties used to model the HAZ in Figure 12. The yield strength of the 0.127 mm wide HAZ layer immediately adjacent to the weld is 1241 MPa, characteristic of the as−quenched martensite found in the grain coarsened HAZ. Between this high hardness layer and the plate, the HAZ is modelled as seven discrete layers of increasing width and decreasing strength. These models realistically represent both peak hardness and total HAZ width. Further, this model presents a greater challenge for accurate J_{SSY} estimation than occurs in an actual weldment. An actual multi−pass weldment has a discontinuous high strength layer due to re−tempering from multiple passes, rather than the continuous high strength layer modelled here. The effect of HAZ modelling on the constraint correction curve for the $0.32W$ square groove weld is illustrated in Figure 13. Here the HAZ has negligible effect on J_{SSY} because the weld / plate interface is sufficiently remote from the crack tip ($L_{min} \geq 5$ mm). This is consistent with previous observations based on welds with no transition zone between the plate and the weld. However, as the distance between the crack tip and the weld / plate interface reduces, the presence of a high hardness HAZ significantly affects J_{SSY}, as shown in Figure 14 for the $0.16W$ square groove weldment. The expectation from analyses where the HAZ is ignored (Figure 11) is that a weldment with $L_{min} = 2.5$ mm will have J_{SSY} estimation errors below 10% if $J_{SE(B)}/(b\sigma_{flow}) \leq 0.0028$. This prediction is slightly non−conservative, for the 20% undermatched $0.16W$ square groove modelled with a HAZ, which has a J_{SSY} estimation error of 13% at $J_{SE(B)}/(b\sigma_{flow}) = 0.0028$. However, as noted previously, the high strength HAZ is discontinuous in an actual multi−pass weldment, rather than continuous as modelled here. Thus, it does not appear that expectations regarding J_{SSY} estimation accuracy based on the simpler bi−material models are grossly inaccurate for real weldments with a constitutive property gradient across the HAZ.

Effect of constitutive properties

All of the results discussed thus far are generated using the constitutive properties detailed in Table 1. To assess the applicability of these results to mismatch for different constitutive properties, two weldments (45°Single−V and 60°Double−V) containing $a/W = 0.15$ cracks are analyzed using properties characteristic of a lower strength steel ($\pm 20\%$ mismatch about a weld metal having $\sigma_o = 414$ MPa, $n = 6.3$). These two weldments are selected as they have L_{min} values that are near ($L_{min} = 4.8$ mm for 45°Single−V) and below ($L_{min} = 3.8$ mm for 60°Double−V) the 5 mm cutoff above which negligible effects of $\pm 20\%$ mismatch on J_{SSY} are observed. J_{SSY} estimation errors remain within previously established error bounds for both weldments considered, as illustrated in Figure 15. Thus, the the effect of $\pm 20\%$ mismatch on J_{SSY} estimation accuracy discussed previously appears approximately correct irrespective of the baseline σ_o and n values used in the finite element analysis.

Effect of extreme overmatch

All results previously presented are for $\pm 20\%$ mismatch. However, certain construction practices cause considerably greater overmatch. This is investigated by performing supplemental analysis for three $a/W = 0.15$ weldments: 70°Double−V, 45°Single−V, and 60°Double−V having L_{min} values of 9.9, 4.8, and 3.8 mm, respectively. The J_{SSY} estimation errors caused by mismatch ranging from 20% under to 100% over are given in Figure 16. The final point on these graphs indicates the last load step at which the stress field near the crack tip in the SE(B) is sufficiently self−similar to the SSY reference solution that J_{SSY} remains independent of the critical distance selected for its calculation. The data in Figure 16 indicate that both increased overmatching and small L_{min} values limit the maximum applied−J for SE(B) − SSY self−similarity. Over the range of applied−J that J_{SSY} can be calculated, the J_{SSY} estimation error for the two Double−V welds associated with 50% and 100% overmatch is approximately the same as for $\pm 20\%$ mismatch. However, J_{SSY} estimation error for the Single−V weld is increased significantly by extreme overmatching. L_{min} parameterizes the departure of J_{SSY} for a weld from that for a homogeneous SE(B) at $\pm 20\%$ mismatch irrespective of joint geometry. This simplification breaks down for greater levels of overmatching where weld geometry becomes important.

SUMMARY AND CONCLUSIONS

Dodds and Anderson provide a framework to quantify finite size effects on cleavage fracture toughness when failure occurs at deformation levels where J no longer uniquely describes the state of stresses and strains in the vicinity of the crack tip. Size effects on cleavage fracture are quantified by defining a value termed J_{SSY}: the J to which an infinite body must be loaded to achieve the same stressed volume, and thereby the same likelihood of cleavage fracture, as in a finite body. In weld metal fracture toughness testing, mismatch between weld metal and baseplate strength can alter deformation patterns, which complicates size effects on cleavage fracture toughness. However, the virtually limitless number of weld joint geometry / crack depth combinations preclude calculation of J_{SSY} for each individual case. This study addresses the accuracy with which J_{SSY} for a welded single edge notch bend, SE(B), specimen can be approximated by previously published results for homogeneous specimens. The case of a crack located on the weld joint centerline is treated. The combined effects of weld groove type, degree of mismatch, and crack depth to specimen width (a/W) ratio are considered by performing plane strain elastic−plastic finite element analyses of SE(B) specimens containing a variety of common weld groove details. These results demonstrate virtually no effect of $\pm 20\%$ mismatch on J_{SSY} if the distance from the crack tip to the weld/plate interface (L_{min}) exceeds 5 mm. If L_{min} falls below 5 mm, there is a deformation (applied−J) dependent value of L_{min} below which reasonably accurate J_{SSY} estimation is possible. At higher levels of overmatch (50% to 100%), it is no longer possible to quantify departure of J_{SSY} for a weldment from that for a homogeneous SE(B) based on L_{min} alone. Weld geometry significantly influences the accuracy with which J_{SSY} for a welded SE(B) can be approximated by J_{SSY} for a homogeneous specimen at these extreme overmatch levels.

ACKNOWLEDGEMENTS

This research was conducted as part of the Surface Ship and Submarine Materials Block under the sponsorship of I. L. Caplan (Carderock Division, Naval Surface Warfare Center (CDNSWC), Code 011.5). The work supports CDNSWC Program Element 62234N, Task Area RS345S50. The work of the first author was conducted at the University of Illinois as part of a training program administered by CDNSWC. Support for R.H. Dodds was provided by CDNSWC under contract number N61533−90−K−0059. Computational support was made possible by CDNSWC Contract No. N61533−90−K−0059.

REFERENCES

(1)   Hutchinson, J.W., *Journal of Mechanics and Physics of Solids*, Vol. 16, pp. 13−31, 1986.

(2)   Rice, J.R., and Rosengren, G.F., *Journal of Mechanics and Physics of Solids*, Vol. 16, pp. 1−12, 1968.

(3)   Dawes, M.G., et al., Proceedings of the *International Conference on Fracture Toughness Testing: Methods, Interpretation, and Application*, 9−10 June 1982, London, England.

(4)   Sumpter, J.D.G., *ASTM STP 995*, pp.415−432, 1989.

(5)   Kirk, M.T., and Dodds, R.H., *Engineering Fracture Mechanics*, Vol. 39, No. 3, pp. 535−551, 1991.

(6)   Sorem, W.A., Dodds, R.H., and Rolfe, S.T., *International Journal of Fracture*, Vol. 47, pp. 105−126, 1991.

(7)   Kirk, M.T., Koppenhoeffer, K.C., and Shih, C.F., to appear in the proceedings of the ASTM conference on *Constraint Effects in Fracture* held in Indianapolis, Indiana, May 1991.

(8)   Dodds, R.H., Anderson, T.L., and Kirk, M.T., *International Journal of Fracture*, Vol. 48, pp. 1−22, 1991.

(9)   Keeney−Walker, J., Bass, B.R., and Landes, J.D., to appear in *ASTM STP 1131*.

(10) Anderson, T.L., and Dodds, R.H., *Journal of Testing and Evaluation*, Vol. 19, pp. 123−134, 1991.

(11) Ritchie, R.O., Knott, J.F., and Rice, J.R., *Journal of the Mechanics and Physics of Solids*, Vol. 21, pp. 395−410, 1973.

(12) Rice, J.R., and Tracey, D.M., in *Numerical and Computer Methods in Structural Mechanics*, S.J. Fenves et al. (eds.), Academic Press, New York, pp.585−623, 1968.

(13) McMeeking, R.M., *Journal of the Mechanics and Physics of Solids*, Vol. 25, pp. 357−381, 1977.

(14) McMeeking, R.M., and Parks, D.M., *ASTM STP 668*, pp.175−194, 1979.

(15) Miglin, M.T., Wade, C.S., and Van Der Sluys, W.A., *ASTM STP 1074*, pp.238−263, 1990.

(16) Herrens, J., and Read, D.T., NISTIR 88−3099, National Institute for Standards and Technology, Boulder, Colorado, December, 1988.

(17) Kirk, M.T., and Dodds, R.H., Jr., University of Illinois, Urbana, Illinois, Structural Research Series Report SRS−568, February 1992.

(18) Barsom, J.M., and Rolfe, S.T., *Fracture and Fatigue Control in Structures − Applications of Fracture Mechanics*, p. 265, Prentice−Hall Inc., Englewood Cliffs, NJ, 1987.

(19) Kirk, M.T., and Dodds, R.H., Jr., Proceedings of the *Eleventh ASME Conference on Offshore Mechanics and Arctic Engineering*, 7−11 June 1992, Calgary, Canada.

(20) Dodds, R.H., and Lopez, L.A., *International Journal for Engineering with Computers*, Vol. 13, pp. 18−26, 1985.

(21) Li, F.Z., Shih, C.F., and Needleman, A., *Engineering Fracture Mechanics*, Vol. 21, pp. 405−421, 1985.

(22) Shih, C.F., Moran, B., and Nakamura, T., *International Journal of Fracture*, Vol. 30, pp. 79−102, 1986.

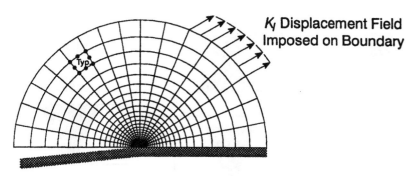

Figure 1:   Small scale yield (*SSY*) model.

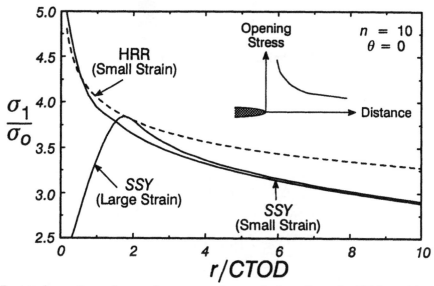

Figure 2:   Comparison of opening mode stress on the crack plane from the HRR and *SSY* solutions.

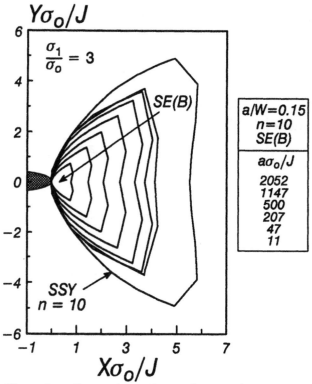

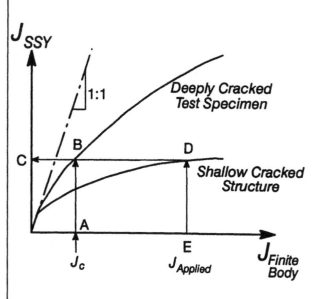

Figure 3:   Comparison of a maximum principal stress contour in *SSY* with those from an *a/W*=0.15 *n*=10 SE(B). SE(B) contours decrease with increasing deformation (i.e. with decreasing $a\sigma_o/J$).

Figure 4:   Conceptual variation of J_{SSY} with J for two finite bodies.

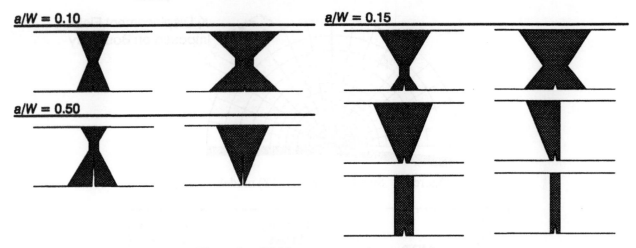

Figure 5:   Weldment geometries analyzed.

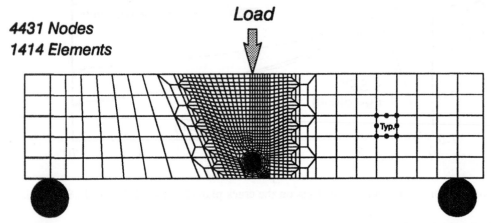

Figure 6:   Finite element model of a SE(B) specimen containing an a/W = 0.15 crack in a single bevel joint.

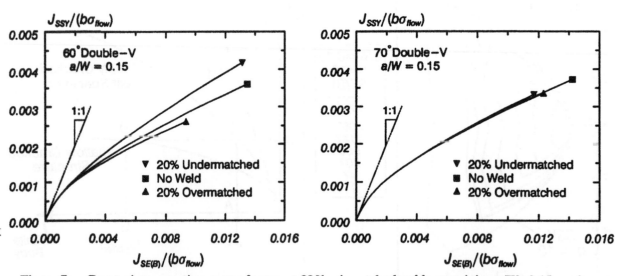

Figure 7:   Constraint correction curves for two ± 20% mismatched welds containing a/W=0.15 cracks.

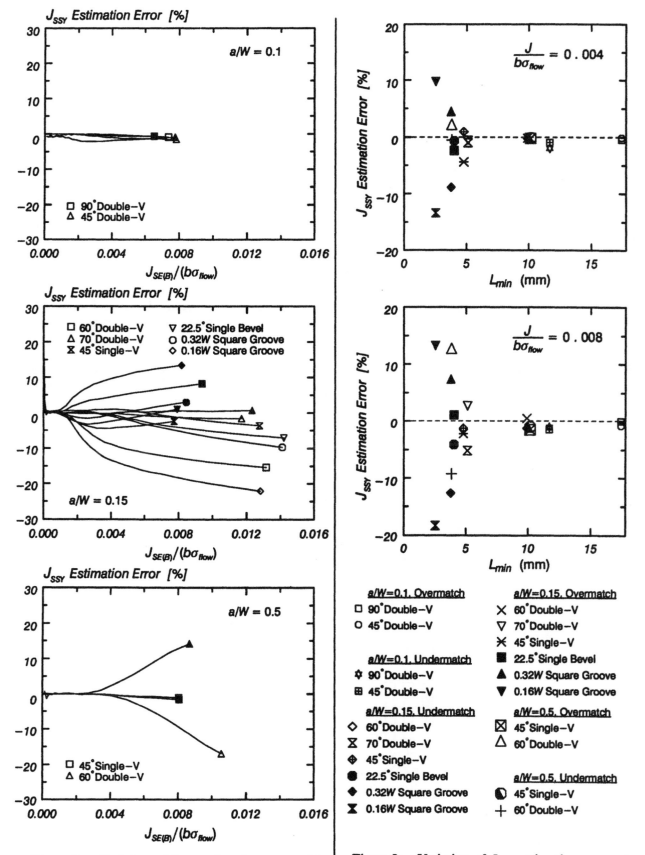

Figure 8: Variation of J_{SSY} estimation error with applied–J for $\pm 20\%$ mismatched welds. Filled symbols represent overmatching, open symbols undermatching.

Figure 9: Variation of J_{SSY} estimation error at two fixed applied–J levels with minimum distance from the crack tip to the weld/plate interface for $\pm 20\%$ mismatched welds.

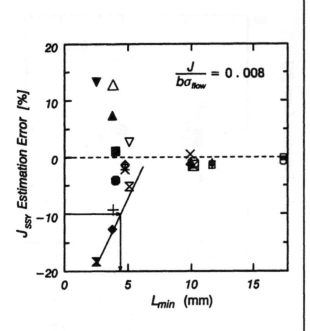

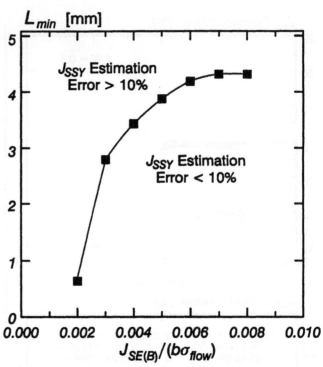

Figure 10: Determination of distance needed from the crack tip to the weld/plate interface to keep J_{SSY} estimation error below 10% for ±20% mismatched welds.

Figure 11: Effect of applied−J on distance needed from the crack−tip to the weld/plate interface to keep J_{SSY} estimation error below 10% for ±20% mismatched welds.

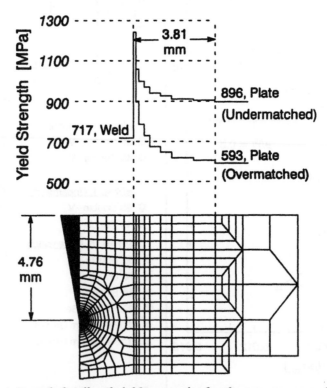

Figure 12: Finite−element mesh detail and yield properties for the square groove HAZ model. Strain hardening coefficient (n) is calculated from yield strength by eqn. [1].

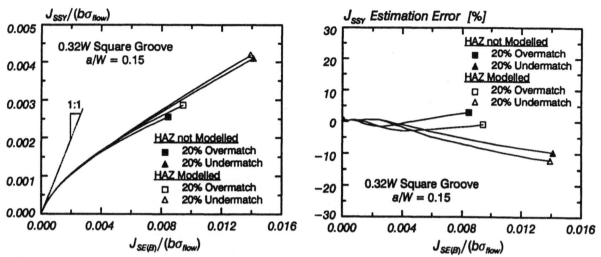

Figure 13: Effect of HAZ modelling on constraint correction curve for the 0.32W square groove weld.

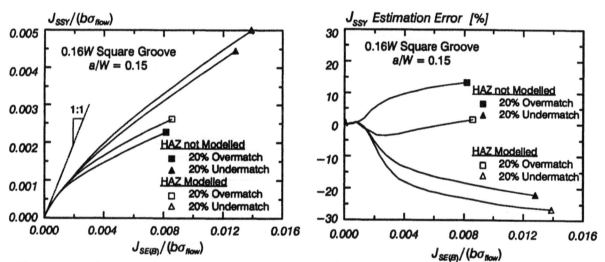

Figure 14: Effect of HAZ modelling on constraint correction curve for the 0.16W square groove weld.

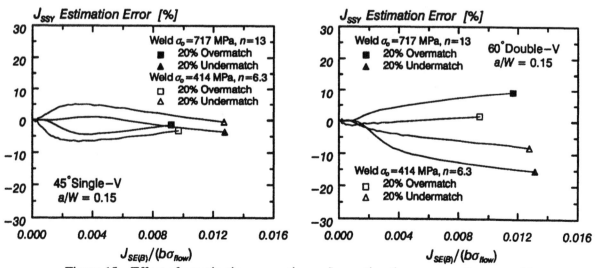

Figure 15: Effect of constitutive properties on J_{SSY} estimation accuracy for two welds.

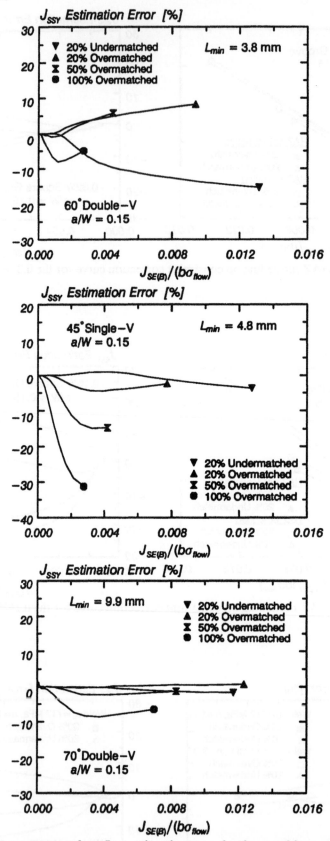

Figure 16: Effect of extreme overmatch on J_{SSY} estimation error for three welds containing a/W=0.15 cracks.

PAPER 35

Emission and reflections of surface waves from shallow surface notches

Prof C Thaulow (NTH/SINTEF), M Hauge (SINTEF Production Engineering) and J Specht (RWTH, Aachen)

SUMMARY

The existence and nature of surface waves initiated from a pop-in have been investigated by finite element calculations. The pop-in was simulated by a sudden release of stresses at the crack tip.

It is concluded that the surface waves propagating along the free surfaces of the crack may be a dominating mechanism in the neighbourhood of a crack tip. Reflected surface waves lead to crack closure and a reduction of the stresses close to the crack tip.

INTRODUCTION

Pop-ins are frequently observed in fracture mechanics testing of weldments. The occurrence of pop-ins is either explained on the basis of the test specimen dimensions, with the plane stress -plane strain relationship and the statistical size effect, or the existence of embrittled microstructural zones.

Once a pop-in has initiated it will extend a limited distance before it is arrested. Sources for the arrest of the crack are:

- the crack propagates into a region with lower stresses
- the crack propagates into a adjacent microstructure with higher toughness
- volume stress waves emitted at crack initiation are reflected back to the running crack

In this paper the attention is focussed on the surface, or Rayleigh waves emitted at crack initiation and the possible interaction between the reflections from these waves and the pop-in arrest.

A recent examination and literature study concluded that surface waves could be of importance with respect to arresting cracks Thaulow and Burget (1). Through the release of energy at crack initiation, surface waves were formed, and the waves will be reflected back from edges and cause crack closure and eventual arrest.

If we suggest that a crack propagates at a speed of 1:50 compared with the volume stress wave v_c (for example a crack speed of 100 m/s and a stress wave speed of 5000 m/s, according to Willoughby and Wood (2), the distance of crack extension before it is influenced by the reflecting wave will be:

$$\frac{\Delta a}{\frac{1}{50} v_c} = \frac{2(W-a_o)-\Delta a}{v_c} \qquad [1]$$

$$\Delta a = 0.02 \ (W - a_o) \qquad [2]$$

The similar crack extension without influence of a reflected surface stress wave will be

$$\frac{\Delta a}{v_{pop}} = \frac{2a_o+\Delta a}{v_R} \qquad [3]$$

$$\Delta a = 2a_o \ \frac{v_{pop}}{v_R - v_{pop}} \qquad [4]$$

where v_{pop} is the speed of the crack extension and v_R is the Rayleigh speed. For $v_{pop} = 1/50 \ v_c$ and $v_R = 0.5 \ v_c$, the equation will be reduced to

$$\Delta a = \frac{a_o}{12} \qquad [5]$$

The crack depth, where the volume - and surface stress waves hit the crack tip at the same time, can then be calculated according to

$$\frac{1}{50}(W-a_o) = \frac{a_o}{12} \qquad [6]$$

$$a_o = \frac{12W}{38} \approx \frac{W}{3} \qquad [7]$$

Hence, for a crack depth of W/3, both the volume and the surface stress wave will reach the extended crack tip at the same time. For crack depths shorter than W/3, the crack extension may be influenced by surface waves reflected from the crack mouth before the volume waves reach back. The influence of surface waves may be increased if the reflection is caused by steps / notches introduced on the notch surface or reflections from the edges of secondary cracks developed before the pop-in takes place. The influence from such surface irregularities will to a large degree depend on the wave length versus the extension of the irregularity.

The effect of reflected surface waves was first accidently observed in crack arrest testing of polymer material, by Shmuely et al.(3). In a reanalysis of shadow optical measurements from crack arrest testing of high strength steels, it was found that the crack velocity decreased remarkably just after the theoretical calculated arrival time of the reflected surface waves, Thaulow & Burget (1).

In a reanalysis of pop-in experiments from Japan, Arimochi (4), we have found that the reflected surface waves will always arrive at the crack tip before the cracks are observed to arrest, Table 1. The specimen with the highest crack velocity, and therefore probably also the highest level of stored energy at crack initiation, is arrested just after the arrival of the surface waves. It must be underlined that there are uncertainties with respect to the interpretation of the data published in the reference.

In a study of dynamic crack growth by shear strain measurements on steel tested statically at - 150°C, a smooth fracture surface appeared after an initial unstable crack growth. Hedner (5). It was believed that the crack growth had occurred slowly in the smooth area, and this was supported from the strain gage measurements of the crack velocity. A reanalysis with the application of eq. [2], strongly indicate that the onset of slow crack growth coincides with the reflections of the first surface wave.

Other indications, supporting the effect of reflected surface waves, are discussed by Thaulow and Burget (1). Hence, unstable cracks have a tendency for auto-arrest, eg. the crack itself supports its own arrest. The experience so far supports this idea on a more global scale, but it may be a dominating mechanism on a finer scale in the neighbourhood of the crack tip. In this area we expect to have the highest frequencies of the wave, hence small surface irregularities can act as effective reflectors, Viktorov (6).

In the present paper, the existence and nature of the surface waves have been examined with FE modelling on a Cray computer. The propagation and reflection of the surface waves, including the radiation pattern over the crack surface, are visualized with video recordings directly from the computer calculations.

Table 1. Results from three point bending CTOD testing of an embrittled weld metal, Arimochi (4). The crack extension at the arrival of the reflected surface wave, $\Delta_{Rayleigh}$, is included in the table.

Thickness (mm)	Temperature (°C)	Δa_{pop} (mm)	t_{pop} (μs)	v_{pop} (ms)	$\Delta a_{Rayleigh}$ (mm)
25	-70	4.6	21	220	4.0
25	-70	5.4	19	280	5.2
25	-100	5.1	20	260	4.8

FEM Models

Three models have been examined. The 2D model is made up by 8-node isoparametric elements with 2 x 2 integration and plane strain formulation, Fig. 1. The size of the model is 12 x 10 mm, with a crack depth of about 4.6 mm. The elements were collapsed at the crack tip. The smallest mesh size was 0.125 mm.

A CTOD bead-in-groove, shallow surface notched specimen was simulated with a 3D model. Because of symmetrix is only one fourth part of the specimen modelled, with a model size of 20 x 10 x 7.5 mm. The crack depth was close to 3.5 mm. The smallest mesh size was about 0.42 mm and the number of elements 3036.

A deep notched CTOD specimen was simulated with a model of 150 x 50 x 25 mm, Fig. 3, and with a crack depth of about 20 mm. The smallest mesh size was about 0.76 mm and the number of elements 1172.

For all the numerical analysis the general purpose finite element computer program ABAQUS version 4.8 was applied, with linear elastic conditions. In the dynamic mode ABAQUS provides both explicit and implicit integration. Implicit integration was used for the 2D- and the shallow crack models, and the more time effective explicit integration for the deep crack model.

The pre- and postprocessing was performed in the programme PATRAN, and the results were visualised and animated in the programme MPGS (Multi - Purpose - Graphic - System).

Source mechanism

The detailed source mechanism of a pop-in is not known, but large literature on acoustic emission monitoring of cracks has indicated some of the characteristics. The relationship between the brittle fracture process and the acoustic emission waveforms has been examined with a special recording system giving a flat response up to 254 MHz, Wadley et al. (7), and by Scruby et al. (8). It was concluded that microfracture events in steel carry energy at frequencies in excess of 10 MHz, i.e. wave lengths in the range of 0.3 mm. Testing of both a quenched and tempered steel at room temperature and a mild steel at 77°K, revealed cleavage micro cracks propagating at an average speed of 450 ms^{-1}, while a speed of 930 ms^{-1} was recorded for cracks in electrolytic iron.

In order to simulate crack extension, the nodes at the crack tip have been loaded in accordance with the following procedure:

- A static calculation with the remote boundary conditions (fixed displacement d$_o$) before the pop-in is performed. The reaction forces at a set of nodes next to the crack tip are registrated. These nodes represent the crack extension during pop-in.

- The remote boundary conditions are replaced by zero displacement in all boundary nodes except the pop-in node set. The boundary conditions for the pop-in node set are represented by concentrated node forces (F$_o$) equal to the calculated reaction forces.

- In a dynamic analysis, the boundary conditions are now changed by reducing the concentrated forces to zero according to a predefined decay function.
  A linear decay function was applied. The decay time corresponded to the crack growth rate.

- The dynamic analysis is continued as long as needed in order to find the response of the Rayleigh wave.

The simulated source mechanism, with the static displacements before the initiation of the cracks, are illustrated in Fig. 4. The displacements give a qualitative measure of the CTOD value, which then will be in the range of 0.045 mm (3D shallow crack) to 0.016 mm (3D deep crack).

In order to verify that the source mechanism was realistic, dynamic analysis was performed with the remote boundary conditions as the starting conditions Fig. 4a. By subtracting the calculated displacements with the static displacements after the pop-in, exactly the same dynamic behaviour was obtained.

The R-waves are not developed at time zero, a certain incubation time is needed in dependence of the mesh size. A fine mesh will promote the most rapid decay function and emit the highest frequency of the resulting surface wave.

RESULTS

2D model

In general, a R-wave will give rise to an elliptical movement of particles at the free surface, Viktorov (6). Figure 5 shows the resulting X-displacement as a function of time for some selected nodes at the crack flank. The passage of the R-wave and the reflection from the corner, are identified.

The dominating displacement is the decay of the whole crack flank back to zero position, Figure 4. The displacement resulting from the passing of the R-wave will be much smaller and of opposite sign, and will therefore be completely hidden by the decay of the crack flank.

The reflection, however, is easily identifiable, and agrees very well with the theoretically calculated arrival times with an assumed R-wave velocity of 2950 m/s.

The reflected R-wave will contribute to a crack closure effect of about 1 µm. The effect on the J-integral, assuming a stationary crack, has also been calculated, and a marked decrease in the J-integral of 5 - 10 % is recorded, as shown in Figure 6.

Shallow crack. 3D model

The arrival of the direct and reflected surface waves could be identified along the crack surface and around the corner, Fig. 7. Again, the direct surface wave on the crack surface will be hidden by the overall crack opening displacement moving in the opposite direction, but the displacement is clearly visible in the y-direction, Fig. 7. This is in accordance with theoretical calculations, given by Pao et al. (9, 10), where the x- and y- displacements have the same sign.

It can be seen that the reflected surface wave lifts the surface, while the transmitted one lowers the surface. Theoretical calculations have also shown that there will be no shift in phase on the reflected surface wave, Freund (11).

The displacement at the crack tip on arrival of the reflected wave is about 2 µm, with a wave length of about 3 mm, Fig. 8. Only one crack surface is visualised in the figure, hence with both surfaces close to each other, the reflected waves will result in crack closure.

The arrival of the pressure and shear waves, can be identified, but no further effort to determine the radiation pattern is performed within this presentation.

At some selected nodes along the crack surface, the wave pattern under the surface, at the passage of the surface wave, was also measured. The shear stress τ_{xy}, and normal stress σ_y, referred to the normal stress σ_y on the surface ($x = 0$), as presented in Fig. 9. The depth ratio is calculated with a surface wave length of 3.2 mm. The stress components are rapidly decaying with increasing distance from the surface, but the shear stress has its maximum at a distance of 0.21 λ_R under the surface. This is very close to the measured maximum at 0.27 λ_R, Viktorov (6).

Deep crack. 3D model

The shallow crack had a crack depth of the same order as the wave length of the surface wave, hence the conditions for examining the development and radiation pattern of the wave are poor. In order to improve the conditions, a deep crack model was made, Fig. 3. But in order not to increase the computation time too much, the mesh size had to be almost doubled. Since the source depends on the mesh size, the wave length has now increased to about 5.3 mm. Again the surface waves can easily be identified, Fig. 10, but the signal response is more disturbed because of the coarse mesh size.

The source mechanism was then changed to simulate a more localised event. The line source was substituted with an elliptical shaped source in the middle of the specimen. Earlier measurements on the surface wave radiation pattern, with so-called wave-blocks, had shown that the waves radiate in preferred directions, Thaulow and Burget (1). Attention was now focussed on the radiation pattern from the localised source.

The results confirmed that a large fraction of the surface wave energy radiates in a direction opposite the crack extension, and is reflected back as a concentrated beam. Hence, the energy will not dissipate in all directions, and eventually be reflected back over an extended period of time. Both the source mechanism applied in the present model, and previous measurements, show that the surface waves radiates in a preferred direction.

DISCUSSION

The present models are much simplified and only the tendencies and relative effects have been examined. The existence of the reflected surface wave is well documented in the models, but the direct influence at the crack tip stress field has not been examined in detail yet. The only indication is the reduction of the J-integral in the 2D model.

The work so far has concentrated upon documenting the nature of the waves and their development over time. The next step is to examine the important boundary conditions such as the source mechanisms, a propagating crack, edges where reflections take place and the effect of surface irregularities.

Real crack sources may have much shorter wave lengths. Hence surface irregularities and secondary cracks will be of increasing importance. New models have to be developed in order to increase the frequency content of the source. The aim is to come close to the crack tip, and to investigate the effect of surface waves in the near field area.

The release and subsequent interactions of surface waves are not limited to pop-ins. Acoustic emission monitoring of corrosion fatigue cracks in sea water has revealed that the surface waves released from the crack can be very strong, Thaulow and Berge (12), and Eriksen (13). Detailed measurements using broad band transducers showed that the emissions were mainly caused by interactions between the crack surfaces, where surface waves were emitted by the breaking of local bonds between the crack surfaces.

CONCLUSIONS

The existence and nature of surface waves initiated from a sudden release of stresses at the crack tip have been examined with FE models. Both the emission and reflection of surface waves were identified with the following characteristics:

- The surface wave results in a lift of the surface, and there is no shift in phase after reflection at the crack mouth edge.

- The resulting deformation is in the range of 1 - 2 µm on each crack surface, causing a crack closure effect upon hitting the crack tip.

- A reduction of the J-integral at the arrival of the reflected surface wave at the crack tip, in the range of 5 - 10 %, was calculated for the 2D model.

- The shear stress component of the surface wave reaches a maximum at a depth of about 0.21 λ_R under the surface, while the normal stress decays gradually.

- The radiation pattern from an elliptical shaped source shows that the surface waves radiate in a preferred direction, opposite to the direction of crack extension. The wave is reflected back as a concentrated beam.

The surface waves may be a dominating mechanism in the neighbourhood of a crack tip. In this area the frequency content will be high, hence surface irregularities and secondary cracks will be of increasing importance.

REFERENCES

(1)   Thaulow, C., Burget, W.: "The emission of Rayleigh waves from brittle fracture initiation, and the possible effect of the reflected waves on crack arrest". Fatigue and fracture of engineering materials and structures. Vol. 13, no. 4, pp. 327-346, 1990.

(2)   Willoughby, A.A. and Wood, A.M.: "Crack initiation and arrest in 9 % Ni steel weldments at liquified natural gas temperature". Int. Symp. on Storage and Transport of LPG and LNG, Brügge, 1984.

(3)   Shmuely, M. et al.: "Effect of Rayleigh waves in dynamic fracture mechanics". Int. Journal of Fracture, 14, R69-72, 1978.

(4)   Arimochi, K. and Isaka, K.: "A study on pop-in phenomenon in CTOD test for weldment and proposal of assessment method for significance of pop-in". IIW Doc. X-1118-86, 1986.

(5) Hedner, G.: "Study of dynamic crack growth by shear strain measurements".
Symposium "Fracture Mechanics in the Nordic Countries", June 1989, Trondheim, Norway.

(6) Viktorov, I.G.: "Rayleigh and Lamb waves. Physical Theory and Applications".
Plenum Press, New York, 1967.

(7) Wadley, H.N.G., Scruby, C.B., Shrimpton, G.: "Quantitative acoustic emission source characterization during low temperature cleveage and intergranular fracture".
Acta Metallurgica, Vol. 29, 1981, pp. 399-414.

(8) Scruby, C.B., Jones, C., Titchmarsh, J.M., Wadley, H.N.G.: "Relationship between microstructure and acoustic emission in Mn-Mo-Ni A533B steel".
Metal Science, June 1981, pp. 241-261.

(9) Pao, Y-H., Gajewski, R.R., Ceranoglu, A.N.: "Acoustic emission and transient waves in an elastic plate". J. Acoust. Soc. Am., 65, 1979, pp. 96-105.

(10) Ceranoglu, A.N., Pao, Y-H.: "Propagation of elastic pulses and acoustic emission in a plate". J. Appl. Mech., 48, 1981, pp. 125-147.

(11) Freund, L.B.: "Influence of the reflected Rayleigh wave on a propagating edge crack".
Int. J. Fract., 17, 1981, pp. R83-86.

(12) Thaulow, C., Berge, T.: "Acoustic emission monitoring of corrosion fatigue crack growth in offshore steel". NDT Int., 17, 1984, pp. 147-153.

(13) Eriksen, M., Thaulow, C.: "Dominant acoustic emission sources during corrosion fatigue of offshore steels."
4th European Conf. on Non-Destructive Testing. Specialist Symp. on Acoustic Emission, London (edited by C. B. Scruby and R. Hill), Vol. 4, 1987, p.3007.

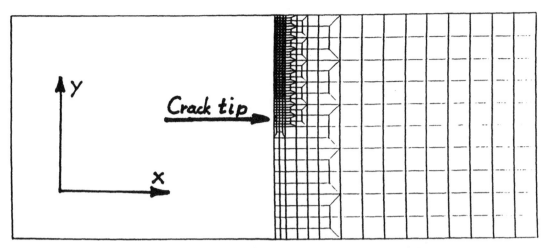

Fig. 1   2-dimensional finite element model.

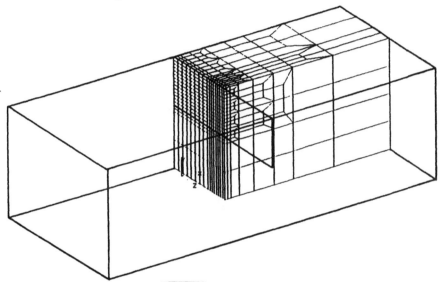

Fig. 2 3D FE model showing the reduction of model by division along symmetry planes.

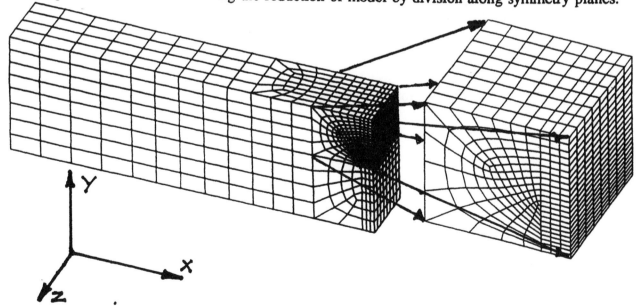

Fig. 3   3-dimensional finite element model.

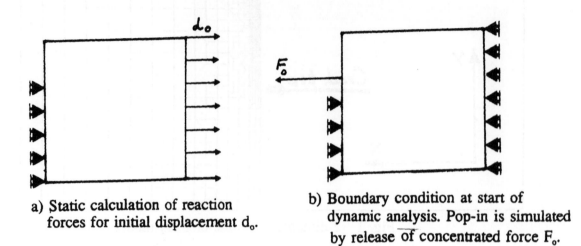

a) Static calculation of reaction
forces for initial displacement d_o.

b) Boundary condition at start of
dynamic analysis. Pop-in is simulated
by release of concentrated force F_o.

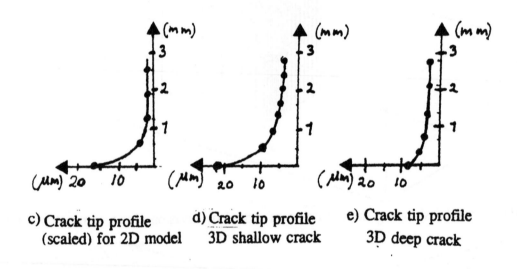

c) Crack tip profile
(scaled) for 2D model

d) Crack tip profile
3D shallow crack

e) Crack tip profile
3D deep crack

Fig. 4 Source mechanism.

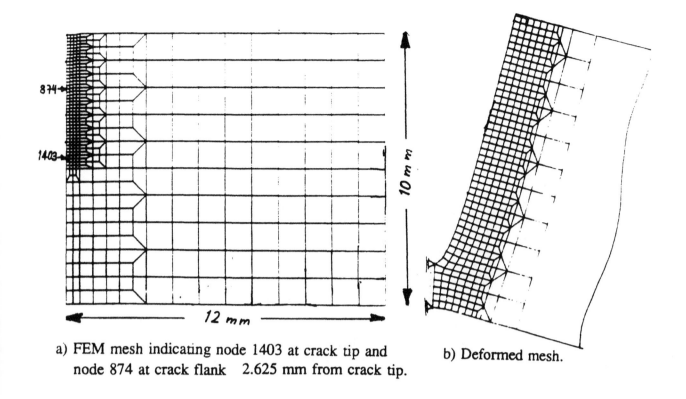

a) FEM mesh indicating node 1403 at crack tip and
node 874 at crack flank   2.625 mm from crack tip.

b) Deformed mesh.

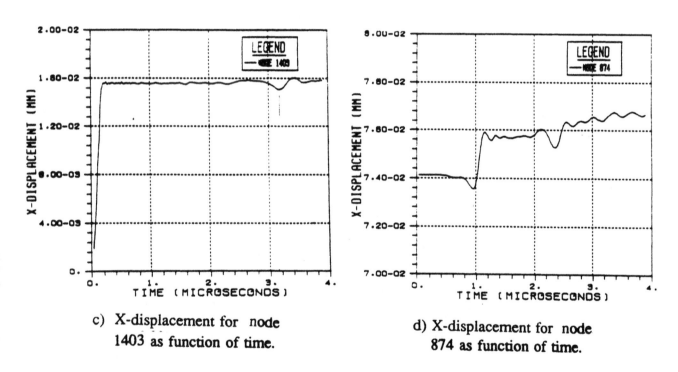

c) X-displacement for node
1403 as function of time.

d) X-displacement for node
874 as function of time.

Fig. 5   2 dimensional model. Displacements in x-direction as a function of time.

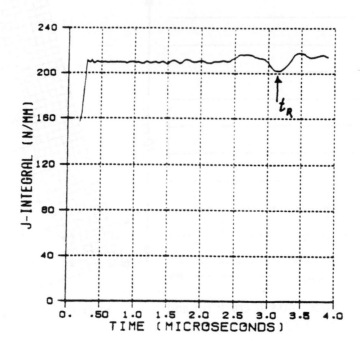

Fig. 6   J-integral at the stationary crack tip.
The arrival of the reflected surface wave is indicated with an arrow.

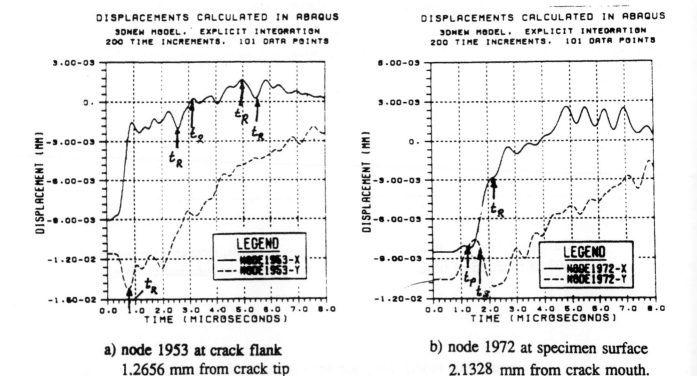

a) node 1953 at crack flank
1.2656 mm from crack tip

b) node 1972 at specimen surface
2.1328 mm from crack mouth.

Fig. 7   3 dimensional model with shallow crack. Displacements in x and y-direction
as function of time for two node positions. The arrival of the direct and reflected
surface waves are indicated with an arrow, t_R. The arrival of the direct pressure
and shear waves are also indicated with t_P and t_S respectively. $v_P = 6000$ m/s,
$v_S = 3200$ m/s, $v_R = 2950$ m/s.

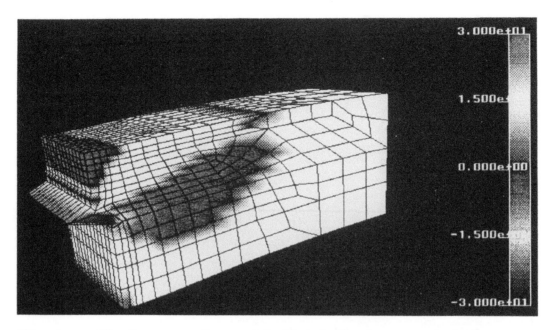

Fig. 8 a   Shallow crack model. Total displacements before crack initiation

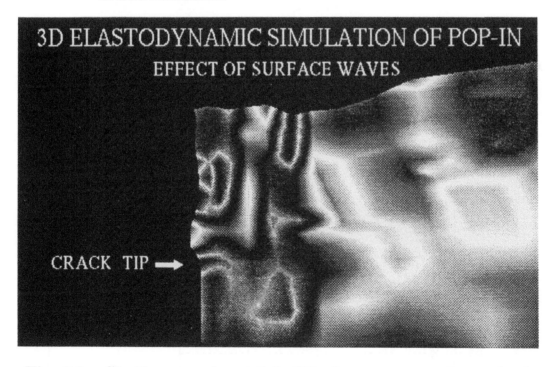

Fig. 8 b   Shallow crack model. Displacements at the arrival of the reflected surface wave at the crack tip. Displacements are magnified.

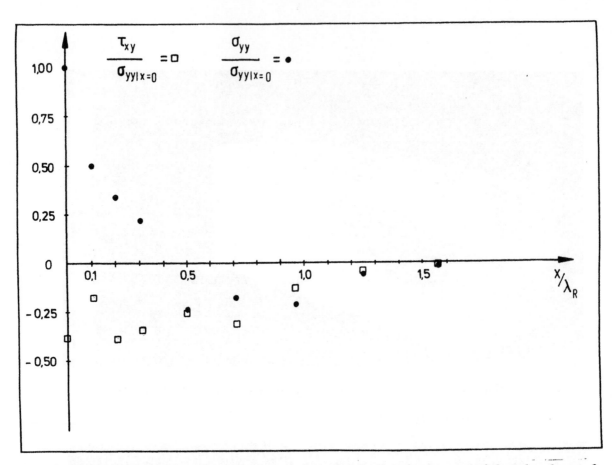

Fig. 9   Shear stress (τ_{xy}) and normal stress (σ_y) distribution in the material under the surface wave.

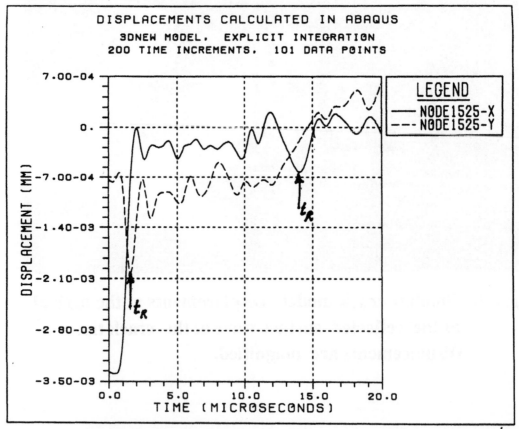

Fig. 10   3 dimensional model with deep crack. Displacements in x and y-direction as function of time for node 1525 located at crack flank 1.55 mm from the crack tip.

Shallow surface cracks in welded T butt plates — a 3D linear elastic finite element study

B Fu, BSc, MSc (University of Newcastle-upon-Tyne), J V Haswell, BSc, MSc, PhD, MIMechE (British Gas Plc) and
P Bettess, BSc, MSc, PhD, ACGI, BIC, MICE, MBCS, MRINA (University of Newcastle-upon-Tyne)

SUMMARY

Stress intensity factor solutions for surface cracks in welded structures are necessary for engineering fracture and fatigue assessments. Solutions are usually obtained by superposition of weld magnification factors derived from the analysis of two-dimensional (2D) edge cracks and planar three-dimensional (3D) surface crack solutions. This paper describes an investigation of the differences between the weld magnification factors for 2D edge cracks and realistic 3D crack shapes.

INTRODUCTION

Shallow surface cracks occur in the welded joints of many engineering structures, and the practical analysis of such cracks is of major importance in fatigue and fracture assessments. The weld geometry introduces a localised notch region, which causes high stress gradients and local plasticity at the weld toe. The linear elastic behaviour of short cracks is dominated by the local stress concentration, and elastic plastic behaviour by local plasticity.

The weld notch effect has been widely investigated in many experimental, analytical and numerical studies, such as those by Maddox (1), Smith and Hurworth (2), Thurbeck and Burdekin (3), Niu and Glinka (4), Straalen et al (5), Dijkstra et al (6), Bell (7) and Pang (8). Most quantitative studies have been based on edge cracks in fillet welded joints, and have been carried out to evaluate the increase in linear elastic stress intensity factor (SIF) due to the weld notch. The results are usually presented in terms of weld magnification factors. In general, published results show that the weld magnification factor reduces to unity as the crack depth increases to approximately 20% of the plate thickness.

Although surface cracks are practical engineering defects, their 3D geometry is difficult to model. This has restricted comprehensive analysis to simple surface cracks in plane plates and cylinders. Consequently, the behaviour of such cracks is well understood, and the SIFs obtained are widely accepted as realistic predictions for engineering analyses.

Methodologies for the assessment of surface defects in welded joints, such as those proposed in the British Standard Published Document PD 6493 (9), are usually based on the superposition of simple plate and cylinder crack SIF solutions and weld magnification factors. These methodologies assume that the weld notch effect, determined from the 2D analysis of an edge crack in a welded joint, can be applied as a simple multiple to the complete 3D crack front SIF distribution.

This paper describes an investigation of the differences between the weld magnification factors for 2D edge cracks and realistic 3D crack shapes. A detailed linear elastic fracture mechanics (LEFM) analysis of relatively shallow surface cracks in a welded T-butt plate has been carried out using 3D finite element (FE) analysis. The effect of the local weld notch stress concentration on the crack front SIF distribution has been investigated. The results have been used to extend and improve existing crack assessment methodologies.

WELD MAGNIFICATION FACTOR

The effect of the weld notch on the SIF value of cracks in welded joints is generally represented in terms of a magnification factor, M_k, defined as:-

$$M_k = \frac{K_I^T}{K_I^p} \qquad\qquad [1]$$

in which the superscripts "T" and "p" refer to the welded T-butt joint and plane plate respectively, and K_I is the mode-I SIF, ie. the SIF of opening mode crack.

The use of such M_k factors in conjunction with the Raju and Newman's SIF equations for semi-elliptical surface cracks in plates (10,11) is recommended in PD 6493 (9) for the assessment of flaws in welded joints.

M_k values are generally obtained from the detailed FE analysis of cracks in welded joints. Most published data refers to the 2D analysis of edge cracks at welded joints, with some limited 3D data. This data is relevant to the deepest point on the crack front only. SIF values of the crack front at the free surface are usually calculated according to the stress concentration factor (SCF), K_c, at the weld notch, which is dependent on local weld notch parameters, such as weld toe radius, weld angle etc. M_k factors at the deepest point a and free surface c of the crack front, as shown in Fig.1, are therefore calculated as:-

$$M_{k(a)} = C(\frac{a}{T})^\beta \qquad (1.0 \leq M_k \leq K_c) \qquad\qquad [2a]$$

$$M_{k(c)} = \alpha K_c \qquad (\alpha \sim \frac{a}{T} \ and \ \alpha \leq 1.0). \qquad\qquad [2b]$$

The current work uses detailed 3D FE analysis to investigate the variation of M_k around the complete crack front.

FINITE ELEMENT MODELS

Detailed 3D linear elastic FE analyses of semi-elliptical cracks in plane and welded T-butt plates subjected to tension and bending have been carried out. The geometric details of the welded T-butt plate with a weld toe crack are given in Fig.1. Cracks with aspect ratios a/c=0.2, 0.4 and 1.0, with depth/thickness ratios a/T=0.05, 0.10, 0.15, 0.20 and 0.40 have been modelled at a sharp weld toe. In addition, edge cracked welded T-butt plates were also analysed to provide results for the limiting crack aspect ratio a/c=0.

FE models constructed using 20-node isoparametric elements were generated using the PA-TRAN graphics software (12). Semi-elliptical cracks were modelled by mapping semi-circular crack front meshes onto a semi-ellipse. The crack tip was modelled using a focused mesh of collapsed elements, with 1/4 point node shifting to generate the $1/\sqrt{r}$ singularity, as discussed by Barsoum (13).

Stress analysis was carried out using the ABAQUS FE package (14). J-integral values were calculated using an equivalent domain integral procedure employing Parks' virtual crack extension (VCE) method (15), and converted to SIF values assuming the 2D J-SIF relationship:-

$$K_I^2 = \begin{cases} \frac{EJ}{(1-\nu^2)} & (\text{ for plane strain }) \\ EJ & (\text{ for plane stress }) \end{cases} \tag{3}$$

in which "E" is Young's modulus, "J" is the J-integral data and "ν" is Poisson's ratio.

FINITE ELEMENT RESULTS AND ANALYSIS

Stress Intensity Factors

SIF results obtained from the FE analyses are presented in non-dimensional form as:-

$$\tilde{K}_{It}(\frac{a}{c}, \frac{a}{T}, \phi) = \frac{K_{It}(\frac{a}{c}, \frac{a}{T}, \phi)}{\sigma_t \frac{\sqrt{a\pi}}{\Phi}} \tag{4a}$$

$$\tilde{K}_{Ib}(\frac{a}{c}, \frac{a}{T}, \phi) = \frac{K_{Ib}(\frac{a}{c}, \frac{a}{T}, \phi)}{\sigma_t \frac{\sqrt{a\pi}}{\Phi}} \tag{4b}$$

where subscripts "t" and "b" refer to tension and bending conditions respectively, σ_t and σ_b are tension and bending stresses respectively, and Φ is the complete elliptic integral of the second kind:-

$$\Phi = \int_0^{\frac{\pi}{2}} \{\cos^2\theta + (\frac{a}{c})^2 \sin^2\theta\}^{\frac{1}{2}} d\theta. \tag{5}$$

Results obtained from the analysis of plane plates agree closely with Raju and Newman's results (10,11).

Crack front SIF distributions for a surface crack of aspect ratios a/c of 1.0 and 0.2 subjected to tension and bending are plotted in Fig.2. The effects of the local stress concentration and the additional stiffness of the welded attachment demonstrated are contradictory and can be summarised as follows:-

1. The SIF distribution is increased along all sections of the crack front located within the region of the weld stress concentration. This includes the whole crack front of shallow cracks, and shallow sections of the crack front of deeper surface cracks. The increase in SIF decreases to zero as the depth of the crack front increases.

2. The welded attachment of the welded T-butt plate provides additional structural stiffness, which reduces the SIF value of finite length cracks. M_k values for the deepest point on the crack front calculated from 3D analysis are therefore lower than values obtained from the 2D analyses of cracks of equivalent depth. The effect decreases as the crack length increases, and is zero for cracks of infinite length.

SIF values for the deepest point of the crack front for all crack geometries analysed are plotted in Fig.3. This shows the variation of SIF with crack aspect ratio and depth, and the effect of the welded attachment.

Weld Magnification Factors

M_k values for the deepest point of the crack front were derived from the results of 2D and 3D analyses, as shown in Fig.3.

The effect of the local stress concentration induced by the weld was considered by comparing the M_k values obtained at different positions on the same crack front. The crack front M_k results obtained for the crack shape $a/c = 0.2$ are presented in terms of relative depth of the crack front, y/T, in Fig.4. The results for different crack depths can be represented on a common curve of the form:-

$$M_k(\frac{a}{c} = 0.2) = f(\frac{y}{T}).$$ [6]

This implies that the effect of the local stress concentration is approximately common to all cracks of the same aspect ratio, and the variation in M_k around the crack front is approximately dependent on the relative crack front depth.

Curves fitted to the crack front M_k vs. relative crack front depth y/T are shown in Fig.6 for tension and bending loads. The relationships derived using statistical regression analysis are:-

$$M_{kt}(\frac{a}{c} = 0.2, \frac{a}{T}, \phi) = 0.8946 + 1.1749\{1.0 + 36.2471(\frac{a}{T})\sin\phi\}^{-1.4626}$$ [7a]

and

$$M_{kb}(\frac{a}{c} = 0.2, \frac{a}{T}, \phi) = 0.7377 + 1.7837\{1.0 + 136.6248(\frac{a}{T})\sin\phi\}^{-0.6283}.$$ [7b]

This form of relationship between M_k and relative crack front depth y/T can be derived for each crack aspect ratio.

Results plotted in Figs.4 and 5 indicate that the variation in M_k due to relative crack front depth y/T and crack aspect ratio a/c can be treated separately. This provides the basis of an engineering model for the prediction of 3D M_k factors. It is proposed that the M_k equations derived from 2D analyses are generalised in terms of relative rather than maximum crack depth a/T, and that the limiting value of the M_k with depth is a variable dependent on a/c, rather than unity as shown for 2D cases. This variable can be determined from interpolation between results for the limiting crack aspect ratios $a/c=0.0$ and 1.0. The proposed form of the expression for M_k is therefore:-

$$M_k(\frac{a}{c}, \frac{a}{T}, \phi) = f_1(\frac{a}{c}) + f_2(\frac{a}{T}, \phi)$$ [8a]

$$f_1(1.0) \leq f_1(\frac{a}{c}) \leq f_1(0.0) = 1.0 \qquad (0.0 \leq \frac{a}{c} \leq 1.0)$$ [8b]

$$f_2(\frac{a}{T}, \phi) = C'\{(\frac{a}{T})\sin\phi\}^{-\beta'} \qquad (0.0 < \frac{a}{T} \leq 0.4; \quad 0 \leq \phi \leq \frac{\pi}{2}).$$ [8c]

The constants $"C'"$ and $"\beta'"$ can be determined from 2D studies taking into account specific weld geometry parameters. The function $f_1(a/c)$ can be derived from limited 3D results for specific welded joints by assuming a constant reduction in M_k for all points along the crack front.

The relationship between M_k and relative crack front depth means that M_k values for very shallow crack can be derived from analyses of deeper cracks of the same aspect ratio. This removes the need to model very shallow cracks, which present practical modelling problem associated with mesh refinement and crack front element distortion.

SIF predictions obtained using equation [7] for a surface crack of constant aspect ratio $a/c=0.2$ and various depths and subjected to tension and bending loads are compared with Raju and Newman's SIF values for plane plates in Fig.6. The increase in SIF due to the high M_k factors in shallow regions of the crack front is demonstrated. This indicates that surface cracks in such a notch region will grow along the surface in preference to through-thickness, which is consistent with experimental crack growth data for welded structures.

CONCLUSIONS

The variation of the weld magnification factor M_k along the crack front has been investigated through the 3D FE analysis of surface cracks in welded plates. Conclusions drawn from the results are summarised below.

1. The results indicate that two contradictory effects influence the crack front M_k distribution, i) the local weld notch stress concentration, and ii) the stiffness of the welded attachment.

2. The SIF value is increased along all sections of the crack front located in the region affected by the weld stress concentration, and is maximum at the surface. The effect decreases to zero as the depth of the crack front increases. Expressions relating the M_k factors to the relative depth of a semi-elliptical crack front have been derived from the results of 3D FE analyses using regression analysis.

3. The welded attachment provides additional structural stiffness which reduces the SIF of finite length cracks. The effect decreases to zero as the crack length increases. The reduction in M_k value due to the stiffness of the welded attachment is a variable dependent on crack aspect ratio, and can be obtained by interpolation from M_k values for crack aspect ratios $a/c=0.0$ and 1.0.

4. An engineering model which extends existing 2D M_k values to 3D M_k values for semi-elliptical cracks of the form:-

$$M_k\{\frac{a}{c}, \frac{a}{T}, \phi\} = f_1(\frac{a}{c}) + f_2(\frac{a}{T}, \phi)$$

is proposed based on the results of 3D FE analyses.

ACKNOWLEDGEMENT

This study was sponsored by British Gas Plc, Engineering Research Station. The authors wish to thank British Gas for permission to publish this paper. Technical discussions with Dr C Ruiz, University of Oxford, and Dr P Hopkins, British Gas Plc, Engineering Research Station, are also acknowledged.

REFERENCES

1. Maddox S J, An Analysis of Fatigue Cracks in Fillet Welded Joints, Int J of Fracture, Vol 11, pp 221-243, 1975.

2. Smith I J and Hurworth S J, The Effect of Geometry Changes upon the Predicted Fatigue Strength of Welded Joints, TWI Report 244/1984, 1984.

3. Thurlbeck S D and Burdekin F M, Effects of Geometry and Loading Variables on the Fatigue Design Curve for Tubular Joints, IIW Doc XIII-1428-91, 1991.

4. Niu X and Glinka G, Stress Intensity Factors for Semi-Elliptical Surface Cracks in Welded Joints, Int J of Fracture, Vol 40, pp 255-270, 1989.

5. van Straalen I J, Dijkstra O D and Snijder H H, Stress Intensity Factors and Fatigue Crack Growth of Semi-Elliptical Surface Cracks at Weld Toes, Proc of Int Conf on Weld Failures, pp 367-376, TWI, London, 1988.

6. Dijkstra O D, Snijder H H and van Straalen I J, Fatigue Crack Growth Calculations Using Stress Intensity Factors for Weld Toe Geometries, Proc of 8th Int Conf on Offshore Mech and Arctic Engng, Vol 3, pp 137-143, 1989.

7. Bell R, Stress Intensity Factors for Weld Toe Cracks in Welded T Plate Joints, DSS Contract No OST84-00125, Faculty of Engineering, Carleton University, Ottawa, 1987.

8. Pang H L J, A Review of Stress Intensity Factors for Semi-Elliptical Surface Cracks in a Plate and Fillet Welded Joint, TWI Report 426/1990, 1990.

9. British Standard Published Document PD 6493: Method for the Derivation of Acceptance Levels for Flaws in Fusion Welded Joints, 1991

10. Newman J C and Raju I S, Stress Intensity Factors for a Wide Range of Semi-Elliptical Surface Cracks in Finite Thickness Plates, Engng Fracture Mech, Vol 11, pp 817-829, 1979.

11. Raju I S and Newman J C, An Empirical Stress Intensity Factor Equation for the Surface Crack, Engng Fracture Mech, Vol 15, pp 185-192, 1981.

12. PDA Engineering, PATRAN Plus User Manual, 1988.

13. Barsoum R S, On the Use of Isoparametric Finite Elements in Linear Elastic Fracture Mechanics, Int J of Num Meth in Engng, Vol 10, pp 25-27, 1976.

14. Hibbitt, Karlsson and Sorensen Inc, ABAQUS Users Manual Version 4.7, 1988.

15. Parks D M, The Virtual Crack Extension Method for Non-Linear Material Behaviour, Comp Meth for Appl Mech and Engng, Vol 12, pp 353-364, 1977.

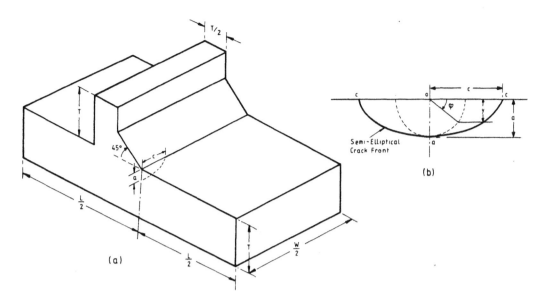

Figure 1   Geometries of (a) a Cracked T-Butt Plate and (b) a Semi-Elliptical Surface Crack

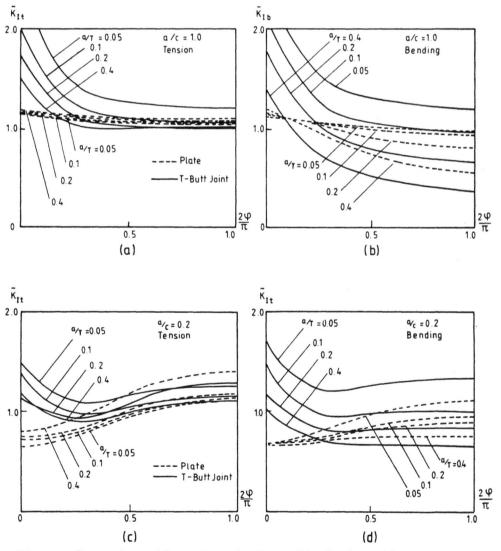

Figure 2   Comparison of Stress Intensity Factor Distributions of Surface Cracks in
a Plate and a Welded T-Butt Joint Subjected to Tension and Bending

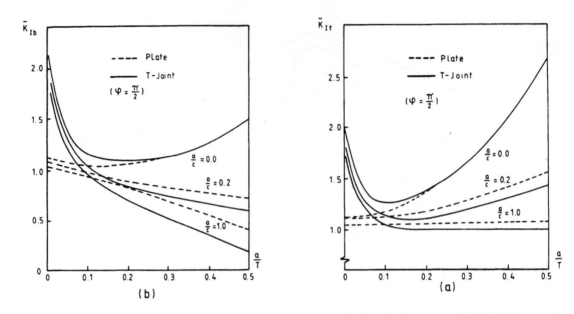

Figure 3 Comparison of Stress Intensity Factors of Cracks in Plates and Welded T-Butt Joints Subjected to Tension and Bending ($\phi = \frac{\pi}{2}$, SIFs at the Deepest Point of Crack Front)

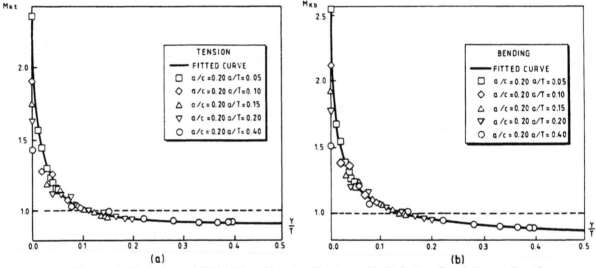

Figure 4 Variation of Weld Magnification Factor with Relative Crack Front Depth

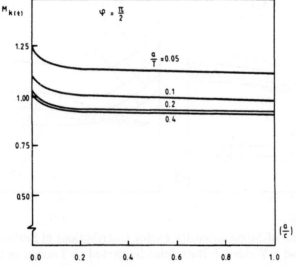

Figure 5 Variation of M_k with Crack Aspect Ratio

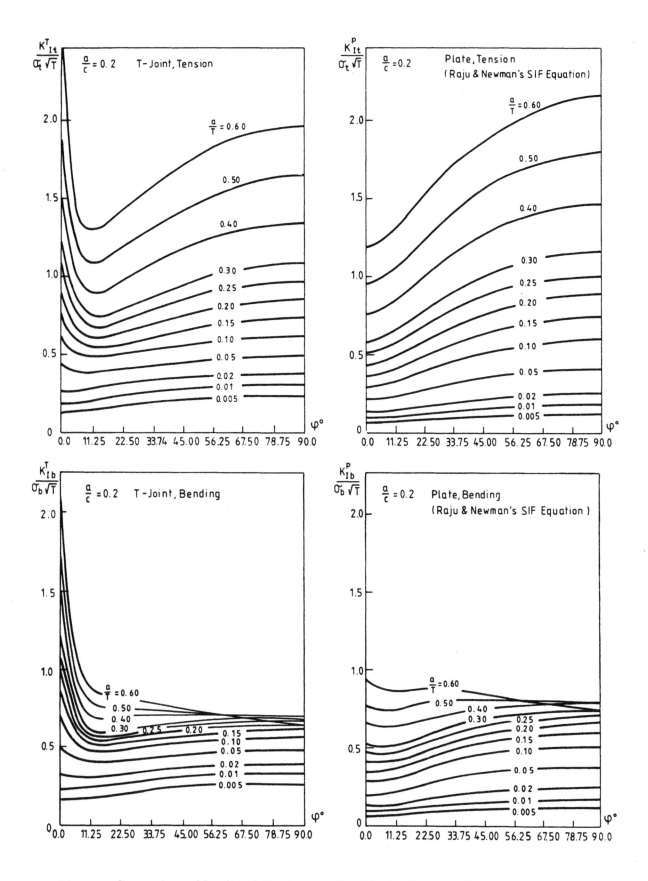

Figure 6 Comparison of Predicted Crack Front SIF Distributions for Plane and Welded T-Butt Plates with a Semi-Elliptical Surface Crack ($\frac{a}{c} = 0.2$)

SESSION D: Fracture assessments

Prediction of overmatching effects on fracture of cracked stainless steel welds (Paper 4)
C ERIPRET and P HORNET

An experimental investigation of the transferability of surface crack growth characteristics (Paper 5)
J FALESKOG, F NILSSON and H ÖBERG

The influence of microstructural variation on crack growth (Paper 9)
V GENSHEIMER DEGIORGI, P MATIC, G C KIRBY III and D P HARVEY II

The deformation and failure at −30 °C of cracked broad flange beams of ordinary structural steel in bending (Paper 10)
K ERIKSSON, F NILSSON and H ÖBERG

Experimental and analytical investigations into the mechanical behaviour of pipe with short through-wall cracks (Paper 32)
Ph GILLES, D MOULIN and D GOUTERON

Application of K_J rule to different specimens and to a cracked elbow (Paper 18)
C COUTEROT, Y WADLER and P TAUPIN

TWI

PAPER 4

Prediction of overmatching effects on fracture of cracked stainless steel welds

C Eripret (Electricité de France) and P Hornet (formerly at Ecole Centrale de Nantes, France)

SUMMARY

This paper discusses the problem of the accuracy of fracture prediction models when considering crack initiation in a weld in an engineering structure, and how that relates to crack extension in a small scale specimen. Weld metal mismatching consequences may greatly affect the crack growth resistance of such a structure, which makes integrity assessment becoming more complicated. Generally, the fracture mechanics assessment procedure assumes that the crack is located in an homogeneous material of uniform tensile and toughness properties, adopting either the weld metal properties, or the lowest tensile properties and the lowest toughness value of the base metal or the weld metal. However, this solution may lead to overconservatism in some cases and overestimates the critical size of allowable flaws. This paper presents the preliminary results of an original study which aims at determining a fracture criterion that could be applied to welds in conditions where classical fracture mechanics concepts fail. Elastic plastic finite element analyses have been performed for various specimens containing strongly overmatched cracked welds. It has been shown that overmatching can provide a substantial benefit in terms of toughness when considering short cracks in the weld metal. It is recommended also to examine cracks located in the fusion line where difference of strain carrying capacities of base and weld metal generates local concentration of shear stresses, which may affect drastically the weld toughness properties.

INTRODUCTION

In integrity assessment analyses of components containing cracks, the classical procedure is to obtain the crack driving force expressed in terms of J-integral. Then, one has to compare it with the material's crack growth resistance curve described through a J-resistance curve determined on the basis of small specimen testing. This J-R curve is commonly accepted as an intrinsic property of the material tested. However, the transfer of toughness properties from laboratory specimens to engineering structure may be problematic in some cases, especially when a crack is in the vicinity of a weld. In that case, the material surrounding the crack is inhomogeneous regarding to its mechanical and fracture properties; base metal, heat affected zone and weld metal exhibit different yield stress, ultimate stress, hardening coefficient, and resistance to ductile tearing. Each material plays a different role in the extension of the plasticity from the crack tip to the structure, and mismatching effects (i.e. undermatching or overmatching) will influence whether the plastic zone remains contained in the crack tip area or spreads out of the weld into the parent material, which leads to large scale yielding (1)(2)(3).

Integrity analyses for assessing the risk significance of cracks in overmatched welds postulate that the crack is located in an homogeneous material. For large cracks, one assumes that this homogeneous material has the same mechanical properties as the weld metal. Stress singularity at the crack tip is very important owing to the large size of the crack, thus yielding occurs mainly in the crack tip area. Plasticity is contained and it can be reasonably assumed that the parent material has no influence on the welded structure resistance. For medium cracks, another plastic zone may appear at the fusion line in the base metal above the crack tip. As long as loading is increasing, these two plastic zones grow until they join together, which may generate local unloading of the crack tip zone. In that case, it is not easy to determine which mechanical and toughness properties must be used for safety analyses and one often chooses the lowest yield stress and the lowest toughness value of parent and weld metal. Lastly, when considering short cracks in very overmatched welds, one can observe that plasticity only occurs in the base metal because of its lower yield strength and the crack in the weld may be not dangerous at all. These different configurations and the parameters that control the plasticity extension deserve to be investigated to better understand crack initiation in a welded component and to improve the accuracy of integrity assessment methods.

OBJECTIVES OF THIS STUDY

This paper presents the preliminary results of a numerical study of the fracture behaviour of overmatched stainless steel welds. The objective of this study is to highlight the problems arising from the application of classical integrity assessment procedures to cracked welds, as well as to determine which fracture criterion is relevant to describe the extension of cracks in inhomogeneous components.

To do so, experiments have been carried out at 300°C on Compact Tension (CT) specimens made of base metal and weld metal in order to determine the toughness of those materials, expressed in terms of J0.2. Through finite element analyses, we have simulated those tests and we have also applied various fracture mechanics criteria such as CTOD, δ5, and finally the so-called local approach of ductile tearing (i.e. the Beremin's model (4) based on the pioneering works of Rice and Tracey). The critical value of each fracture criterion corresponding to crack initiation (conventionnally J0.2) has been determined. Each of these criteria provides a procedure to predict the fracture behaviour of different specimen configurations on the basis of CT specimens tests. Therefore, these different failure criteria have been applied to predict the failure of other specimens such as centre cracked panels (CCP) and single edge notched bend specimens (SENB) as well as one CT specimen containing a crack right on the fusion line.

The conclusions obtained in terms of resistance to ductile tearing have been shown to vary quite widely, depending on which fracture criterion is used for integrity assessments. These results, that will be soon validated by experiments, will provide a better understanding of weld metal fracture behaviour as well as a basis for improving the classical fracture assessment procedure.

MATERIAL DATA

The stainless steel weld seam presented in this study is a 316 L type weld manufactured by connecting two stumps of primary cast stainless steel pipe (CF8 M) using a TIG welding process. No heat treatment has been applied to the weld. The pipes were 700 mm diameter and 70 mm thick. The chemical composition is presented in Table 1.

	C	S	P	Si	Mn	Ni	Cr	Mo	N	O
Base Metal	0,015	0,003	0,022	1,15	1,01	9,74	20,01	0,39	0,081	0,0063
Weld metal	0,008	0,011	0,018	0,51	1,49	12,70	19,20	2,60	0,051	0,0072

Table 1 : Chemical analyses of stainless steel base metal and weld (% weight).

The mechanical properties at 300°C of the base and weld metal are given in Table 2. The tensile curves plotted in Figure 1 show that the tensile behaviour of the tested materials are very different, the weld metal yield strength is three times higher than that of the parent material.

	0.2 % Y.S. (MPa)	Tensile strength (MPa)	Elongation (%)	Area reduction (%)	Young's modulus (MPa)
Base metal	133	366	40	73	176500
Weld metal	412	474	16,4	54	176500

Table 2 : Tensile properties at 300°C of stainless steel base metal and weld.

20% side grooved ASTM E1152-87 standard Compact Tension specimens 25 mm thick have been tested to determine the toughness properties of base metal and weld metal. The weld CT specimen has been positionned along the centreline of the through-thickness of the weld seam. Each of the CT specimens has been fatigue pre-cracked in order to get an a/w ratio of approximately 0.6. Toughness properties have been estimated following ASTM E813-88 procedure, and are presented in Table 3.

	Base metal	Weld metal
J0.2 (kJ/m2)	434	208

Table 3 : Toughness properties at 300°C of stainless steel base metal and weld.

NUMERICAL STUDY

Elastic-plastic finite element analyses have been performed assuming plane strain conditions. For all the cases, eight-node isoparametric elements were used. The computations were performed using an incremental Von Mises rule with isotropic hardening. As the crack tip cells were subjected to very large elongation and shear stresses, it was necessary to use an updated Lagrangian numerical scheme at each load step. The stress strain curves introduced in the finite element code as a set of numerical data, are the true curves drawn from tensile tests previously presented in Figure 1. However, for higher strains, those curves have been extrapolated using a Ramberg-Osgood power-law hardening fit which parameters were found to be equal to :

n = 1.94 and α = 28.3   for the base metal,

n = 14.66 and α = 1.286   for the weld metal.

Homogeneous Compact Tension specimens

The tests performed on CT specimens mentioned previously have been analysed in order to check the ability of the finite element code to simulate this standard ASTM E 1152-87 toughness test. A comprehensive benchmark (5) has been carried out using three different finite element codes developed in the French nuclear industry (CASTEM, SYSTUS, and ALIBABA). The comparison between experiments and finite element analyses showed that a satisfactory agreement between the three FEA and experimental results could be obtained for load-displacement curve as well as for the crack-driving force evolution. Moreover, local values of stress and strain fields in the crack tip area were derived from the above FEA with 5% discrepancy in a loading regime where the plastic strains are greater than 40%. On the basis of this benchmark, we concluded that a plane strain FEA proved to be able to model a standard ASTM CT specimen tests. Furthermore, the use of local fracture criteria (based on the cavity growth rate) can be standardised and reliably applied to investigate toughness tests on stainless steels. Then, through the combination of experimental toughness values and finite element analyses, we derived the critical values of different fracture criteria, as presented in Table 4.

	ASTM E813-87 J0.2 (kJ/m2)	CTOD (mm)	δ5 (mm)	(R/R0)c
Base metal	434	1,12	0,77	5,93
Weld metal	208	0,26	0,23	3,03

Table 4 : Critical values of different fracture criteria for the parent material and the weld metal of a stainless steel weld.

Centre Cracked Pannels

The centre cracked panel analyses were carried out considering a plate width (2W) of 200 mm, containing a central 40 mm wide band of weld metal. The height of the specimen is 500 mm. Two crack sizes of 7.2 mm and 50 mm (2a) located in the middle of the weld have been investigated, in order to evidence the crack size effect on plasticity development and its influence on the specimen resistance to ductile tearing. The path-independent J-integral was calculated numerically as well as evaluated through an estimation procedure similar to the ASTM procedure (6), which derives a global energetic J value from the load-displacement curve:

$$J = Jel + Jpl \text{ , where } Jpl = \frac{\eta pl * A}{b * B} \qquad 1]$$

ηpl is equal to 1, A is corresponding to the plastic area under the load-displacement curve, b is the ligament, and B is the net thickness of the specimen.

A comparison between these two values of J is plotted in Figures 2 and 3. This comparison shows that the difference between a theoretical value of J, calculated numerically, and the "experimental" value that was derived from a CCP test. The latter is not really a J-integral value but rather the amount of energy provided to the specimen, that includes the energy for crack initiation and propagation, but also the plastic strain energy which is dissipated in the base metal. This plastic strain energy does not make the crack propagate and must not be included in the toughness value derived from the test. It may be concluded from those results that great care is needed when using an experimental assessment procedure based on ASTM to determine toughness properties of welded components. This experimental procedure is relevant to homogeneous structure testing, involving limited plastic zone extension, but may not be adapted for mismatched welds.

The second part of this numerical analysis concerns the application of the Rice and Tracey local fracture criterion to predict crack initiation in the CCP specimens. The CTOD and $\delta 5$ criteria have bot been applied because they are not directly transferable from a CT specimen, for which the crack mouth is free, to a CCP specimen. The J integral derived from the ASTM procedure which is corresponding to the critical value of $(R/R0)c$ has been found to be about 600 kJ/m2 for the larger crack (a/w = 0.25) and should be higher than 750 kJ/m2 for a/W=0,036 (in fact we stopped the computation at 750 kJ/m2 when the whole specimen section made of base metal was yielded, while the singularity at the crack tip had only generated a contained plastic zone). It should be remembered that, on condition that the Rice and Tracey failure criterion is relevant to the weld metal fracture initiation, these values of J would be measured experimentally by the assessment procedure. This point will be studied carefully when performing these tests.

Single Edge Notched Bending specimen

The SENB specimen analysis was carried out considering a plate width W of 100 mm, containing once again a 40 mm wide band of central weld metal. The height of the specimen is 500 mm and the crack size is 25 mm in order to get an a/w ratio of 0.25. By comparing this case with the second CCP specimen studied previously, we revealed the influence of loading (bending instead of tension) on the mismatch effects. The comparison between the numerical path-integral J and that derived from the ASTM procedure is plotted on Figure 4. The same trends than previously exhibited can be noticed. One can conclude that the ASTM procedure includes a certain amount of plastic strain energy dissipated in the base metal, which does not play any role in crack initiation. This energetic quantity can be considerd as an "apparent toughness" of the structure, that is measured experimentally but which does not represent the real loading which the crack tip is subjected.

The application of different local criteria (CTOD, $\delta 5$, and local approach) in order to predict crack initiation in the SENB specimen is presented in Table 5, where the apparent toughness J derived from each failure criterion is given. It will be noticed that the CTOD criterion leads to more or less the same value of apparent toughness J, while the two others $\delta 5$ and $(R/R0)c$ indicates higher values and therefore an artificial increase in the weld resistance to ductile tearing. Here again, performing experiments is the only way to determine which analysis is representative of the real behaviour of the material.

Fracture criterion	CTOD = 0.26 mm	$\delta 5$ = 0.23 mm	$(R/R0)c$ = 3.03
ASTM J-value (kJ/m2)	195	370	530

Table 5 : Apparent toughness of a SENB specimen at 300°C according to different fracture criteria predictions.

Composite CT specimen with an interface crack in the fusion line

This composite CT specimen is made of one half base metal and one half weld metal. It contains an interface crack corresponding to an a/w ratio of 0.6. For this analysis, we derived the crack driving force only from the ASTM J-estimation procedure from the load-displacement curve. The different fracture criterion have been applied considering the two CT halves independently, which means as shown in Figure 5 that we consider that crack initiation occur as soon as one of the specimen halves has reached its own critical value. For instance, when applying the CTOD criterion, we consider the evolution of the 1/2 CTOD corresponding to each part of the specimen, which is measured in regard to the fusion line. Then, the fracture criterion becomes :

1/2 CTOD (base metal part) < 1.19/2 or 1/2 CTOD (weld metal part) < 0.26/2 where the critical values come from Table 4. The application of the δ5 fracture criterion is based on the same procedure than that of the CTOD criterion. Lastly, the application of the Beremin's model consists in computing the cavities growth rate at the crack tip in each part of the composite CT specimen, and to compare it to the corresponding critical value. The results obtained are gathered in Table 6.

Fracture criterion	CTOD	δ5	(R/R0)c
ASTM J-value (kJ/m2)	195	205	185

Table 6 : Apparent toughness of a composite CT specimen at 300°C according to different fracture criteria predictions.

It will be observed that the apparent toughness values obtained are slightly lower than those of the base metal. This surprising conclusion has to be investigated experimentally to proove whether the differences of strain carrying capacities of the base and weld metal, that generates additionnal shear stresses along the fusion line, are high enough to cause this apparent weakening of the welded component. Moreover, the fracture process that initiates the crack propagation may be far from being ductile tearing, but rather a shear localisation mechanism or even decohesion between the two materials. In that case, the fracture criteria used in this study are no more representative of the real fracture mechanisms.

The second point to emphasise concerns the crack initiation and propagation. The three criteria have been predicting that fracture should occur in the base metal. This prediction is consistent with the different observations made in (7) and (8) that showed that cracks tend to propagate in the softer material, for instance the base metal. Once again, experiments will be essential in order to conclude on the ability of these fracture criterion to describe the fracture behaviour of interface cracks.

CONSEQUENCES FOR INTEGRITY ASSESSMENT ANALYSES

The "classical" structural integrity assessment procedures rely on the concept of J-resistance curve. One essential point of this procedure is that the crack driving force J applied to the structure is compared to the material's J-R curve. This J-R curve is supposed to be an intrinsic property of the material, and is determined experimentally through small specimens testing. The present study demonstrates that some problems may result in the experimental determination of the material toughness, and emphasizes that one should know accurately the mismatching effects that can be expected before using the standard testing methods developed for homogeneous material. If these preliminary results are confirmed by experiments, it will show that what one measures when performing a standard specimen test on an inhomogeneous structure may be not its real resistance to ductile tearing, but rather what we can call an "apparent toughness" that includes an amount of mismatching effects, that play no role in the fracture process. Consequently, extreme caution is needed when using estimated fracture toughness measurements derived from standard testing procedures.

Another important point that should be mentioned is the influence of the crack size on the structure response, and especially on the development of plasticity. As referenced in (2), the apparent toughness as may be measured on a deeply cracked specimen should be more or less identical to the toughness of the weld metal. In that case, the singularity arising from the presence of the crack in the overmatched weld is sufficient to generate a contained plastic zone ahead of the crack, while the base metal parts of the specimen remain quasi elastic. The linear elastic fracture mechanics conditions are satisfied and the global as well as local response of the welded structure is controlled by the tensile and toughness properties of the weld metal. On the opposite, when considering a medium or a shallow crack, the influence of the base metal increases as the ratio a/w is decreasing (Figures 2 and 3). Then, it is necessary to cope with interactions between the plastic zone developed at the crack tip and the other one that appears at the interface between the base metal and the weld right above the crack tip. When this second plastic zone appears in the base metal, it may produce an unloading of the crack tip zone which finally leads to an increasing of the apparent toughness, as revealed in the above finite element analyses. The toughness measurement from the area under the load displacement curve, called "apparent toughness", provides higher values than the numerical path-independent J-integral because local unloadings of the crack tip zone are not accounted for. The energy balance, which the standard ASTM E 1152-87 testing procedure refers to, includes also the plastic strain energy dissipated in the base metal.

CONCLUSIONS

This paper gathers the first results of a programme studying the effect of overmatching on fracture resistance of welded structures. Elastic plastic finite elements analyses have been carried out for different specimen geometries in order to characterise the effects of mismatching on plasticity development and its consequences on toughness measurement. Firstly, for shallow cracks, some problems may result in using the standard test methods and the associated J estimation formula recommanded by ASTM. The values determined through experiments correspond to what can be called an "apparent toughness", but which may not exactly correspond to the resistance of the structure to ductile tearing. These testing procedures should be used cautiously when considering welded components. On the second hand, the tensile and toughness properties that should be used in the integrity assessment analyses are not easy to be standardized, because of the crack size effects on the plasticity development in the structure. Keeping to the lowest value of yield stress and the lowest value of toughness for the base metal and the weld could be a conservative solution, but exhibits in some cases too important safety margins. Finally, it is neccessary to perform tests considering various configurations in order to better understand the behaviour of mismatched welds and to determine which relevant fracture criterion can be reliably used to carry out structural integrity assessment of welded engineering structures.

REFERENCES

(1)   Smith E. - "Comments on the problem of crack extension from a defect in a weld" - Pressure Vessel and Piping Conference - Vol. 215 pp 67-71 - San Diego, June 23-27, 1991.

(2)   Dong P. and Gordon J.R. - "The effect of under and overmatching on fracture prediction models" - Welding '90 Conference - pp.363/370 .

(3)   Schwalbe K.H. - "Effect of weld metal mismatching on toughness requirements : simple analytical considerations using the engineering treatment model" - Communication at the ESIS / Technical Committee 1 specialists' meeting - Vienna - April 1-2 - 1991.

(4)   Beremin F.M. - "Experimental and numerical study of the different stages in ductile rupture : Application to crack initiation and sTable crack growth" - Three dimensionnal constitutive relations and ductile fracture - North-Holland publishing Compagny - pp. 185/205 - 1981

(5)   Franco C. and al. - "Benchmark on computation simulation of a CT specimen fracture experiment" - Pressure Vessel and Piping Conference 92 - New Orleans - June 22-25 1992.

(6)   Roos E. and al. - "A Procedure for the experimental assessment of the J-integral by means of specimens of different geometries" - Int. J. Pres. Ves. & Piping - Vol.23 pp.81-93 - 1986.

(7)   Tschegg E. and al. - "Cracks at the ferrite-austenite interface" - Acta Metall. Mater. - Vol.38 N°3 pp.469/478 - 1990.

(8)   Koçak and al. - "Interfacial and subinterfacial cracks in the copper-ferrite system" - Engineering Fracture Mechanics - Vol.39 N°4 pp739/750 - 1991.

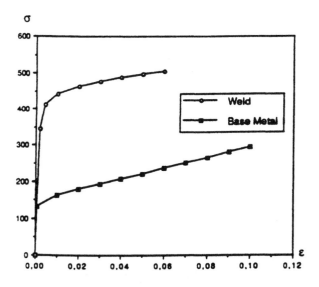

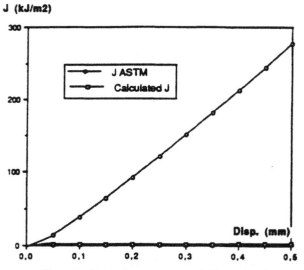

Figure 1 : True stress - true strain relation of the base metal and the weld

Figure 2 : Comparison of calculated J-integral and values derived from ASTM evaluation procedure - CCP specimen - a/w = 1/36

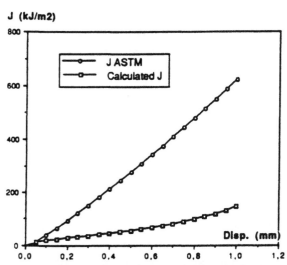

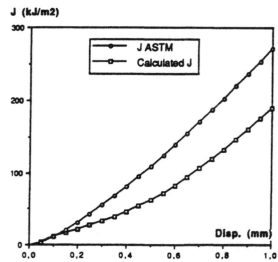

Figure 3 : Comparison of calculated J-integral and J-value derived from ASTM estimation procedure - CCP specimen - a/w = 0.25

Figure 4 : Comparison of calculated J-integral and value derived from ASTM estimation procedure - SENB specimen - a/w = 0.25

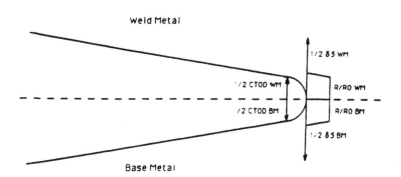

Figure 5 Fracture prediction of a composite CT specimen - Calculation of fracture criteria in each half of the specimen

PAPER 5

An experimental investigation of the transferability of surface crack growth characteristics

J Faleskog, MSc, F Nilsson, Prof, PhD and H Öberg, MSc (Royal Institute of Technology, Norway)

SUMMARY

The goal of the study was to compare J_R-curves from testing on relatively small conventional specimens of a pressure vessel steel with results from large slightly curved surface cracked (SCT) plates subjected to different loading histories. Six SCT experiments were carried out at 20°C and three at 60°C. Unstable crack growth occurred very shortly after crack growth initiation in four of the tests at 20°C, while some amount of stable crack growth was observed in the remaining two and in all of the small specimens. The analysis showed that initiation occurred at roughly the same value of J for all the SCT specimens. For all three SCT experiments carried out at 60°C ductile crack growth was observed. The J_R-curves from the SCT tests coincided qualitatively with each other as well as with the results from CT testing. It seems that the J_R-philosophy can be used if the temperature is well above the transition temperature. The initiation point is well predicted while some deviations can occur at larger crack growth increments if the loading is mainly of tensile character. This also holds for the initiation event at the lower temperature while for the further growth the situation is less clear.

INTRODUCTION

The main goal of fracture mechanics is to enable prediction of initiation and propagation of growth of existing cracks of known configurations in structures of arbitrary shape. It should be possible to base a prediction of the fracture behaviour of a structure on results from experiments performed on specimens of the same material but with a geometry that differs from the one under consideration. If we can rely on transferability, results from experiments conducted on small laboratory specimens can be used to predict the fracture behaviour of a large structure for which an assessment is desired. In the early development of non-linear fracture mechanics results from experiments on fairly similar specimens were correlated successfully. The expectations based on the relative success with these experimental investigations seem now to have been overly optimistic and many problems are still not entirely resolved. A trend of increased interest in these questions can now be observed.

The object of the present investigation has been to study some aspects of this general question. In most fracture mechanics theories it is assumed that the state at a point of the crack front can be characterized by a scalar parameter. For non-linear problems several different suggestions for a suitable fracture parameter have been put forward, but the so called J-integral is the most widely used. In the present study particular attention is paid to the possible geometry invariance of the J_c value and the J_R-curve. Experiments have been performed both on typical laboratory specimens and on large surface cracked plates. Some preliminary results on the part of the investigation performed at the relatively low temperature 20°C have been previously reported, Faleskog et al (1), Nilsson et al (2). In the present article the experimental program has been extended to include tests at 60°C.

THE *J*-INTEGRAL

For three-dimensional situations the J-value related to a point s on the crack front can be formally defined (cf Carpenter et al (3)) by the expression [1]. The coordinate system is assumed to be oriented so that the x_1-axis lies in the crack plane and is normal to the crack front at the considered point. The

x_3-axis is tangential to the front and the x_2-axis is normal to the crack plane.

$$J(s, C) = \int_C (w\, \delta_{1j} - \sigma_{ij} u_{i,1}) n_j\, dC - \int_{A_s} \frac{\partial(\sigma_{i3} u_{i,1})}{\partial x_3} dA \qquad [1]$$

Here C is a curve enclosing the crack front in a plane perpendicular to the front and A_s is the area enclosed by C. σ_{ij} is the stress tensor and u_i the displacement vector. w is the deformation work density defined as

$$w = \int \sigma_{ij}\, d\varepsilon_{ij} \qquad [2]$$

If the state of loading is nearly one of proportional loading approximate independence of the integration region prevails and the integral has a finite, non-trivial value even for incrementally deforming elasto-plastic materials. This suggests the following criterion for initiation of growth at the point s

$$J(s, P) = J_c \qquad [3]$$

Here J_c is assumed to be a material property and P is a measure of the applied loading.

For a stationary crack front the J-integral is usually reasonably independent of the integration region if the global loading conditions are reasonably near proportional loading. For a *growing* crack tip in a material that behaves according to an incremental plastic constitutive law, the loading state is very far from proportional in the crack front region. In fact, the order of the singularity for a growing crack in such a material is of such order that an integration path taken infinitesimally close to the tip would yield a zero value for the J-integral. In order to obtain integration region independent values the curve C in [1] must be taken so remotely from the tip that proportional loading conditions prevail. A common approach is instead to neglect the non-proportionally deforming region and calculate the J-integral assuming the tip has been stationary at its instantaneous position during the loading process. This has been shown to be a reasonable approximation for at least some two-dimensional situations (cf Nilsson (4)). In the present study this will be assumed to be valid also for three-dimensional cases considered. In this way the problem of simulating the crack growth is avoided.

In order to predict further crack growth after initiation it is assumed that the growth is governed by

$$J(s; P, a) = J_R(\Delta a(s)) \qquad [4]$$

where J_R is a material function only depending on the crack growth increment Δa. In fully three-dimensional cases it is assumed that [4] holds locally at every point along the crack border.

The object of the present investigation was to conduct experiments carefully on widely different types of specimens. The crack growth was observed and the J-values corresponding to each growth increment were calculated. The validities of the hypotheses [3] and [4] were tested by comparing the resulting $J_R(\Delta a(s))$ values for the different specimens.

MATERIAL AND EXPERIMENTAL ARRANGEMENTS

The experimental work in this project has been performed on a pressure vessel steel (2 1/4 Cr 1 Mo) taken from a decommissioned chemical reactor. The material composition and the conventional mechanical properties at the test temperatures (20°C and 60°C) are given in Table 1 and 2, respectively. 20°C is about in the middle of the material's transition region according to impact testing while at

60°C the material is expected to exhibit fully ductile behaviour.

Table 1. Material composition and properties for 2 1/4 Cr 1 Mo

C, %	Si, %	Mn, %	P,%	S, %	Cr, %	Mo, %
0.143	0.23	0.48	0.01	0.016	2.12	0.93

Table 2. Mechanical properties of 2 1/4 Cr 1Mo

Temperature, °C	Yield stress, MPa	Tensile strength, MPa	Elongation, %	Impact Energy, Joule
20	280	530	32	48
60	258	500	34	145

The specimens were conventional fracture specimens, CT- and SEN(B) specimens, and large plates with a surface crack (SCT, see Fig. 1 for dimensions). Some of the conventional specimens were side-grooved in order to promote plane strain conditions. The total thickness B of the small specimens was 25 mm. The loading of the specimens was done in accordance with ASTM E-813 (5) with partial unloadings made in order to facilitate the determination of crack growth by the compliance technique.

For the CT- and SEN(B)-experiments the evaluation of J was based on the measured force-displacement curve according to

$$J = \frac{K_I^2(1-v^2)}{E} + \frac{\eta}{B_N(W-a_0)} \int_0^\delta P d\delta_{pl}$$

[5]

K_I is the stress-intensity factor for the actual load, P the applied force and B_N the net thickness. δ_{pl} denotes the plastic part of the displacement and η is a non-dimensional geometry constant depending on the specimen type. W is the characteristic dimension of the specimen and a_0 the initial crack length.

The SCT specimen was a slightly curved 50 mm thick plate where the midpoint was offset a distance e (see Fig. 3) compared to the mid-points of the ends of the specimen. The specimen ends were welded to thick plates with a slant angle θ_0 with respect to the normal to the plane of the plates and an initial bending moment was thus accomplished. Thus, the specimens were subjected to two loading systems, where the applied tension could be regarded as a primary load which induced both tensile and bending stresses. The initial displacement could be considered as introducing a secondary bending load. By varying the offset e and the angle θ_0, different M-P-histories could be obtained where P is the membrane load and M the bending moment as referred to the cracked section.

The initial crack was produced by first machining a semi-circular slit perpendicular to the surface of the specimen. After fatigue loading in bending the cracks assumed approximately semi-elliptical shapes. The depth and length of the initial cracks together with the specimen offset and the initial bending stress are given Table 3.

One of the objectives was to study the influence of different loading histories on the crack growth behaviour. Nine experiments were performed with different offsets and initial bending moments M_i at the two different temperatures according to Table 3. M_i is here normalized with respect to the moment M_Y needed for initiating plastic deformation. M_Y is calculated without any account taken for the presence of the crack. The yield stress σ_Y is here taken as the 0.2 % offset stress. A minus sign means that the moment is oriented so that tensile stresses occur on the non-cracked side of the specimen.

Table 3. Crack dimensions, specimen offset and initial bending moment.

Temperature, °C	Experiment, no	Crack length l, mm	Crack depth a, mm	Offset e, mm	Initial bending moment M_i/M_Y
20	1	98.4	23.9	0.	0.
	2	92.8	22.1	0.	0.025
	3	88.8	20.2	5.0	1.10
	4	89.8	20.2	2.5	0.56
	5	98.2	21.7	30.0	-0.36
	6	100.	22.3	30.0	1.05
60	7	98.4	23.9	2.5	0.53
	8	92.8	22.1	30.0	-0.21
	9	88.8	20.2	30.0	1.06

During the testing the specimens were instrumented with strain-gauges at many different locations for measurement of loads and crack depth. The crack depth measurements were made by partial unloadings and strain measurements in the crack vicinity. In addition a technique with subsequent colourings of the crack surfaces was used. By this technique two to three distinct markings of the successive crack front positions could be obtained in addition to the one provided by the position of the crack before the fatigue loading used to finally break the specimen. At the higher temperature the specimens were heated by infrared heaters and the temperature was measured by thermocouples. When steady state conditions were reached the desired test temperature could be maintained within ±1.0°C .

During the fracture test the specimen was loaded with increasing tensile force upon which the partial unloadings were superposed. The loading was taken to such a level that the estimated crack growth was about 2-2.5 mm provided that total fracture had not occurred earlier. After unloading the specimens were dismounted and subjected to fatigue loading in bending until each specimen was entirely broken.

FINITE ELEMENT ANALYSIS OF SCT EXPERIMENTS

Since no reliable approximate method was available for the analysis of the SCT specimen, finite element analyses had to be performed. The general task program system ABAQUS (6) was chosen for this purpose. In the calculations the material was assumed to be incrementally elastic-plastic and to obey the von Mises criterion with its associated flow rule. The hardening was taken as isotropic with a piece-wise linear approximation of the uniaxial test curve. A special element mesh generating program for the complicated models needed was developed. This program allowed for easy generation of models with successive refinements of the mesh division in the vicinity of the crack border. Some typical mesh configurations are shown in Fig. 2.

All calculations were performed taking non-linear geometry effects into account. This proved necessary because of the experimental arrangements where the bending loads were significantly affected by the finite deformation. Because of symmetry only one quarter of the specimen needed to be modelled. On the symmetry planes the appropriate boundary conditions were imposed. At the loaded boundary the displacements and rotations were assumed to be prescribed so that a clamped and rigidly moving boundary plane was simulated. The J values were calculated by using the routine for such evaluations available in ABAQUS which is based on a domain integral method. It can be shown that this method is equivalent to the definition [1] provided that the conditions for path-independence of J are satisfied.

EXPERIMENTAL RESULTS

Small specimen testing at 20°C and 60°C

Some representative results from the testing of conventional specimens are shown in Fig. 3. In Fig. 3a the J_R-Δa curves from the tests (CT and SEN(B)) at 20°C are shown. Here, solid points are from specimens with side-grooves while the open ones are from specimens without grooves. The same coding is also used in Fig. 3b for the CT-testing at 60°C. At 20°C there is a considerable difference between specimens with or without side grooves, respectively. At 60°C it is seen that for small amounts of growth, there is no marked difference, while for larger amounts of growth, the J-values for the grooved specimens tend to fall slightly below those of the non-grooved specimens. The reason why this effect is more marked at the lower temperature is not known. Another general observation is also that the J values are somewhat higher at 60°C than at 20°C which is to be expected. No differences between SEN(B) and CT specimens was noticed at the lower temperature.

An evaluation of J_{Ic} according to ASTM E-813-87 (5) was attempted for the tests with side-grooves. Since points with a J-value higher than $J_{max}=\sigma_Y B/15$ (here 630-660 kN/m) may not be used in such an evaluation, it is evident that a strict ASTM E-813 evaluation was not possible. Disregarding this limit and still evaluating the side-grooved tests in the spirit of ASTM E-813-87 resulted in the mean J_R-curves shown in Fig. 3. The intersection of the J_R-curve and the 0.2 mm offset line gives J_Q-values of about 400 kN/m at 20°C and 700 kN/m at 60°C. Since the J_Q-value is determined by the intersection with the offset line, the apparent Δa-value at initiation becomes about 0.7 mm and 1.3 mm, respectively, and can thus hardly be true initiation values. Fractographical examination performed on specimens used in the tests at 20°C indicated that initiation occurred much earlier than the evaluation by the ASTM-procedure suggested. From this it was concluded that the initiation level ought not to be higher than 250 kN/m. For the tests at 60°C careful colour markings pointed to a true initiation value not higher than 250 kN/m also at this temperature.

SCT-testing at 20°C

A lot of data were collected during the testing of the SCT specimens. These data have been document-ed in a data record book (7) together with the results of a rather extensive fractographical examination. For the different tests the P-M relations shown in Fig. 4 resulted. P_l and M_l denote the limit load and limit moment, respectively, for the uncracked section based on the yield stress. It is noted that the nominal loading was strongly non-proportional. It can also be seen that after a rather extensive plastic deformation unstable crack growth occurred at or very shortly after crack growth initiation in four (no. 1, 2, 5 and 6) of the experiments. In two tests (no. 3 and 4 in Fig. 4) significant stable crack growth occurred. The point of crack growth initiation was in these cases esti-mated by considering the unloading compliance measured by the strain gauges in the vicinity of the surface crack.

The fractographical examination revealed that the early stages of the crack growth initiation and pro-pagation exhibited the same features for all SCT as well as conventional specimens. A clear blunting or stretch zone could be observed as well as a zone of typically ductile dimpled propagation that cor-responded to at least 0.1-0.2 mm of growth. In the specimens that experienced rapid crack growth shortly after initiation this ductile growth was followed by a cleavage failure. Before the cleavage growth occurred no significant differences could be observed between these specimens and those that experienced significant amounts of ductile growth.

In Fig. 6 the J-distribution along the crack front at the estimated initiation load is shown as obtained from the FEM calculations. The non-dimensional variable s denotes the position along the front according to the definition in Fig. 6. It can be seen that the maximum J-value for all SCT tests falls within a relatively narrow scatter-band (192±12 kN/m). In the small specimen testing apparent initiation evaluated according to the ASTM principles as was discussed above occurred at a J-level that was about twice as high. The fractographical investigation suggested that initiation occurred much earlier. Thus, the J-levels at initiation for the different types of specimen may not differ very much from each other should the true initiation point be known.

In order to investigate if the instability behaviour could be explained by the J_R philosophy a limited study of crack growth was made. Two experiments (no. 3 and 6) were selected and J-calculations were performed for cracks with the same length (l) as the original one but with the depths (a) extended. For experiment 3 which exhibited stable crack growth the extension was chosen to coincide with a colour marking. J was then calculated for the load at initiation and after the given crack growth, respectively, and the results are shown in Fig. 7. As can be seen, apart from some uncertainty in the Δa position the results agree quite well with the mean curve from the two-dimensional testing. For experiment 6 which failed in an unstable manner, the J-calculations were performed for constant end displacement. It is seen that the slope of the J-Δa curve in this case is very close to zero. The conclusion is obviously that the J_R-concept could for these experiments not be used to explain the unstable crack growth.

No systematic effect of the loading history on the critical J for initiation could be detected in the results of the SCT experiments even though the nominal loading in some of the tests significantly deviated from a proportional one. In a full analysis of this kind the possible effects of history on J seem to be automatically accounted for and there is no need of distinguishing between primary and secondary stresses.

A complete explanation of why unstable crack growth occurred very shortly after initiation in four of the SCT experiments and in none of the conventional specimens (in all about 25) is not available presently. The results shown in Fig. 7 showed that dJ/da at constant loading for the unstable SCT-specimens was orders of magnitude smaller than the slope of the small specimen J_R-Δa curves. Thus, the instability theory based on the J_R concept was not able to explain this difference in behaviour.

Some possible causes for the behaviour may be mentioned. It must obviously be connected to the geometric differences between the specimens in some way. One possible measure of these differences may be the constraint factor. There are problems with the quantification of this effect and it can be suspected the measures commonly used are very dependent on the mesh used to model the structure. Despite the theoretical shortcomings of the concept, some constraint effect seems to be confirmed by the different J_R-curves for specimens with and without side-grooves, respectively, as well as by the different behaviour of the small specimens and the SCT specimens, respectively. Obviously this effect is not wholly systematic since stable crack growth occurred in two SCT experiments and this suggests that statistical effects may be influential. According to the weakest link concept the probability of a cleavage fracture should increase with the length of the crack front. It should further not matter how a given amount of crack front is distributed among different specimens. The total amount of crack front of all the conventional tests was about the same as the total amount of the SCT tests. The event of an unstable crack growth directly or shortly after initiation should thus be about equally probable for the two sets of experiments which is clearly not consistent with the experimental outcome. Thus we conclude that the interaction of geometrical (constraint) and statistical effects is the probable cause of the observed behaviour.

SCT-testing at 60°C

In all three of the SCT experiments at 60°C the cracks grew stably during the entire test. The different tests resulted in the calculated *P-M* relations shown in Fig. 5. The loading histories almost coincided for experiments 8 and 9 at high load levels. These had approximately the same geometrical data while the main difference between the two tests was the amount of the initial bending moment. Apparently the large difference in secondary stress had little influence on the load carrying behaviour of these specimens. Fractographical examination revealed that the growth was of a wholly ductile character. In all cases the maximum growth occurred at the deepest point of the original crack and then decreased gradually to almost vanish at the intersection of the crack front with the free surface.

All SCT-experiments were evaluated by FEM in the manner described above. The calculated *M* and *P* values agreed well with the experimental results. This was also the case for the overall force-displacement relation. In this series of tests it was not possible to pinpoint the true initiation point in the same way as for the tests at 20°C since the growth in all cases was stable. In Fig. 8 the distribution of *J* along the crack front at three subsequent colour markings is shown for one specimen (no. 8).

In Fig. 9 the *J*-values calculated by FEM at the points $s=1$. and $s=0.4$ at the crack front are given as functions of the amount of crack growth as measured through the colour markings. Also shown in the figure is the mean J_R-curve from the CT-tests. There are certain differences between the results of the different experiments. The tendency is that they tend to converge for small amounts of growth thus supporting the hypothesis that initiation occurs at a common *J* value. According to an extrapolation of the results in Fig. 8 the critical *J*-value seems to be in the interval 200-250 kN/m. It can also be noted that the results at $s=0.4$ for all experiments and especially no. 8 and 9 fall fairly close to each other. The results at $s=1$. for no. 7 also fall near the three sets just discussed. Test no. 7 experienced much less bending than tests no. 8 and 9 for which the values for $s=1$ are near to the CT results. The first four of the sets could thus be said to represent a more tensile loading situation than the last three where the bending component of the loading was larger. The observation that tensile loading tends to elevate the J_R-curve is consistent with previous observations in the literature for two-dimensional tests.

CONCLUSIONS

The results indicate that crack growth initiation can be well correlated by the *J* integral both in the transition interval and at the upper shelf. Admittedly, this conclusion is dependent on some rather uncertain conjectures about the exact point of growth initiation. Clearly the ASTM procedure is inadequate for materials that behave as the actual one. Unfortunately no other standardized procedures seem to exist which can deal with situations of the present kind and development is urgently needed.

In the transition interval (20°C) the continued propagation can be either of the cleavage or of the ductile type. Which one of the two possibilities that will actually occur is to some extent random in nature and it is reasonable to assume that amount of constraint will affect the relative proportion between the two mechanisms. If ductile propagation results, it appears that the J_R-curve for this continued growth will be similar to that of the small specimens.

At 60°C it was observed that somewhat different J_R-curves were obtained in the different SCT tests. It appears that the most tensile loaded test exhibited the highest J_R-values and that an increased bending moment tended to lower the J_R-curves. This is consistent with earlier observations on laboratory type specimens. The effect seems to be mostly accentuated when the crack has grown somewhat. Otherwise history effects seem to be small. Thus we can conclude that the J_R-philosophy can be used if ductile growth is ensured. The deviations that occur for the tensile cases will result in conservative predictions provided data from bend tests are used.

ACKNOWLEDGEMENT

This work has been supported by a grant from the Swedish Nuclear Inspectorate (SKI). The authors want to express their gratitude for this support. Most of the numerical calculations were performed using the CRAY YMP resources provided by the National Supercomputer Center (NSC). This support is gratefully acknowledged.

REFERENCES

(1) J. Faleskog, K. Zaremba, H. Öberg & F. Nilsson, "An investigation of two- and three-dimensional elasto-plastic crack growth initiation experiments", in *Defect Assessment in Components-Fundamentals and Applications*, ESIS/EGF9, ed. by J. G. Blauel and K. H. Schwalbe, Mechanical Engineering Publications, London, 1991, 433-444.

(2) F. Nilsson, J. Faleskog, K. Zaremba & H. Öberg, "Elastic-plastic fracture mechanics for pressure vessel design", *Fatigue & Fracture of Engineering Materials & Structures*, **15**, 1992, 73-89.

(3) W. C. Carpenter, D. T. Read & R. H. Dodds, "Comparison of several path independent integrals including plasticity effects", *Int. J. of Fract.*, **31**, 1986, 303-323.

(4) F. Nilsson, "Numerical investigation of *J*-characterization of growing crack tips", *Nuclear Engineering and Design*, **133**, 1992, 457-463.

(5) ASTM E-813, "Standard test method for J_{Ic}, a measure of fracture toughness", Section 3 of the 1988 General Book of ASTM Standards, American Society for Testing and Materials, Philadelphia.

(6) ABAQUS, Users Manual 4.8, Hibbit, Karlsson and Sorensen Inc, Providence, Rhode Island, 1988.

(7) J. Faleskog, F. Nilsson & H. Öberg, *"Fracture experiments on surface cracked plates-data record book"*, Report 130, Department of Solid Mechanics, Royal Institute of Technology, Stockholm, Sweden, 1990.

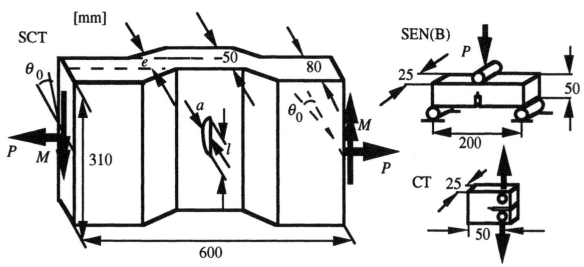

Figure 1. Specimens used in the investigation

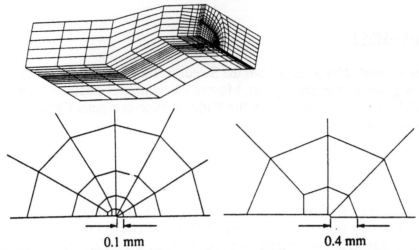

0.1 mm   0.4 mm

Figure 2. Examples of FEM models and mesh arrangements around the crack front

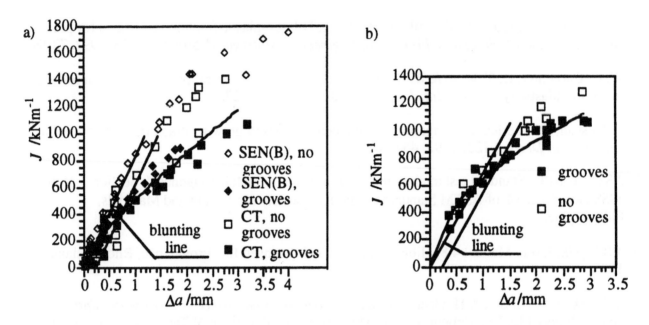

Figure 3. *J* as a function of crack growth for conventional specimens, a) 20°C, b) 60°C

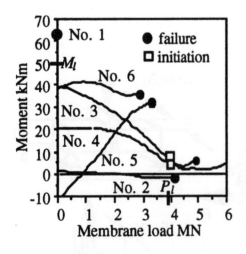

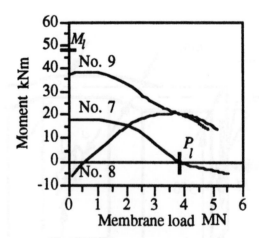

Figure 4. Moment versus normal load for SCT-tests at 20°C

Figure 5. Moment versus normal load for SCT-tests at 60°C

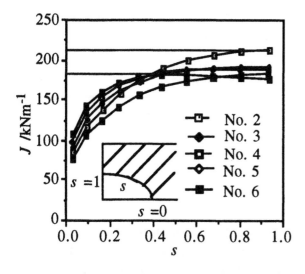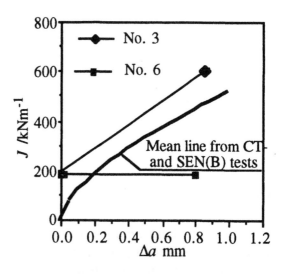

Figure 6. *J* at estimated initiation point along the crack front for SCT tests at 20°C

Figure 7. *J* at the deepest point of crack as function of crack growth for two SCT experiments at 20°C

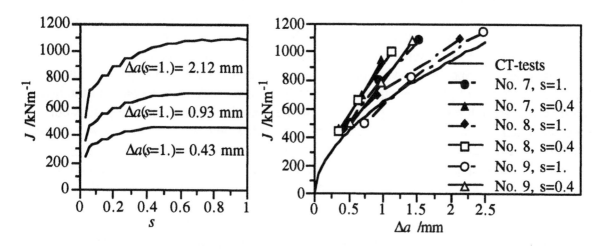

Figure 8. *J* along the crack front for SCT test no. 8 at 60°C for three different crack front positions

Figure 9. *J* at two locations at the crack front as functions of crack growth for SCT experiments at 60°C

The influence of microstructural variation on crack growth

V Gensheimer DeGiorgi, BSc, MEng, PhD, P Matic, BSc, PhD, G C Kirby III, BSc, MS, PhD and
D P Harvey II, BSc, MS (Naval Research Laboratory, USA)

SUMMARY

Welding, casting and heat treatment of structural components are common processes which produce material zones with pronounced microstructural gradients. Crack inititaion and subsequent crack growth through these zones is affected by the local microstructure. The purpose of this investigation was to create a controlled environment in which to observe and model these effects in laboratory specimens, employing a local approach to fracture, through a set of relevant experiments and computational analyses.

Panel specimens of 1045 steel were machined with a small edge notch and heat treated by differential quenching to produce transverse microstructural gradients in the panels. Hardness measurements made on the panel surfaces indicated the desired inhomogeneous material characteristics across the panel widths and the relative brittle to ductile character of the microstructural gradient. Specimens were pre-cracked to an crack lenght (a) to panel width (w) ratio of 0.2. In one group of panel specimens, the notch and pre-crack were located on the brittle side to study crack growth into the comparatively ductile material. In another group of panel specimens, the notch and pre-crack were located in the ductile side to study crack growth into the comparatively more brittle region. Stable, ductile tearing was produced by tensile loading. Load, displacement and crack tip position were measured. The effect of crack growth from ductile into more brittle microstructure versus brittle into more ductile microstructure was evident.

A corresponding set of homogeneous tensile specimens were produced which spanned the same microstructural range as found in the panel specimens. The tensile specimens were loaded to fracture. The role of the microstructure on specimen yield strength and ductility was evident. Computational simulations were performed to establish the large strain elastic-plastic constitutive parameters associated with the microstructure in the specimen. The local material fracture toughness, expressed as the absorbed energy density to fracture, was calculated for each material hardness at the point of fracture initiation in the corresponding specimen.

With the knowledge of material response and local fracture toughness versus microstructure obtained from the tensile specimens, computational simulations of the panel responses were performed to model crack initiation and growth. The ABAQUS (1) finite element program with a debonding algorithm to model crack growth was used in the evaluation.

INTRODUCTION

This study investigates nonuniform material effects on stable crack growth in edge notched bars. The material nonuniformity was produced by microstructural changes generated by heat treatment under controlled conditions. While strongly analogous to heat affected zones in welds the choice of this system over an actual weld ensured a more controlled microstructure and the absence of additional features such as fusion lines, porosity, lack of penetration defects or lack of fusion defects.

The objective of this investigation was to experimentally measure and computationally simulate crack growth through relatively brittle material into progressively ductile material and the complimentary case of crack growth through ductile material into brittle material. Material selection of 1045 steel was based on its inherent ductility as manufactured, the sensitivity of the microstructure to specific heat treatment regimes and the range of constitutive responses associated with different microstructures. Experiments were performed on flat tensile specimens of corresponding microstructures for tensile characterization and on Mode I cracks in edge notched panels for crack growth studies. All finite element analyses were performed using the commercial finite element code ABAQUS. The crack growth analysis was accomplished using a debonding algorithm in ABAQUS in conjunction with a local fracture criterion.

MATERIAL MODEL AND FRACTURE CRITERIA

The material constitutive model used is based on an incremental rate independent plasticity theory available in the ABAQUS finite element program (1,2). Total strains in the multiaxial strain state are obtained by the integration of the linearly decomposed rate of deformation tensor. Elastic strains are assumed to remain infinitesimal. A von Mises yield function is used. The uniaxial Cauchy stress - logarithmic strain constitutive response of the material is formally input to the ABAQUS program.

The local fracture criterion used is absorbed energy. Absorbed energy is measured as the strain energy in the material. The absorbed energy per unit mass, w, of the material, is:

$$w = \lim_{\Delta V \to 0} \left(\frac{\Delta W}{\rho \Delta V} \right)$$
[1]

where W is the energy and V is volume. The terms of stress components, σ_{ij}, and strain components, ε_{ij}, and the mass density, ρ, the absorbed energy or strain energy is:

$$w = \int_0^{\varepsilon_{ij}} \frac{\sigma_{ij}}{\rho} d\varepsilon_{ij}$$
[2]

The measure of absorbed energy chosen, the strain energy, incorporates the contributions of both stress and strain quantities to the material history. The value of strain energy density corresponding to local fracture of the material is:

$$w_c = \int_0^{(\varepsilon_{ij})_c} \frac{\sigma_{ij}}{\rho} d\varepsilon_{ij}$$
[3]

where w_c is the strain energy required for fracture for a defined stress-strain history. The value of w_c is generally path dependent, although a representative value may be practical for engineering applications.

For ductile metals, the mass density values only slightly, even over large ranges of deformations. For this reason, it is possible to define an energy per unit volume density:

$$w = \lim_{\Delta V \to 0} \frac{\Delta W}{\Delta V} \tag{4}$$

or

$$w = \int_0^{\varepsilon_{ij}} \sigma_{ij} de_{ij} \tag{5}$$

with an associated critical value:

$$w_c = \int_0^{(\varepsilon_{ij})_c} \sigma_{ij} de_{ij} \tag{6}$$

The energy per unit mass is a fundamental quantity however, the energy per unit volume is equally appropriate for use with constant volume deformation processes.

For the case of a uniaxial representation of Cauchy stress versus logarithmic strain material response, the critical absorbed energy as represented by the critical value of strain energy density, corresponds to the area under the uniaxial stress-strain curve:

$$w = \int_0^{\varepsilon_c} \sigma d\varepsilon \tag{7}$$

For multiaxial state of stress, each of the six stress-strain pairs must be evaluated and summed. One or more individual terms, but not all six terms, in the multiaxial expression can be negative. The total, w_c, must be positive.

Prediction of fracture inititaion by a local criterion requires the identification of local maximum energy densities and comparison of these maxima with a critical energy density value. The location of the energy density maxima may vary during the loading history. Additional deformation can be sustained without fracture as long as:

$$w < w_c \tag{8}$$

Local fracture occurs when the absorbed energy attains the critical strain energy density value at some point within the material:

$$w = w_c \tag{9}$$

EXPERIMENTAL

Quenching Procedure For Nonuniform Microstructure

The material used for generating nonuniform microstructures was 1045 steel rectangular bars subjected to a differential quench. For quenching, the steel had a nominal width of 50.8 mm (2.0 in), a nominal length of 254 mm (10.0 in) and a nominal thickness of 12.5 mm (0.5 in). A 2.54 mm (0.1 in) notch was machined into each bar at a location halfway along the length.

The differential quench was performed by heating the bars to their austenitic temperature of 800 °C for 90 minutes. The bars were removed from the oven and allowed to air cool for 45, 60 or 90 seconds before half of the width was immersed in water. Air cooling allowed for sufficient time for the diffusion of carbon. Therefore, no martensite was formed and microscopic examination of the steel indicated the presence of pearlite dominating the microstructure.

Microstructure gradients were obtained by immersing one side and allowing the other side to air cool during quenching. Immersion of the notch side during the quenching process produced a material microstructure that was relatively brittle near the notched edge and increased in ductility across the width to the opposite edge. The complementary quenching process produced a ductile material near the notch and transitioned into a more brittle material.

After the quenching process, the bars were split into halves with a nominal thickness of 3.81 mm (0.15 in) so that the net force distributed over the cross section would be within the load capacity of the testing machine. The resulting edge notched specimens (Fig. 1) were stress relieved at 600 °C for 10 minutes. Examination of the hardness values from the original exterior and interior faces of each specimen exhibited no evidence of a through thickness hardness gradient nor did the resulting split halves exhibit any out of plane warping that might be associated with relief of residual stresses. In order to map the hardness values over the surface of the plate, a 66 mm x 50.8 mm (2.6 in x 2.0 in) sheet of paper ruled with a 5 mm (0.2 in) square grid was overlaid on each specimen. Rockwell B hardness measurements were obtained at each grid intersection on the front and back of the specimen surface. The corresponding measurements from the front and back were averaged and a surface spline was fitted through the averaged values for interpolation at specific points as dictated by the finite element model. Figure 2 shows hardness contours for one of the specimens; typical hardness values ranged from 108 near the immersed edge to 82 near the air cooled edge.

Constitutive Characterization

Load displacement data for each of the six hardness values corresponding to the six specimens were reduced to approximate Cauchy stress and logarithmic strain using standard averaging techniques. The strain energy density required to produce fracture for each specimen hardness value was determined by integration of each stress-strain curve to the point of fracture.

The six stress-strain curves were used to generate a stress-strain-hardness surface using a least squares technique in the computer program Mathematica (3). The resulting surface was used to generate individual stress-strain curves. Stress-strain response data was generated for hardness values from 80 to 110 on the Rockwell B scale.

Crack Growth Experiments

A 2.54 mm (0.1 in) precrack was fatigued into each specimen in order to sharpen the notch tip. Precracking was performed using a servohydraulic testing machine in load control mode with the load cycling between ± 26.7 kN (± 6000 lbs) at 5 Hz. To maintain a straight crack front, the specimen was turned about its loading axis every 10 000 cycles. The progress of the crack growth was monitored with a travelling microscope. Measurements of the front and back crack lengths were within 0.051 mm (0.002 in) for the final precrack. Post mortem examination of the fracture surface revealed that the precrack was straight throughout the thickness of the specimen and perpendicular to the front and back surfaces.

For the crack growth experiments, each specimen was gripped with hydraulic wedge grips.

Measurements of load, displacement and crack length were made at discrete time intervals. The load was measured using a 22.3 kN/volt (5000 lbs/volt) load cell. The displacement was measured using a 50.8 mm (2.0 in) extensometer that straddled the anticipated crack path and was located 10 mm (0.4 in) from the notched edge. All tests were performed in displacement control.

Measurements of the crack length were made using a travelling optical microscope. The resolution of crack length measurements was 2.54 μm (0.0001 in). Throughout the test, the region surrounding the crack tip was observed through the microscope.

In the linear region of the load-displacement response, crack tip opening occurred as the load was increased. After the crack tip was fully opened, crack tip blunting occurred with subsequent loading until crack propagation began. Once crack propagation initiated, the load was increased such that the change in crack length was less than 0.25 mm (0.01 in) for the first 4 or 5 load increments. Thereafter, the load was increased until the crack tip moved out of the field of view of the microscope. The loading history was operator controlled.

Figure 3 compares the load-displacement response for the brittle to ductile and ductile to brittle crack growth sets. Specimens from each set exhibit comparable load displacement response. For the brittle to ductile crack growth case a comparatively rapid decrease in load is seen once crack propagation occurs. For the case of crack growth from ductile to brittle material, the load increases as crack propagation occurs. However, as the crack advances into the comparatively more brittle material the load begins to decrease rapidly until final failure.

Figure 4 compares the crack length as a function of displacement for both sets. The behavior of specimens from each set exhibit comparable features. For crack growth from brittle to ductile material the curves are concave downward, indicating that crack growth is more rapid through the brittle material than the ductile material. For crack growth from ductile to brittle behavior the curves are concave upward which indicates the rate of crack growth begins to accelerate as the crack tip advances into the more brittle material.

COMPUTATIONAL

Two edge notched panel computational analyses were performed using ABAQUS; crack growth from brittle to ductile material and crack growth form ductile to brittle material. A two-dimensional finite element mesh (Fig. 5) was generated using PATRAN (4). The model consisted of 5026 nodes comprising 1650 elements. On the basis of post mortem identification of inclined fracture surfaces, plane stress conditions were determined to be dominant. Therefore, eight noded plane stress elements, CPS8, were used in the mesh. Quadraic interface elements, INTER3, were used to connect opposite sides of the crack face prior to crack advance along the projected growth path. The mesh was finer near the original notch, the precrack and in the adjacent area ahead of the precrack into which it propagated. Elements adjacent to and in the path of the crack plane were 0.077 times the original crack length. The full specimen was modeled to account for the absence of material symmetry on either side of the crack plane.

The notch and pre-crack of the specimen were modeled as an initial crack of total length 7.88 mm (0.20 in). This geometric simplification was made on the basis of previous studies (5) which showed no significant effect at the crack tip. Interface elements of type INTER3 were used to bond the nodes of elements ahead of the crack tip together for a distance of 11.8 mm (0.3 in) ahead of the crack tip. This defined the path to be followed by the crack. Geometric nonlinearity was incorporated in the solution method to account for large strains and rotations.

Material parameters were assigned to each element from the set of parameters determined for each hardness value between 80 and 110 on the Rockwell B scale. A computer program was written to determine the hardness value for each individual finite element mesh. This was accomplished by interpolating over the measured hardness values, obtained from the edge notch specimens prior to testing, to each finite element centroid.

One end of the specimen was fixed with zero displacements corresponding to the fixed grip of the test machine. Increasing displacements were applied on the other end consistent with the moving grip of the test machine.

Crack growth was governed by a debonding algorithm which implemented a local fracture criterion to trigger nodal release between the elements immediately ahead of the current crack. The crack path was specified in the finite element model by interface elements. A user subroutine was used to specify a debond parameter associated with the crack tip. For this investigation, the debond parameter was defined as the difference between the strain energy density, w, at the crack tip and a critical value, w_c, of the strain energy density. The value of w_c at the current crack tip node was taken as the average of the critical w_c values associated with the material parameters in the four elements surrounding the current crack tip. The energy density at the current crack tip was evaluated from the integration point values in the surrounding elements at each solution increment.

When the value of w equaled or exceeded the value of w_c debonding occurs. Nodal forces are released in the element ahead of the crack tip. The load increment for the debonding process was scaled by the proportion of the absorbed energy increment required to reach the critical value. The algorithm allowed for the possibility of more than one element of crack growth at constant load.

DISCUSSION OF RESULTS

The objective of the finite element analyses was to assess the ability of the computations to predict the specimen responses and discriminate between the two microstructural gradients over a significant amount of crack growth.

The brittle to ductile crack growth case had a total gage displacement of 0.484 mm $(1.91 \times 10^{-2}$ in). Twelve elements debonded during the analysis, increasing the crack length from 5.08 mm (0.20 in) to 10.2 mm (0.40 in). The load versus gage displacement is shown in Fig. 6a. The magnitude of the reaction force is comparable to the experiment. The maximum force was reached and proceeded to decrease as observed in the experimental data. The crack length versus gage displacement for the brittle to ductile crack growth case is shown in Fig. 7a. The crack initiation is predicted somewhat prematurely by the analysis. The growth rate is very similar to the experimental post-initiation crack growth rate.

The ductile to brittle crack growth case had a total gage displacement of 0.924 mm $(3.64 \times 10^{-2}$ in). Fifteen elements debonded during this analysis. The crack length increased from 5.08 mm (0.20 in) to 12.7 mm (0.50 in). The load versus gage displacement is shown in Fig. 6b. The magnitude of the reaction force is slightly greater than observed in the experiment. The predicted force continued to slowly increase, however, as was observed in the experiment. The crack length versus gage displacement for the ductile to brittle crack growth case is shown in Fig. 7b. The crack initiation is predicted somewhat prematurely. The growth rate is again very similar to the experimental post-initiation crack growth rate for the case considered.

Factors which influence the accuracy of these results include the accuracy of the constitutive parameters and the accuracy of the w_c values for the material at each hardness. Additional modeling of the tensile specimen response can be used to extract more accurate stress and strain and fracture parameters (6). Furthermore, the use of a constant w_c value for each material hardness value does not account for observed strain history dependence of w_c (7). Differences between w_c for material in the immediately vicinity of the initial crack tip and material further ahead of the crack which fractures as part of subsequent crack propagation can be expected due to the differences in strain history. Parametric studies have shown the computationally determined crack growth rate to be sensitive to these differences (8). Both of these effects are currently under investigation.

CONCLUSIONS

An experimental and computational study of crack growth in edge notched 1045 steel specimens with a quenched induced microstructural gradient has been performed. The effect of this gradient on crack growth emanating from the notch placed on the brittle side of the panel versus that of the notch on the ductile side of the panel is clearly demonstrated from the experiments.

Computational simulations of both cases were performed using a local fracture criterion and a debonding algorithm. Comparison of both computational analyses shows that the essential features between reaction force versus displacement histories and crack growth rates are predicted. Crack growth initiation is predicted somewhat prematurely, while subsequent crack growth rates are accurately predicted. The application of these methods to the geometrically more complex, but conceptually identical, problems involving welding, casting and heat treating is possible on the basis of these results.

REFERENCES

1. Hibbitt, Karlsson and Sorensen, Inc., ABAQUS User's Manual, Providence, RI, 1989.

2. Hibbitt, Karlsson and Sorensen, Inc., ABAQUS Theory Manual, Providence, RI, 1989.

3. Wolfram, S., Mathematica: A system for doing mathematics, Addison-Wesley, New York, 1988.

4. PDA Engineering, PATRAN User's Manual, Santa Ana, CA, 1987.

5. DeGiorgi, V. G., Kirby, G. C. III and Jolles, M. I., Prediction of Classical Fracture Initiation Toughness, Engineering Fracture Mechanics, Vol. 33, 1989, pp. 773-785.

6. Matic, P., Kirby, G. C. III and Jolles, M. I., The relation of specimen size and geometry effects to unique constitutive parameters for ductile materials, Proc. Royal Soc. London A 417, pp 309-333, 1988.

7. Matic, P., Kirby, G. C. III, Jolles, M. I. and Father, P. R., Ductile alloy constitutive response by correlation of iterative finite element simulation with laboratory video images, Engineering Fracture Mechanics, Vol. 40, pp 395-419, 1991.

8. DeGiorgi, V. G. and Matic, P., A computational investigation of local material strength and toughness on crack growth, Engineering Fracture Mechanics, Vol. 37, pp 1039-1058, 1990.

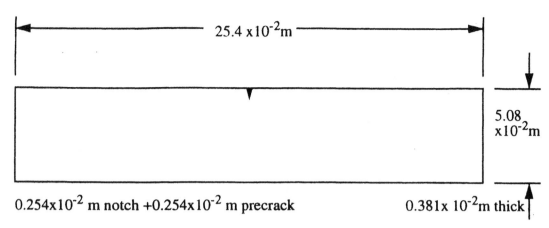

Figure 1 Single edge notched panel geometry.

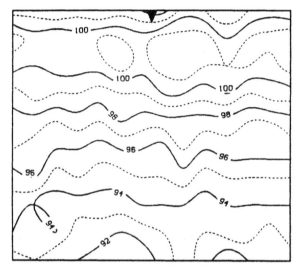

Figure 2 Typical Rockwell B hardness map for case of edge notch on brittle side of bar.

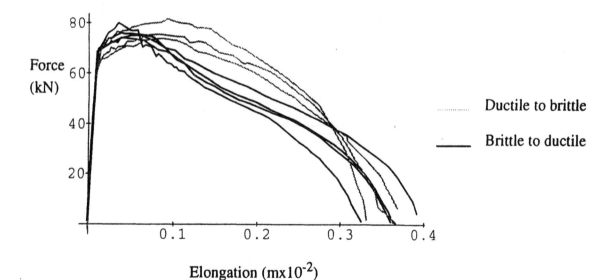

Elongation (mx10^{-2})

Figure 3 Comparison of reaction force (kN) versus specimen elongation (mx10^{-2}) for crack growth from ductile to brittle side of panel and brittle to ductile side of panel.

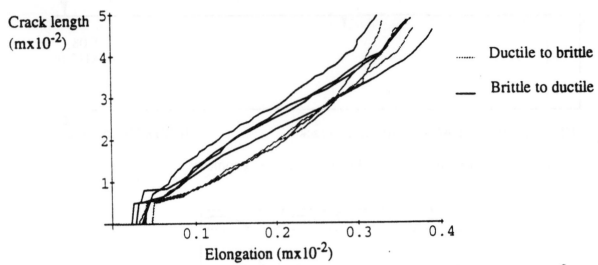

Figure 4 Comparison of crack length (mx10^{-2}) versus specimen elongation (mx10^{-2}) for crack growth from ductile to brittle side of panel and brittle to ductile side of panel.

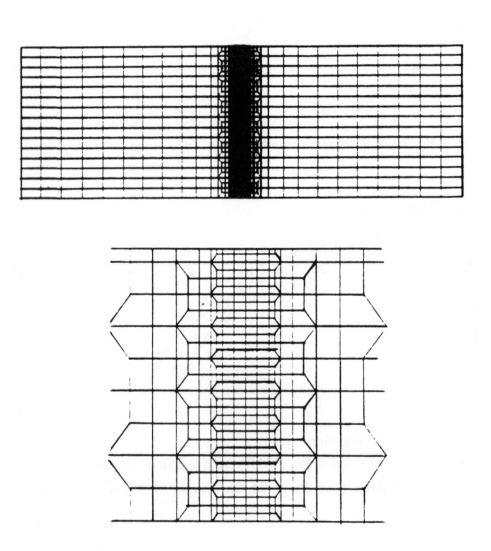

Figure 5 Finite element mesh for edge notched panel.

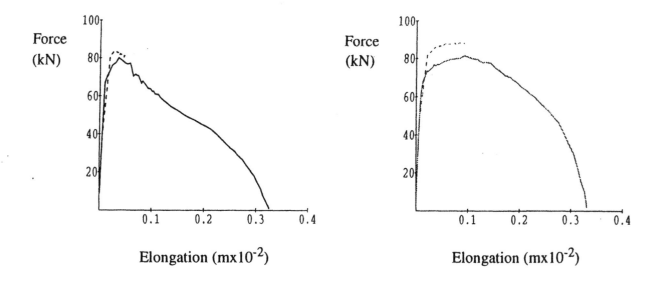

Figure 6 Predicted and measured reaction force versus displacement for crack
growth from (a) brittle to ductile and (b) ductile to brittle material.

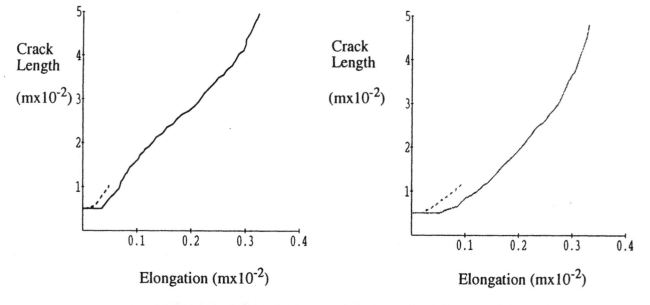

Figure 7 Predicted and measured crack length versus displacement for crack
growth from (a) brittle to ductile and (b) ductile to brittle material.

TWI

PAPER 10

The deformation and failure at −30 °C of cracked broad flange beams of ordinary structural steel in bending

K Eriksson, PhD (University of Luleå), F Nilsson, Prof, PhD and H Öberg, MSc (Royal Institute of Technology)

SUMMARY

The deformation and failure in bending of precracked broad flange beams of type HEB 400 and similar was studied. 6 m long beam elements were slowly loaded to fracture or plastic deformation at the testing temperature -30°C.

The J contour integral for the full scale beam element was calculated with the finite element method for two typical crack lengths. J as a function of crack length was approximated with the R6-method using analytical estimates of the fully plastic bending moment for thin-walled beam cross-sections.

Critical values of J and CTOD were obtained with standard Compact and SENB specimens machined from the beam flanges. Although scatter is considerable the fracture toughness ranges of the laboratory specimens and of the full scale tests are very much in coincidence.

By means of a residual strength diagram the borderline between "long" and "short" cracks is illustrated. Here "long" cracks are those which caused fracture and "short" cracks just plastic deformation but no crack growth.

INTRODUCTION

Reliable fracture mechanics assessments of large scale structures requires transferability of fracture properties obtained with laboratory specimens. This kind of transferability has more or less been taken for granted because toughness data obtained with not too different types of specimens correlate reasonably.

The original objective of the present work was primarily to perform testing of full scale structural elements and simply observe whether failure was brittle or ductile for a given crack, load and temperature corresponding to the worst case estimated in practice. These results are used for practical assessments of railway bridges in service.

A second objective emerged from the primary results, namely to study the transferability of fracture data obtained with laboratory specimens. In the present paper emphasis is put on the latter.

Because of the history of the project, the materials studied are very inhomogeneous. Scatter in mechanical properties is considerable and the materials are not ideally suited for a scientific investigation. Some of the materials are taken from structures that have been in service for up to 70 years. On the other hand many structures today are made of steels that are not manufactured to current standards and whose properties are virtually unknown. This is a rather typical situation for practical assessment work. Further it is found that the fracture toughness scatter of recently produced beam steel is not significantly less than that of older steels. It is therefore hoped that the present work can to some extent contribute to the handling of safty considerations in general.

MATERIAL

The present investigation comprises five wide flange rolled beams, one HEB 400 and four DIP 42 1/2. The cross section dimensions are given in Table I. The first beam, No 1, is recently produced, of steel St37 Class D, and used as a reference. The remainder, No:s 2-5, are taken from a railway bridge that was in service 1919-1987. Of the latter, one beam, No 3, was replaced in 1962. Thus, three beams has been in service for some 70 years, one for 25 years and one not.

Conventional material properties

Tensile and notch toughness testing results are shown in Table II. The tensile testing results are typical for the class of material considered although the ratio between upper and lower yield strength is above average in some cases.

The notch toughness figures given are mean values of three tests. The beam No 1 steel satisfies the class D requirement (minimum 27J at -20°C) very well. The notch toughness of the remaining beams is very low, or some 20% of the required minimum.

Chemical composition

The chemical composition of the beam steels is given in Table III. All steels are carbon steels with low carbon content and low to normal content of residuals and impurities, notably P and S. The nitrogen content of steel No 3 is however somewhat above normal. The steels No 1 and No 3 are silicon killed while the others are rimmed.

Killed steels are in general more homogeneous and have better notch toughness. This is true for steel No 1 but not for steel No 3. The inhomogeneity of the latter has been described elsewhere (1).

EXPERIMENTAL

6 m long beam elements were precracked in a flange with an edge crack. The cracks were machined and fatigued at room temperature. The fatigued part was at least 15 mm and the stress intensity range did not exceed 30 MPa$\sqrt{m}$. In order to achieve symmetry the opposite flange was notched to the same depth. Fatigue crack growth from the notch was successfully suppressed with a hole drilled at the tip of the notch. The crack lengths are given in Table II.

The beams were cooled and held at the testing temperature -30°C for at least one hour prior to testing and then slowly loaded in four point bending to fracture or just plastic deformation. The crack was on the tensile side halfway between the inner loading points of the beam. Loading was repeatedly relaxed during the test to enable stable crack growth to be detected. In addition to the load and load point displacement, some fifteen strains, displacements, temperatures, crack mouth opening displacements, etc were recorded for each beam.

As well as the full scale tests a series of standard Compact and Three point bend specimens were tested in accordance with ASTM E813 (2). The aim of these tests was to obtain fracture toughness (J_c) results for comparison with those of the large specimens. The standard specimens were machined from the flanges of the large specimens. The plane of the crack was always perpendicular to the rolling direction of the parent material and the tip of the crack at the same distance from the flange edge as in the large specimens. The nominal specimen thickness B_b was equal to the full flange thickness t. A subset of the standard specimens were side-grooved. Specimen dimensions are given in Table IV.

Experimental observations

Beams No:s 1, 2 and 4 failed from a fracture that was entirely brittle from an engineering point of view. Beam No 3 begun to skew under loading enough to render further loading meaningless. After loading to a J = 41 kN/m at which no fracture occurred the beam was unloaded. In beam No 5 fracture did not occur during the first loading. The crack length was increased somewhat through additional fatiguing. Neither did any fracture occur during the second loading. A longer crack than before was then made in the previously compressed flange. The beam was turned upside down and during the third loading fracture occurred at a lower load than before.

No indication of stable crack growth was found, neither from compliance calculations nor from direct observations of the fracture surfaces. These were mostly perpendicular to the flange plane and showed Chevron markings but no shear lips. The load displacement relationships (for the entire beams) were linear up to the point of failure. The load-CMOD relationships were however slightly curved prior to failure. Also strain close to the crack tip was non-linear. The final fracture was preceeded by plastic deformation just enough to invalidate LEFM analysis. Notably the thickness criterion was never fulfilled.

FINITE ELEMENT MODELLING

The beams were modelled with aid of the FEM-code ABAQUS (3). The elements used in the model were plate elements according to the thin plate approximation with five degrees of freedom per node. Such elements can sustain membrane forces and transverse bending moments but the influence of shear force on the deflection is neglected. Because of double symmetry, only a quarter of the beam was modelled. The beam end boundary conditions were chosen to comply with pure bending deformation.

In total the model comprised 157 elements and 685 nodes. Around the crack tip so called quarter-point elements were used. The material model was elasto-plastic and obeyed the von Mises yield criterion and its associated flow rule. The hardening was chosen according to tensile test data at the testing temperature (-30°C). The J-integral was calculated with the built-in routines of ABAQUS for the crack lengths a = 38 and 70 mm.

For calculation of the nominal bending stress σ_0 and the limit bending moment M_f, engineering type beam theory has been used. The stress σ_0 has been calculated without consideration of the crack, as customary, while the presence of the crack has been taken into account in the determination of M_f.

Purely elastic calculation results differ less than 8% (5% for the shorter crack) from those of the double edge cracked strip in tension (cf. Tada *et al* (4)). The minor differences that are observed could well arise from the influence of the web.

The results of the non-linear calculations are shown in Fig. 1. The two cases are fairly close. For comparison the function g_{R6} (see below) is also shown.

DETERMINATION OF THE J-INTEGRAL

In the present investigation J-estimates were required for beam geometries that differed only slightly from each other. In order to handle such cases we write

$$J_e = K_I^2/E$$

$$J = J_e \, g(L_r)^{-2}$$

[1]

where K_I is the stress intensity factor and E Young's modulus. J_e is the elastic value of the J-integral for a given nominal stress σ_0 and g a correction function depending on

$$L_r = M/M_f$$

[2]

Here M is the bending moment and M_f a measure of the limit bending moment of a cracked beam cross section assuming the material to be perfectly plastic with a yield stress σ_y. The geometry dependence of the function g should be rather weak for small changes of geometry. Changes in crack length are accounted for by the effect on K_I and M_f only.

The R6-method (5) contains a suggestion for g that is intended to be conservative for all geometries

$$g_{R6}(L_r) = (1-0.14L_r^2)(0.3+0.7\exp(-0.65L_r^6))$$

[3]

The applicability of this relation in the interval considered is no doubt acceptable, Fig. 1.

When evaluating laboratory specimen experiments other methods for J-determination have been used. The J-calculation can be written in the form

$$J = \frac{K_I^2 (1-v^2)}{E} + \frac{\eta}{B_n(W-a_0)} \int P d\Delta_p \qquad [4]$$

Here B_n is the net thickness of the specimen, a_0 the initial crack length, Δ_p the plastic part of the displacement and η a dimensionless constant depending on geometry.

The limits to the possible validity of a critical value of J is usually stated in the form

$$l \, \sigma_y/J > \alpha_1 \quad \text{and} \quad t \, \sigma_y/J > \alpha_2 \qquad [5]$$

l is the smallest in-plane dimension of the body and t the thickness. α_1 and α_2 are dimensionless constants. In ASTM E813 (2) α_2 is taken as 25 and for bend specimens α_1 is mostly assumed to be the same. For tensile geometries Shih *et al* (6) suggested that a value of 200 for α_1 is more relevant. The corresponding validity limit for the K_I-criterion is according to (7)

$$t \, (\sigma_y/K_I)^2 > 2.5 \quad , \text{or, using} \quad J = K_I^2 \, (1-v^2)/E$$

$$[6]$$

$$t \, \sigma_y/J > \alpha_3 = 2.5 \, E/(\sigma_y \, (1-v^2) \approx 1900$$

The numerical value for α_3 was obtained by using typical data for the materials in the present investigation.

EVALUATION OF EXPERIMENTS

The critical values of J for the different full-scale tests were calculated from [1] based on the load at crack initiation P_c estimated directly from the load-displacement record and the crack length measured on the fracture surface. The results of the different full-scale tests are summarized in Fig. 2. The thickness requirement for J-characterization, [5], is satisfied for all full-scale tests. The corresponding ratio for the in-plane dimension is typically around three times larger. The suggested validity limit of [6] is thus satisfied.

The critical values of J obtained with laboratory specimens taken from the full-scale tests are given in Table IV and summarized in Fig. 3. The J-values were evaluated according to [4]. The thickness requirement for J-testing is satisfied with a wide margin for steels No:s 2, 4 and 5, with a few exceptions for steel No 1 and by 50% of the tests for steel No 3. For the first steels some tests are in fact not far from being valid according to the plane strain fracture toughness testing norm ASTM E399 (7). In a strict sense the results are not valid according to ASTM E813 since no stable crack growth occurred. The critical values of J are thus designed J_c and not J_{Ic}.

For comparison CTOD according to (8) has been determined. The results are given in Table IV.

DISCUSSION

The beams tested can be divided into two groups, one brittle and one ductile. Brittle and ductile are here used in their engineering sense. A brittle steel is at a temperature below its brittle-to-ductile transition temperature and a ductile material above it. In this sense beams No:s 1 and 3 are ductile and beams No:s 2, 4 and 5 brittle at the testing temperature.

Both ductile beams were made of killed steel but, most likely because of a different manufacturing technique, beam No 1 is more homogeneous than beam No 3. All brittle beams were made of rimmed steel.

The fracture toughness of the two groups are of a different order of magnitude. The fracture toughness of the ductile steels is typically greater than 100 kN/m while that of the brittle materials may be as low as 10 kN/m. No evidence of stable crack growth before fracture was found, not even for the ductile materials and certainly not for the brittle ones. The maximum fracture toughness obtained for a full scale ductile beam was 93 kN/m.

From each beam some ten laboratory fracture toughness testing specimens were made. No characteristic effect of specimen type or size was found for plane specimens. In the laboratory specimens taken from the ductile beams some amount of stable crack growth before fracture has been observed but not in specimens from brittle beams, Table II.

The scatter in fracture toughness is considerable (COV around 0.5) and almost of the same order for both groups. Only for the most brittle steel is scatter below average (COV around 0.25).

The fracture toughness of the large-scale tests was normaly within the scatterband of the corresponding laboratory specimens. In two cases the fracture toughness of the full scale beam is close to the lower end of the laboratory specimen scatterband. The crack size is of the same order and the crack tip is positioned at an identical material fibre in both cases. There is thus no particular reason for different scatter in either case.

The fracture toughness of beam No 2 is much higher than of the corresponding laboratory specimens. Considered in isolation this is a very peculiar result, but in view of all results and assuming that scatter is about the same in full-scale and laboratory specimens we might argue that the number of small specimens was insufficient to produce one single result embracing the fracture toughness of the beam.

At present it is not possible to decide whether the behaviour of beam No 5 is an effect of scatter or crack size. In the first loading the fracture zone might have been located in tougher material but absence of fracture in the second loading requires an unlikely high toughness for this beam. In the third loading the fracture zone encountered more typical material.

The effect of side-grooving is quite different in brittle and ductile materials. Further the effects in ductile homogeneous and inhomogeneous materials are different. In the brittle materials side-grooving did not give different fracture toughness or less scatter than plane specimens.

In (1) it was found that the notch toughness varied considerably across the flange thickness in beam No 3. Near the surface the notch toughness was typically 25-85 J but only some 7-8 J in the core material. This is rather typical for older structural steel. For common thicknesses a Charpy specimen which is machined on all sides represents the core material more than the surface material and may give very low notch toughness while full thickness fracture toughness specimens indicate a much higher effective toughness. Because of these findings, it was decided to use full thickness laboratory specimens only in this investigation.

In ductile homogeneous materials there is a smaller amount of stable crack growth and perhaps a weak concentration towards the lower part of the fracture toughness scatterband for side-grooved specimens. In ductile inhomogeneous materials we have broadly speaking ductile surface parts and a brittle core. The ductile surface material is cut off by the side-grooves and the effective toughness is thus determined by the brittle core material. Also plane laboratory specimens from beam No 3 showed a small amount of stable crack growth but the side-grooved specimens did not.

It can be seen in Fig. 2 that fracture did not occur for shorter cracks although a beam was loaded to a J that exceeded J_c for a longer crack in a fairly similar material. In Fig. 4 these results are replotted in a residual strength diagram. Here the data points are given in terms of bending strength versus crack length. Of the three tests in which no fracture occurred two were still within the nominal elastic region but close to the elastic plastic border and one in the elastic-plastic region. For all beams in which fracture occurred the crack length was greater than 40 mm. Compare the results of beams No:s 1 and 5 (loading 2) or No:s 2 and 3 with fairly equal loading respectively for which fracture occurred only for the longer crack.

CONCLUSIONS

1. Fracture toughness obtained with full-thickness laboratory specimens can be used to predict the behaviour of a full-scale structure.

2. The fracture toughness scatter of both laboratory specimens and of full-scale structures is considerable. The scatterbands are however very much coinciding if full-thickness plane specimens are used.

3. Although some of the structural steels investigated were brittle, fracture did not occur in full-scale testing for edge cracks shorter than 40 mm.

ACKNOWLEDGEMENT

This work was funded by the Swedish National Rail Administration. The authors want to express their gratitude for this support.

REFERENCES

(1) K. Eriksson Charpy notch toughness and fracture toughness of inhomogeneous structural steel. To be published.
(2) *Standard test method for J_{Ic}, a measure of fracture toughness, ASTM E-813* (1986) General Book of ASTM Standards, Sect. 3 American Society for Testing and Materials, Philadelphia, 1986.
(3) *ABAQUS, User's manual*, version 4.8 (1990) Hibbit, Karlsson and Sorensen Inc., Providence, R.I.
(4) H. Tada, P. Paris and G.R. Irwin (1973) *The stress analysis of cracks handbook*. Del Research Corporation, Hellertown, Pa.
(5) I. Milne, R.A. Ainsworth, A.R. Dowling and A.T. Stewart (1986) *Assessment of the integrity of structures containing defects*. CEGB Report R/H/R6-Rev. 3, Central Electricity Generating Board.
(6) C.F. Shih and M.D. German (1981) Requirements for a one parameter characterization of crack tip fields by the HRR singularity. *Int J of Fract*, **17**, 27-43.
(7) *Standard test method for plane-strain fracture toughness of metallic materials, ASTM E-399*, General Book of ASTM Standards, Sect. 3 American Society for Testing and Materials, Philadelphia, 1986.
(8) BS 5762. Methods for crack opening displacement (COD) testing. British Standard Institution London, 1979.

Table I.   Beam cross-section dimensions.

Beam No	Type	Height H (mm)	Width W (mm)	Web thickness d (mm)	Flange thickness t (mm)
1	HEB400	400	300	13.5	24
2-5	DIP 42 1/2	425	300	14	26

Table II.   Evaluation of full scale beam experiments, testing temp -30°C.

Beam No			1	2	3	4	5		
Year			1989	1919	1962	1919	1919		
Desoxidation			killed	rimmed	killed	rimmed	rimmed		
R_{eL}		(MPa)	295	297	300	278	290		
R_{eH}		(MPa)	329	332	359	330	310		
R_m		(MPa)	470	405	480	412	446		
A_5		(%)	33	33	33	33	30		
KV	(J)	-20°C	62	4.1	8.2	4.3	7.4		
a		(mm)	70	70	38	92	36	41	93
J_e		(kN/mm)	51	28	33	20	29	57	9
J_c		(kN/mm)	93	35	>41	23	>50	>100	>12
Failure			brittle	brittle	-	brittle	-	-	brittle
$\sigma_0 = M/W$		(MPa)	220	152	218	110	231	263	80
M/M_f			.66	.45	.65	.35	.71	.81	.24
$t\, R_{eL}/J_c$			76	220	-	314	-	-	377

Table III.   Chemical composition (weight %)

Beam No	C	Si	Mn	P	S	Cr	Ni	Cu	N
1	.090	.29	.65	.021	.040	.031	.04	.11	.008
2	.034	<.01	.58	.030	.018	.013	.04	.009	.009
3	.102	.404	.40	.025	.014	.006	.03	.037	.013
4	.054	<.01	.73	.030	.016	.016	.04	.030	.008
5	.042	<.01	.71	.027	.014	.018	.04	.029	.010

Table IV.   Critical J-values from laboratory specimens

Beam No	Specimen type	W mm	B_b m m	B_n mm	Δa mm	J_c kN/m	CTOD mm	$\dfrac{B_n \sigma_y}{J}$
	CT	50	22.5	-		64		104
	CT	50	23.4	18.5	0	125		44
	CT	50	22.0	-	0.96	375		17
	CT	50	22.0	18.0	0	115		46
	CT	140	23.4	18.5	0.30	175		31
	CT	140	23.4	18.5	0.60	295		19
1	SENB	40	20.0	-	0	100	0.18	59
	SENB	40	20.0	16.0	0	121	0.20	39
	SENB	40	20.0	16.0	0	165	0.28	29
	SENB	50	22.5	-	0.50	356	0.65	19
	SENB	50	22.5	-	0.33	275	0.51	24
	SENB	50	22.5	-	0.22	192	0.40	35
	SENB	50	22.5	18.0	0.12	160	0.29	33
	SENB	50	22.5	18.0	0.17	200	0.34	27

Table IV   Cont.

	CT	50	22.9	-	0	12.0		567
	CT	50	23.0	-	0	9.6		712
	CT	50	23.0	-	0	9.7		704
	CT	50	23.1	-	0	9.6		715
2	CT	50	23.0	18.3	0	13.0		418
	CT	50	23.0	18.3	0	17.6		309
	CT	140	23.0	-	0	14.8		462
	SENB	50	23.0	-	0	10.6	0.021	644
	SENB	50	23.0	-	0	15.2	0.035	404
	CT	50	23.1	-	0.61	450		15
	CT	50	22.0	-	0.18	316		21
	CT	50	18.2	-	0	119		46
	CT	50	18.4	-	0.70	500		11
	CT	50	23.1	-	0.22	300		23
	CT	50	23.1	-	0.10	135		51
3	CT	50	23.1	-	0	110		63
	CT	50	23.1	18.0	0	45		120
	CT	50	23.0	18.4	0	33		167
	CT	50	23.1	18.3	0	43		128
	CT	140	23.0	-	0.70	≈ 500		14
	CT	140	23.0	-	0.31	≥ 270		26
	SENB	50	23.1	-	0.87	560	1.03	12
	SENB	50	23.0	-	0.92	360	0.75	19
	CT	50	23.0	-	0	31.0		206
	CT	50	23.0	-	0	18.5		346
	CT	50	23.0	-	0	15.0		426
	CT	50	23.0	-	0	15.2		420
	CT	50	23.0	18.0	0	79.0		63
4	CT	50	23.0	18.0	0	59.0		85
	CT	140	23.0	-	0	29.5		217
	CT	140	23.0	-	0	34.5		185
	SENB	50	23.0	-	0	30.0	0.07	213
	SENB	48.6	22.0	-	0	90.0	0.20	68
	SENB	48.7	22.0	-	0	11.0	0.03	556
	CT	140	23.0	-	0	59		114
	CT	140	23.0	-	0	34		196
	CT	50	22.0	-	0	30		213
	CT	50	22.0	-	0	22		290
5	CT	50	22.0	18.0	0	77		68
	CT	50	22.0	18.0	0	33		158
	SENB	50	22.5	-	0	61	0.18	107
	SENB	48.9	22.0	-	0	94	0.21	68
	SENB	50	22.0	18.0	0	30	0.06	174
	SENB	50	22.0	18.0	0	20	0.05	261

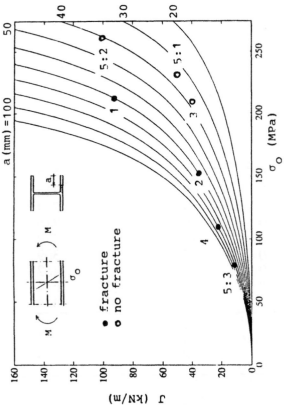

Fig 1. FEM calculation of the function g for a broad flange beam

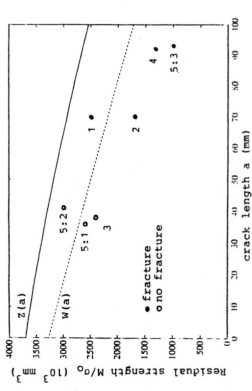

Fig 2. J versus σ_o for beam DIP 42 ½ and experimental results

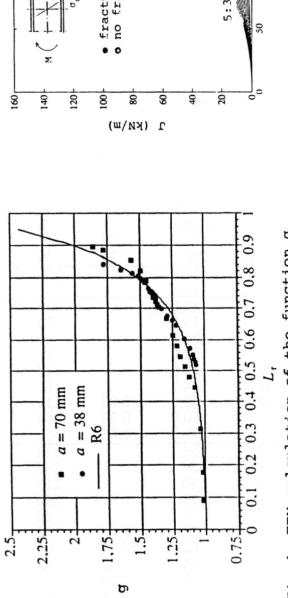

x   CT, SENB specimen
*   side-grooved
●   beam fracture
○   beam, no fracture

x/y σ_o/σ_y and crack length

a) ductile materials

b) brittle materials

Fig 3. Summary of fracture toughness testing, full-scale beams and laboratory specimens

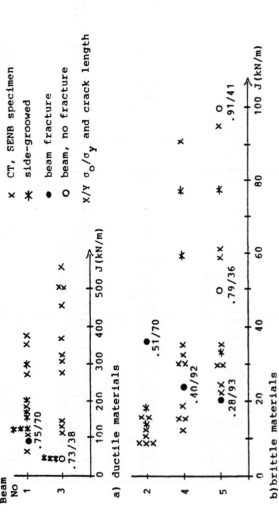

Fig 4. Residual strength versus crack length
W(a) elastic section modulus
Z(a) plastic section modulus

TWI

PAPER 32

Experimental and analytical investigations into the mechanical behaviour of pipe with short through-wall cracks

Ph Gilles, CEng (Framatome, France), D Moulin, CEng (CEA-SACLAY, France) and D Gouteron, CEng (Framatome, France)

SUMMARY

The paper aims to reach a better understanding of the mechanical behaviour of through-wall cracked pipes. Since few results are available for pipes with short through-wall cracks, a 100 mm diameter pipe containing a circumferential crack whose total length is about 15 degrees, was investigated. The material is a very ductile austenitic steel and the pipe is subjected to pure bending.

The analysis combines two parallel investigations : a carefully instrumented four-point bending test has been conducted followed by a detailed Finite Element computation of the test. The purpose of such an approach is to obtain a good simulation of the test, thus allowing the variations of the fracture parameters to be computed during the loading of the pipe. This is of particular importance for checking the validity of the J values derived from the experimental Moment-Pipe-Rotation Curve and to compare them to those obtained on small-scale specimens. Comparisons of experimental and computational values are made on the crack opening displacements, the deformation of the pipe section, the pipe rotations close to the cracked section as well as at the end of the pipe.

Emphasis is laid on a short crack configuration for the following reasons. Few Finite Element results are available for through-wall cracks whose total angle is less than 30 degrees. Previous investigations on limit moment for such pipes have evidenced a different mechanical behaviour than for large crack pipes. Furthermore, analytical studies and Finite Element computations conducted for two-dimensional crack problems have shown that the strain hardening influence on J values is very dissimilar depending on whether the crack length is small or not compared with the remaining ligament. In order to check the validity of simplified methods the J computations are compared with the tabulated Finite Element results of the GE-EPRI handbook.

INTRODUCTION

The peculiar behaviour of short through-wall cracked pipes loaded in pure bending is investigated by three methods. First, an experimental program of 30 tests performed on unpressurised pipes, made of the same austenitic steel is presented. The results obtained on pipes with large cracks are compared with those relative to pipes with no crack or a crack less than 45 degrees of total angle. For the 15 degree pipe, the moment and the driving force J variations with the rotation are reported. Next, a detailed Finite Element computational analysis of this pipe permits comparison with the experimental results and provides additional information on the pipe behaviour. Finally, the differences between the J-load curves derived experimentally, by the Finite Element analysis and by the GE-EPRI simplified method are discussed.

AIM OF THE EXPERIMENTS

The main purpose of the tests is to determine the conditions of crack initiation and crack propagation in austenitic steel pipes subjected to bending.

Larger size tests are used to obtain data on pipes representative of actual industrial configurations. Smaller size tests are used to obtain numerous and detailed results in order to check analytical methods more extensively.

The effect of various parameters is documented : crack length, crack tip radius, tube geometry, temperature and type of weldment.
During these tests benefit was taken from experimental results previously obtained by several investigators : Zahoor (1), and Wilkowski (2) in the U.S.A., Bruckner (3) in Germany and Shibata (4) in Japan.

DESCRIPTION OF THE TESTS

The experiments are conducted at room temperature. The pipes are unpressurised and loaded for four point bending. The pipes have an initial through-wall circumferential crack located in the tensile part of the middle cross section.
Table 1 gives the test matrix for the 30 experiments performed at the CEA-SACLAY. Detailed informations on the experiments can be found in papers of MOULIN (5, 6, 7).

Figure 1 gives a schematic presentation of the experimental device and the specimen dimensions. The displacement controlled loading is increased at a constant rate (quasistatic loading). The instrumentation allows a continuous record of the following parameters :
- applied load,
- load line displacement,
- crack mouth opening,
- d.c. potential drop for characterisation of the crack initiation and propagation
- total rotation of the pipe (i.e. rotation between two symmetrically located cross sections) near to and far from the cracked section, respectively named PHI2 and PHI1,
- pipe diameter (horizontal and vertical) close to the cracked section.

During the tests several unloadings are performed to mark different crack fronts during the propagation. At the end of the experiment, the diameters are measured, to reveal the terminal ovality of the pipe. The crack length is measured after breaking the residual cracked section. All the pipes considered in this paper are fatigue precracked. The final crack is almost symmetrical with respect to the loading plane.
The material of each pipe has been characterised. Tensile tests were performed on cylindrical specimens. J-Resistance curves are drawn from CT specimens. The specimens are extracted from the tubes used in the pipe tests. For instance CT specimens of reduced thickness (B = 8 mm) are obtained from D = 105,36 mm pipes. The material characteristics obtained are reported in Table 1. Since the difference in tensile properties reported is less than 10%, the table is limited to the short crack pipes. Young's modulus values are quite low for this type of steel.

The dimensions of the CEA 27 pipe, analysed in this paper, are 104.80 mm for the external diameter, 8.315 mm for the thickness and 14.3 degree for the final crack size.

TABLE 1. Tests matrix of austenitic pipes

Test Identifier	Total crack angle[*] degree	Yield stress MPa	Ultimate stress MPa	Young Modulus MPa
CEA 9	0	231	537	145700
CEA 27	15	252	517	139600
CEA 12	30	228	529	157500
CEA 23	45	249	547	159300
CEA 22	60	230	527	141600

[*] Measured on outside diameter

EXPERIMENTAL RESULTS

Global behaviour of the cracked pipes

The basic result is in the moment-rotation curve, the rotation being measured in two pipe sections (see PHI2 and PHI1 locations on Fig.1). Figure 2 shows such a curve for the CEA 27 experiment : the moment is still increasing at the maximum rotation. No crack initiation was observed. Of course, the rotation is larger when measured far from the cracked section (PHI1), but the difference between the two curves becomes almost constant for large rotations, which reveals that the non linearity of the pipe behavior is mainly due to yielding close to the cracked section. It has been verified that the PHI1 over PHI2 ratio is constant.

Fracture analysis of the experimental results

The variations of the crack driving force J under the prescribed rotation, are estimated by a single specimen method. This method based on the scale function approach, has been close-checked and presented by MOULIN (5). It relies on the following equation :

$$J = - \frac{1}{Rt} \frac{h'(\theta)}{h(\theta)} \int M \, d\,PHI \qquad [1]$$

where - R is the pipe mean radius
    - t is the wall thickness
    - θ is the total angle defining the through-wall crack
    - $\dfrac{h'(\theta)}{h(\theta)}$ is the scale function, obtained by a static limit analysis of the cracked tube.

$$\frac{h'(\theta)}{h(\theta)} = -\frac{1}{4} \cdot \frac{\sin\frac{\theta}{4} + \cos\frac{\theta}{2}}{\cos\frac{\theta}{4} - \frac{1}{2}\sin\frac{\theta}{2}} \qquad [2]$$

Figure 3 shows, for all the fatigue precracked pipes, the variation of J as a function of the reference stress S_R derived by limit analysis

$$S_R = \frac{M_{exp}}{4 R^2 t \left(\cos\frac{\theta}{4} - \frac{1}{2}\sin\frac{\theta}{2}\right)} \qquad [3]$$

That kind of representation allows to compare the respective merits of the J-based fracture criterion and of the net section collapse criterion.

One can see that for $\theta > 30$ degrees, all the curves come close. For these curves there exists a limit stress, almost independant of θ, whereas J changes are quite large when nearing the maximum bending moment : the curve is nearly vertical. This means that the limit load criterion is not sensitive to the crack length for ductile pipes with a large through-wall crack. A large scatter is observed on J values at the initiation of the ductile tearing. The values corresponding to small cracks seem to be the highest ones. The initiation being defined by the change of slope of the electric potential drop - pipe rotation curve, the range of the corresponding J value obtained in the CEA tests is : 1000 - 2500 kJ/m^2.

Figure 4 gives the variation of the bending moment with the current crack angle θ, accounting for the propagation, in two tests performed on large crack pipes. The limit moment variation, drawn on this figure is obtained through the formula :

$$M_L = 4 R^2 t \left[\cos\frac{\theta}{4} - \frac{1}{2}\sin\frac{\theta}{2}\right] \cdot \sigma_f \qquad [4]$$

where σ_f is the flow stress defined by the average of the yield and ultimate stresses. After the maximum, the two curves and the theoretical curve become close. Thus during the tearing process, the limit stress is almost constant and equal to the flow stress. For ductile pipes with large through-wall cracks under bending, the tearing crack propagation may be successfully modeled using the net section collapse criterion, provided that the crack doesn't initiate too soon before the maximum moment (i.e the J_{IC} value is not too low).

For pipes with cracks shorter than 45 degrees, and for J_{IC} below than 500 kJ/m^2 it is not possible to define a limit stress S_{Rmax} (see Fig. 3). In such configurations, the J criterion seems more appropriate.

Tearing resistance characterisation

The J-Resistance curves drawn from tubes and Compact Tension (CT) specimens made in the same material are shown on Fig. 5. The CT specimens have a reduced thickness (B = 8 mm, w = 40 mm).
No valid J_{IC} value has been obtained using the ASTM E 813 method (8) : the two blunting lines do not intersect the J-Resistance curve of the CT results. The CT specimen curves and those derived from the pipe test are almost linear in the 0 - 4 mm Δa range.
The slope of the CT specimen line is higher by a factor of 4 than the straight line relative to the pipe.

Therefore the CT specimens cannot be used to characterise the material toughness : the present results are not valid, and the pipe curvature and thickness do not allow use of larger CT specimens.

For crack initiation and the propagation of short cracks, the fracture behaviour of such pipes is difficult to characterise and the experimental J-estimation procedure needs to be checked carefully.

TEST SIMULATION BY THE FINITE ELEMENT TECHNIQUE

Definition of the mesh

Defining a Finite Element mesh for a three dimensional cracked structure is a difficult problem. In the present study, we aimed to fulfill the following requirements :

(a)- The mesh should be suitably defined to simulate the behaviour of the tested pipe
(b)- Thickness effects have to be considered
(c)- The strain distributions in the portion of pipe close to the cracked section should be accurately computed. Informations about the spreading of plasticity in the uncraked ligament are useful to develop simplified J-estimation schemes like the one developed by GILLES & BRUST (9)
(d)- The representation of the near-tip stress-strain fields have to be very good.

The tested pipe has been designed to avoid any influence of the crack on the stress distribution at the loading point. The ratio of the length (named L2) along with the bending moment is constant over the external diameter is about 5.25, which is quite large (see appendix of (10)). If pipes are loaded beyond the maximum moment or up to the maximum deflection allowed by the loading device, yielding may occur in the pipe arms. Therefore the pipes have been reinforced between the loading points. This increases the pipe stiffness and consequently the bending capacity of the test bench. When four-point bending induces shear forces at the loading points, modelling the bending by applying a moment at the end of the pipe shortened to the L2 value, is not appropriate. This was checked-on an uncracked pipe with the same dimensions and material characteristics as the CEA 27 pipe : the pure bending model gives a 4% higher bending moment than the four-point bending model. Finally we decided to build a model as close as possible to the experiment, however the following differences remain.

Because of symmetry considerations only one quarter of cracked pipe is modeled. For the sake of simplicity, in the Finite Element model the displacement is prescribed at one node and fixed at another one. These two nodes are distant from 400 mm, like between the roller bearings at rest. During the loading the bearings are moving on the pipe, and the experimental arm length increases contrary to the Finite Element one. This has been taken into account for the computation of the experimental value of the moment. Another difference is with the tested pipe, where the reinforced portion of pipe is modeled by increasing the Young's modulus, with the outer diameter unchanged. By mistake, the meshed pipe is sligthly longer than the tested one, however the reinforced part is also lengthened, giving a quite low overestimate of the stiffness (around 0.1%) .

With regard to the requirement (b), we have chosen to use massive elements instead of thick shell elements. The elements are isoparametric and quadratic. Linear elements are unappropriate for problems involving large gradient of stresses through the thickness, which is the case in bending. One layer of elements is probably enough, but the three layers of our model allow to investigate the variation of J through the thickness. For through-wall cracks in pipes the GE-EPRI method developed by KUMAR & al. (11) is based on a shell Finite Element approach accounting for the effect of tranverse shear deformations. Such elements stiffen the structure, which may explain that for short cracks the GE-EPRI document (11) reports negative values for the rotation because of the crack.

The requirement (c) is satisfied by the generation three thin layers of elements close to the crack plane.

It is difficult to set general rules that suit the requirement (d). Of course, refining the mesh around the crack tip is essential. This part of the mesh shown on Fig. 6 is radial close to the crack tip, which gives a better representation of the local singular fields. The good quality of the mesh has been checked in Linear Elastic Fracture Mechanics. The Finite Element code CASTEM (12) we used in this study allows to parametrize the mesh. Therefore the same type of mesh may be generated with other dimensions.

For the 180/16 through-wall crack in a pipe under bending we obtained a Stress Intensity Factor value under plane stress conditions, only 1% higher than the result derived by applying the GE-EPRI formula.
The selected mesh contains 2454 nodes, 412 cubes and 24 prisms.

Elasto-plastic computation

The selected values of the material characteristics are those relative to the test ($\nu = 0.3$ and see Table 1 for the Young's modulus and the yield stress). The rationale stress-strain curve is graduated in 112 points up to a strain of 20.5 %. For the present problem, there is no need to extrapolate the curve.
The Von-Mises'yield criterion is chosen.

The computation is conducted under the small displacement hypothesis, which does not allow a good simulation of the crack opening close to the tip in such a ductile steel. Furthermore the ovality effects are underestimated for large rotations, i.e. when the ratio of the maximum deflection over the lenght L2 is greater than 0.1. Now, the maximum load in the computation gives a ratio of 0.17. The requested level of accuracy on the loads is 2%.

The load is applied in 148 steps : 2 mm steps up to a load line displacement of 24 mm, 10 mm steps up to 50 mm of displacement and 20 mm steps up to 94 mm of displacement. The computation lasted more than 4 hours on a CRAY-XMP machine.

COMPARISON BETWEEN EXPERIMENTAL AND NUMERICAL RESULTS

Global behaviour of the structure

Figure 7 compares the experimental Load-load Line Displacement (LLD) curve to the computed results. This comparison has also been made for the moment rotation (PHI2) and the small displacement curve differs in a same way from the experimental one. The comparison reveals the three following domains. In the elastic regime (i.e. up to 9.8 mm of displacement) the agreement is excellent. In the non linear range between 9.8 mm and 30 mm, the difference is about 2% which is the accuracy requested in the computation. Beyond 30 mm of displacement, the curves are close but the "computed" pipe is stiffer than the tested one. Such a difference has been observed by other scientists and should be attributed to ovality effects which tend to flatten the pipe and therefore reduce the stiffness.

This hypothesis has been ckecked by conducting the same Finite Element analysis under the large displacement assumption up to 34 mm of displacement. The result is shown on Fig. 7 and the large-displacement moment rotation curve is lower than the experimental one which is the case in most of Finite Element analysis. The examination of the Von-Mises equivalent strain distributions in the cracked part of the pipes has evidenced the formation of a plastic hinge around 30 mm LLD. This mechanism appears in the 20 - 30 mm range which corresponds to the 19.2 - 21.1 kN.m in terms of the applied moment.

The limit moment of the cracked section computed for the yield stress equals 18.25 kN.m. At this level the hinge mechanism is far beyond, and opposite to the case for pipes with large through-wall crack where the plastic zone is spreading in the cracked section. In short cracked pipes like the CEA 27 one, the plastic zone develops also in the longitudinal direction and even above the crack face (see the analysis of a cracked bend specimen presented by SOREM et al. (13).

For short cracks it seems that the hinge behaviour is delayed but still exists as proven by the Crack Mouth Opening Displacement versus rotation curve (see Fig.8). At the limit load the rotation PHI2 equals 0.75 degrees and beyond this value the CMOD - Rotation relationship appears to be linear in the computed analysis as well as in the experiment. The agreement between these two curves is very good, if we take into account that the initial value of the experimental CMOD has not been corrected.

Finally, it should be emphasised that the small displacement computed is conducted far beyond the validity limit of this assumption. The deflection of the pipe reaches 90 mm, which is about 0.6 times the length where the moment is onstant. The deflection should have been limited to 50 mm. However the compued evolution of the pipe shape evidences a circular bending behaviour (see Fig. 9).

Fracture mechanics analysis

The J variations as a function of the load are derived from the experiment as previously presented and computed with two methods. From the Finite Element results we computed J using the virtual crack extension method developed by PARKS (14).

The variations of J along the crack front are quite small (less than 5 % except on the pipe inner and outer surface). Therefore only the J value at the node located on the mid surface is considered. The J values are also derived using the GE-EPRI method (11) which is also based on Finite Element small displacement computation.

The three curves are presented on Fig. 10. The agreement between the results obtained numerically is excellent up to a moment of 22.5 kN.m, corresponding to a displacement of 43 mm which is a quite large value.

Beyond this value, the difference may be explained in two ways : although the shell elements used in the GE-EPRI computations are stiffer, the difference is inherent in the GE-EPRI estimation scheme itself. In this method the stress-strain curve has to be filled by a RAMBERG-OSGOOD law. As frequently in the case, we found two different values for the hardening exponent n depending on the strain range considered for the RAMBERG-OSGOOD fit n = 9.3 in the 0-5 % range, n = 6.7 in the 0-10 % range.

We selected the value corresponding with the smallest range, because the strains are small in most of the yielded element if the load is not to high. Beyond the 22.5 kN.m moment this may not be the case and the J values are surely overestimated since J is proportional to M^{n+1}, where M is the bending moment.

The experimental J values are obtained assuming the pipe cross section remains circular. As recorded it in the investigation of the global behaviour of the structure, the pipe becomes oval when the applied displacement exceeds 30 mm and the moment 21 kN.m. The limit moment experimental, on which the method is based decreases when the pipe flattens. This leads to an important overestimation of J. Furthermore the expression of the limit moment used in (3) has been derived for large cracks and as in bars for short cracks, this expression is no longer applies.

CONCLUSION

The close agreement observed between the experimental and numerical results obtained for the mechanical behaviour of a pipe with very short through-wall crack permits the following conclusions. For a crack angle less than 45 degrees the behaviour of a through-wall cracked pipe is quite different and much more complex to model than if the crack is large.

For these pipes as well as uncracked pipes, the ovality noticeably reduces the stiffness of the pipe.

As the bending increases the yielding spreads from the crack tip in almost all the directions, which delays the formation of the hinge mechanism on the cracked section. Therefore the behaviour of such pipes is much more difficult to model : in Finite Element computations the large displacement assumption is recommended beyond the yield limit load. The existing simplified methods give plastic compliances and J values that are inaccurate. The next step in this work should be to develop an improved J-estimation scheme for short cracks.

AKNOWLEDGEMENTS

This work was funded jointly by Electricité de France, Commissariat à l'Energie Atomique and Framatome.

REFERENCES

(1). A. ZAHOOR, M.F. KANNINEN. "A plastic mechanics prediction of fracture insta-
      bility in a circumferentially cracked pipe in bending. Part I-J integral
      analysis". Trans. of ASME. J. of Pressure Vessel Technology.
      November 1981, Vol. 103 p. 352-358

(2). G.M. WILKOWSKI, A. ZAHOOR, M.F. KANNINEN. "A plastic fracture mechanics
      prediction of fracture instability in a circumferentially cracked pipe in
      bending. Part II : experimental verification on a type 304 stainless steel
      pipe". Trans. of ASME J. of Pressure Vessel Technology.
      November 1981, vol. 103, p. 359-365

(3). A. BRUCKNER, R. GRUNMACH, B. KNEIFEL, D. MUNZ, G. THUN. "Fracture of pipes with through-wall circumferential cracks in four-point bending".
P.V.P. vol. 95 p. 123-136
4th National Congress on Pressure Vessel and Piping Technology ASME Portland OREGON. June 19-24. 1983

(4). K. SHIBATA, S. MIYAZONO, T. KAUEKO, N. YOKOYAMA. 1986 "Ductile fracture behaviour of circumferential cracked type 304 stainless steel piping under bending".
Nuclear Eng. and Design, 04 (1986), p. 221-231

(5). D. MOULIN, J. LEBEY, D. ACKER. "Experimental determination of J value on circumferentially cracked stainless steel straight pipes under bending".
Proceeding of the Inst. of Mech. Engineers International Conference on pipe work Engineering and Operation.
London, 1989

(6). D. MOULIN, F. TOUBOUL, N. FOUCHER, J. LEBEY, D. ACKER. "Crack initiation and experimental determination of J in bending for elbows and pipes of austenitic steel".
Proceeding of the Pressure Vessel and Piping Conference ASME, HAWAI, HONOLULU, U.S.A, June 1989

(7). D. MOULIN, F. TOUBOUL, N. FOUCHER, J. LEBEY, D. ACKER. "Experimental evaluation of J in cracked straight and curved pipes under bending".
Transactions of the 10th International Conference on SMIRT, Volume G, Edition HADJIAN , U.S.A., 1989

(8). ASTM E813. J_{IC} a measure of fracture toughness.
Annual book of ASTM Standards 1981

(9). Ph. GILLES and F.W. BRUST. "Approximate fracture methods for pipes - Part I : Theory", "Part II : Application". Nuclear Eng. and Design 127 (1991) pp. 1-31

(10).C. FRANCO, Ph. GILLES. "Three-dimensional elasic-plastic analysis of small circumferential surface cracks on pipes subjected to bending".
ASTM STP 1144 Vol II, American Society for Testing and Materials, Philadelphia, USA, 1992

(11).V. KUMAR et al. "An engineering approach for elastic-plastic fracture analysis, Report NP 3607, EPRI (1984)

(12).CASTEM Finite Element System, June 1988 version. CEA-DMT-SEMT-LAMS-CEN Saclay BP 249 1190, GIF/YVETTE, F.

(13).SOREM, W.A., DODDS, R.H. Jr and ROLFE, S.T., "A comparison of the J-Integral and CTOD parameters for Short Crack Specimen Testing". Elastic-Plastic Fracture Test Methods : the user's Experience (second volume), ASTM STP 1114 J.A. JOYCE et al. American Society for Testing and Materials, Philadelphia, USA, 1991, pp 19-41

(14).D.M. PARKS, The Virtual Crack Extension Method for Non Linear Material Behaviour. Comp. Meth. Appl. Mech. Engng 12, 353-364 (1977)

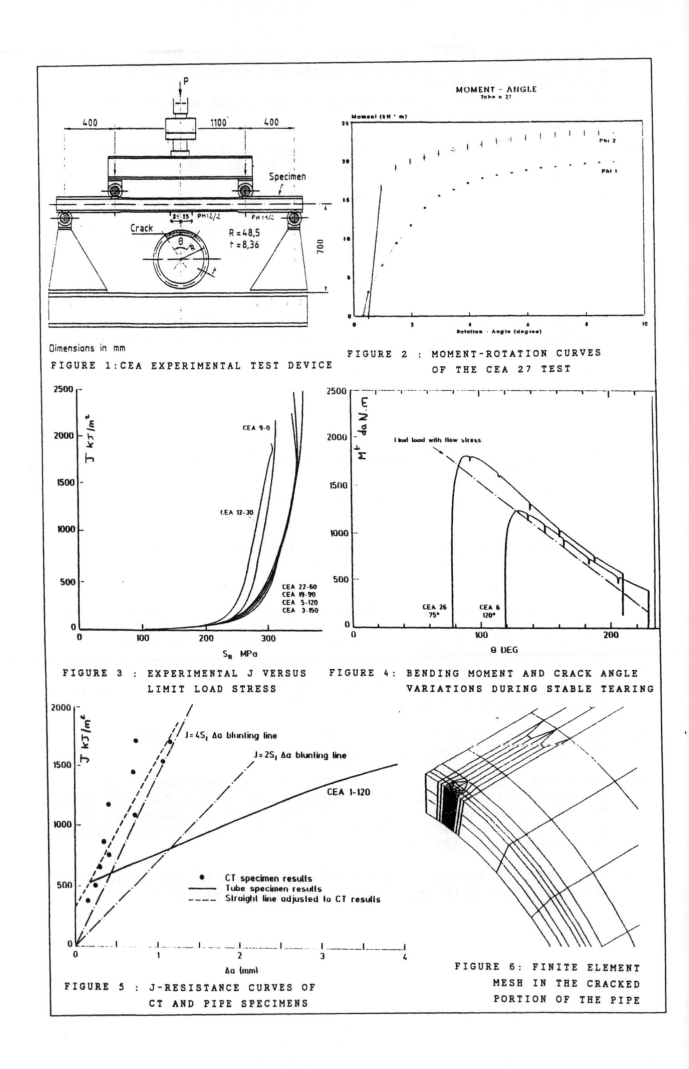

Dimensions in mm

FIGURE 1: CEA EXPERIMENTAL TEST DEVICE

FIGURE 2 : MOMENT-ROTATION CURVES OF THE CEA 27 TEST

FIGURE 3 : EXPERIMENTAL J VERSUS LIMIT LOAD STRESS

FIGURE 4: BENDING MOMENT AND CRACK ANGLE VARIATIONS DURING STABLE TEARING

FIGURE 5 : J-RESISTANCE CURVES OF CT AND PIPE SPECIMENS

FIGURE 6: FINITE ELEMENT MESH IN THE CRACKED PORTION OF THE PIPE

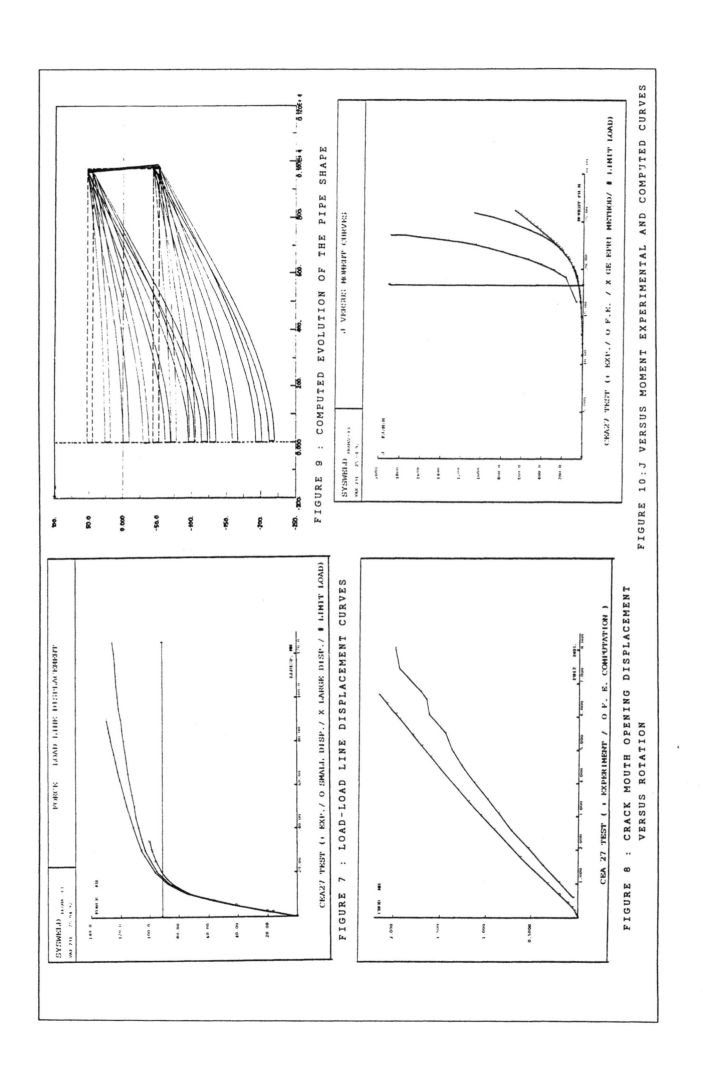

FIGURE 9 : COMPUTED EVOLUTION OF THE PIPE SHAPE

CEA27 TEST (+ EXP./ O F.E. / X GE EPRI METHOD/ ■ LIMIT LOAD)

FIGURE 10 : J VERSUS MOMENT EXPERIMENTAL AND COMPUTED CURVES

CEA27 TEST (+ EXP./ O SMALL DISP./ X LARGE DISP./ ■ LIMIT LOAD)

FIGURE 7 : LOAD-LOAD LINE DISPLACEMENT CURVES

CEA 27 TEST (+ EXPERIMENT / O F. E. COMPUTATION)

FIGURE 8 : CRACK MOUTH OPENING DISPLACEMENT
VERSUS ROTATION

TWI

PAPER 18

Application of K_J rule to different specimens and to a cracked elbow

C Couterot, Ing and Y Wadier, Ing (Electricité de France) and P Taupin, Ing (FRAMATOME, France)

SUMMARY

The behaviour of cast duplex stainless steel elbows is studied under a three-party agreement between Electricité de France, Framatome and the Commissariat à l'Energie Atomique. Within this frame-work, Framatome has developed a simplified fracture analysis method : "K_J rule", based on the CEGB-R6 rule revision 3 option 2. Its margins must yet be estimated. This paper presents the principle of the "K_J rule" and its application to different duplex stainless steel structures : four specimens and a cracked elbow. The J-value margins are evaluated by comparison with 3D-elastic-plastic numerical computations. The results obtained with the "K_J rule" are generally conservative and so we can safely forecast the crack initiation.

INTRODUCTION

The objectives of the present program are to study the thermal ageing embrittlement of cast duplex stainless steels and to predict the fracture of aged components by using engineering method.

3D-elastic-plastic numerical computations on different components containing various sized cracks are very expensive because of engineer time for meshing and computer cost. So Framatome has developed a simplified fracture analysis method, the "K_J rule" and the aim of this paper is to evaluate the J-value margins on several structures.

PRESENTATION OF "K_J RULE" (1)

This simplified method is based on the second option of CEGB-R6 revision 3 rule (2). Its aim is to obtain a conservative value of the elastic-plastic J integral and it is used to predict the crack initiation and propagation in a structure.

Firstly the stress intensity factor K_I is computed using one of these methods : the influence function technique (results of Héliot and al (3)), the EPRI handbook (4) or an elastic finite element computation.

Secondly a non linear correction $\dfrac{1}{K_r}$ is then applied to the above K_I. This K_r value is deduced from L_r using the Failure Assessment Diagram R6 (revision 3 option 2). L_r expression is the ratio of the applied load to the limit load of the cracked component, or the ratio of an equivalent stress σ_e, defined below, to a weakened yield stress $Q \sigma_Y$; so the main point of this method is to determine the L_r value.

According to the situation, two methods are used to compute L_r :

1) **the cracked structure limit load is given** by a handbook (specimen cases), then the ratio L_r is defined by :

$$L_r = \frac{\text{applied load}}{\text{limit load}} = \frac{P}{P_0} \qquad [1],$$

2) **the limit load of the cracked structure is unknown** because of the complex geometry of the component. In this case, the available L_r expression refers to a stress corresponding principally to a local membrane stress; its formula is given in the reference (1) :

$$L_r = \frac{\sigma}{Q \sigma_Y} \qquad [2],$$

in which σ_Y is the yield stress and Q the weakening factor due to the presence of the crack.

If the loading through the wall at the crack plane consists of a membrane stress σ_m and a bending stress σ_b, we propose :

- to keep the above definite Q with its expression for a membrane loading,

- to determine the equivalent stress σ_e with reference to the limit load expression of a plate under membrane plus bending loading.

This proposal leads to the limit condition expression, in which σ_{mL} and σ_{bL} correspond to σ_m and σ_b values at a limit condition :

$$\frac{\sigma_{bL}}{1.5 \, Q \, \sigma_Y} + \left(\frac{\sigma_{mL}}{Q \, \sigma_Y} \right)^2 = 1 \qquad [3].$$

The equivalent stress is defined according to the resolution of this equation with respect of $Q \sigma_Y$ as variable :

$$\sigma_e = \frac{\sigma_b}{3} + \sqrt{\left(\frac{\sigma_b}{3} \right)^2 + \sigma_m^2} \qquad [4].$$

in which expression σ_b is neglected when σ_e is greater than $Q \sigma_Y$.

The principle of the "KJ rule" is explained in the following organization chart with the two options for the computation of L_r.

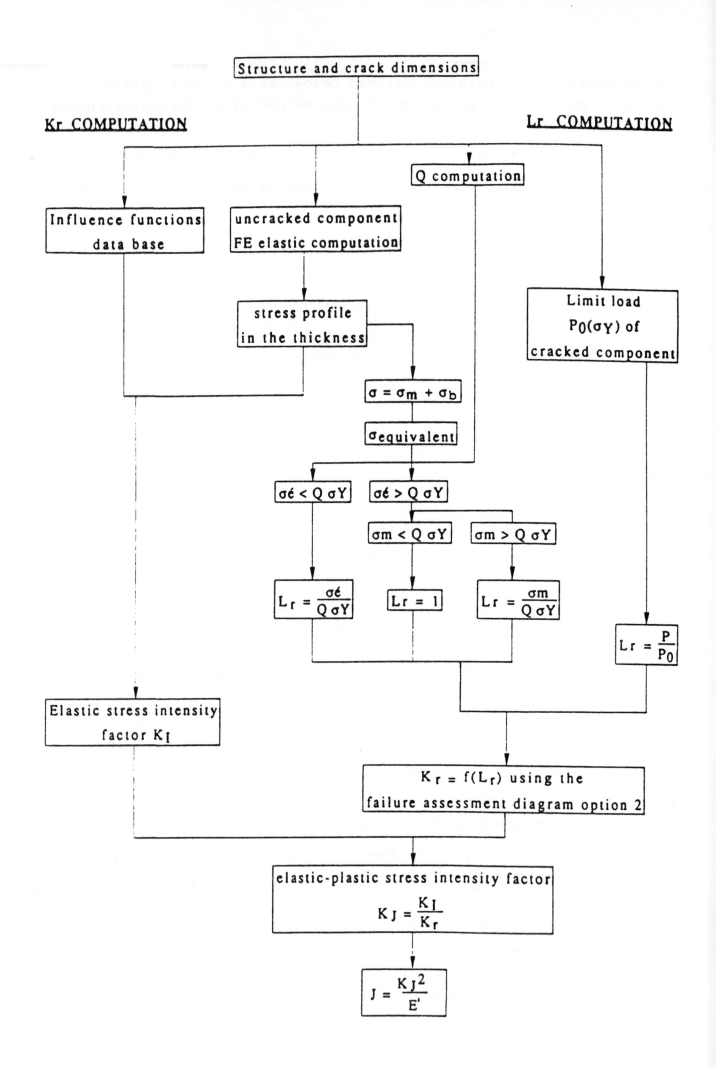

APPLICATION OF THE "K$_J$ RULE" TO FOUR SPECIMENS

Material properties

The cast stainless steel is aged during 1000 hours at 400°C. The
true stress - true strain curve is plotted on figure 1 and its main
properties at 320°C are :

    Young's modulus E = 160000 MPa
    0.2 % offset yield strength σ_Y = 276 MPa.

Description of specimens

Three specimen types are studied :

- a Compact Tension with a deep crack : crack length a = 30 mm and
specimen length b = 50 mm,

- a Center Cracked Plate tension with a short crack : a = 25 mm and
b = 100 mm,

- two Single Edge Cracked Plate Tension with different crack sizes :
a= 10 mm, b = 80 mm and a = 25 mm, b = 40 mm.

Simplified calculation method

Tension specimens are under plane strain conditions.

The limit load P_0 is given by the EPRI handbook (4) or by Miller (5)
as a function of the 0.2 % offset yield strength. The elastic stress
intensity factor K_I comes from the Murakami handbook (6).

The failure assessment diagram (FAD) derived from the R6 revision 3
option 2 is plotted on figure 2.

Comparison of results with a 2-D elastic-plastic finite element analysis

The computation is made with the ALIBABA Finite Element code and a
postprocessor gives the elastic-plastic J-values using the integral
calculation along the crack contour.

Figures 3,4,5 and 6 show J-values evolution for various load P.

The results obtained with the "K$_J$ rule" are conservative for
specimens with a deep crack : CT and SECPT with crack
depth/thickness $\approx$ 0.6.

In the case of short specimens (CCP and SECPT with crack
depth/thickness > 0.25), margins decrease and results are slightly
unconservative for high loading.

APPLICATION TO A CRACKED ELBOW

Material properties

The main properties of the aged cast stainless steel at 300 °C are
E = 176500 MPa and σ_Y = 320.8 MPa. The true stress - true strain
curve is smoothed by a polynomial function.

Crack description

The cast stainless steel elbow loaded with an in plane closure
bending moment is shown in the figure 7. Its dimensions are :

    external radius Re = 287 mm,
    thickness t = 42 mm,
    bending radius R = 900 mm.

It contains a large semi-elliptical surface crack on one of the
external flanks : lengh 2c = 210 mm and depth a = 10.5 mm. It is
fixed at one end with a 450 mm spool and a displacement is imposed
on the other end of a tube 5 m long.

Determination of stress intensity factor along the crack tip

The first stage is the elastic FE computation of the uncracked
component submitted to the same closure bending (100 mm imposed
displacement). The ovalization effects induce a bending stress
σ_b = 370 MPa and a compressive membrane stress σ_m = -22 MPa in the
thickness of the elbow. The equivalent stress is used in the L_r
expression :

$$\sigma_e = \frac{\sigma_b}{3} + 0.5 \left((\frac{\sigma_b}{1.5})^2 + 4 \sigma_m2 \right) 1/2$$

The cracked elbow is meshed with a great refinement in the crack tip
area and contains 2506 parabolic elements and 9998 nodes.The
computation is made with PERMAS Finite Element code. The G-Theta
method (7) introduced in PERMAS code provides the evolution of the
energy release rate G along the crack front; so G-values are
evaluated in the cases of an elastic and an elastic-plastic
computation.

The elastic computation of this cracked elbow is performed to obtain
K_I values along the crack front : the elastic stress intensity
factor is higher at the bottom than in the edges. The crack
initiation will be obtained at the bottom of the crack.

The "K_J rule" is then applied, figure 8, with the bending stress
relieving, according to the value of the ratio $(\sigma_e/Q\ \sigma_Y)$.
The elastic-plastic computation gives J values.

The comparison of simplified results with the elastic-plastic J-
values shows that the rule gives conservative results, about 100 %
higher (figure 9). Therefore we can safely analyze the crack
initiation and estimate the propagation during the bending loading.

CONCLUSION

The "K_J rule" is an easy to use method whose margins must be precised. The comparisons with Finite Element results is conservative for the different specimens (except sometimes for high loading) and the cracked elbow.

The next task will be to confirm these results by tests and further 3D elastic-plastic computations.

REFERENCES

(1) A. Pellissier-Tanon, C. Ensel, D. Guichard, F. Coustillas, H. Churier-Bossenec - Stress classification in industrial fracture mechanics analysis - Paper 7th ICPVP - Düsseldorf - Juin 1992.

(2) I. Milne, R.A. Ainsworth, A.R. Dowling, A.T. Stewart - Assessment of the Integrity of structures containing defects.Central Electricity Generating Board - R/H/R6 - Revision 3 - May 1986.

(3) J. Heliot, R. Labbens, A. Pellissier-Tanon - Semi-elliptical cracks in a cylinder subjected to stress gradients - Fracture Mechanics - ASTM-STP 677, 1980.

(4) An Engineering Approach for Elastic-Plastic Fracture Analysis. Note E.P.R.I. - NP 1931 - Section 3 - July 1981.

(5) A.G. Miller - Review of Limit Loads of Structures Containing Defects. Internal Journal of Pressure Vessels and Piping - Vol. 32 1988.

(6) Y. Murakami - Stress Intensity Factors Handbook. Committee on Fracture Mechanics - 1987.

(7) Y. Wadier, O. Malak - The Theta Method applied to the analysis of 3D Elastic-Plastic Cracked Bodies - SMIRT 10 - Los Angeles, 1989.

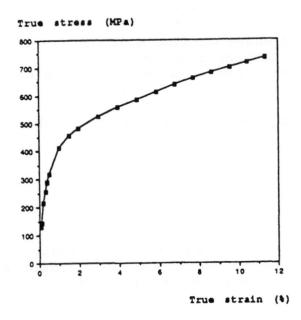

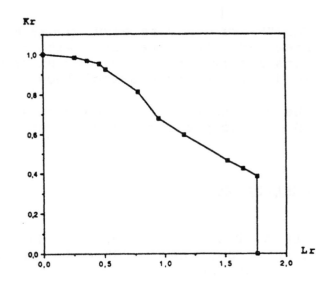

Figure 1   True stress - True strain at 320°

Figure 2   R6 Failure Assessment Diagram

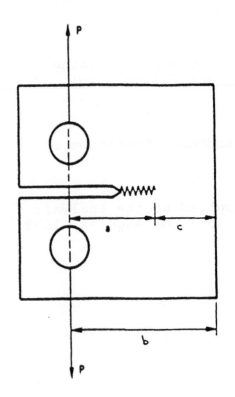

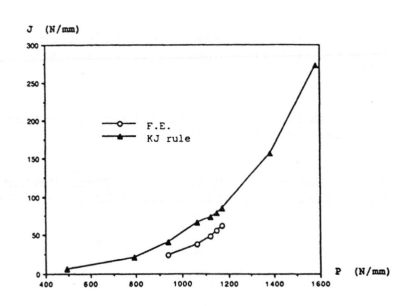

Figure 3   J-values comparison for CT specimen

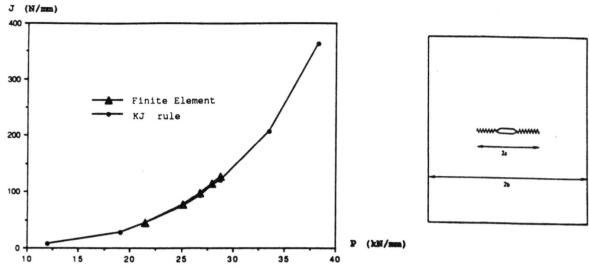

Figure 4   J-values comparison for CCP specimen

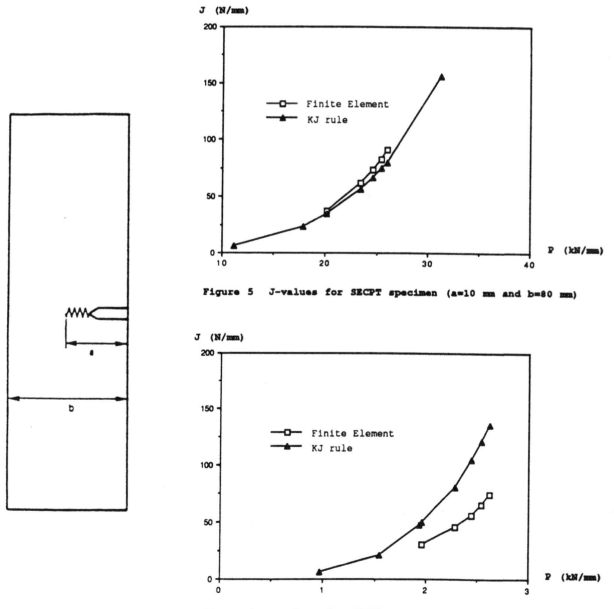

Figure 5   J-values for SECPT specimen (a=10 mm and b=80 mm)

Figure 6   J-values for SECPT specimen (a=25 mm and b=40 mm)

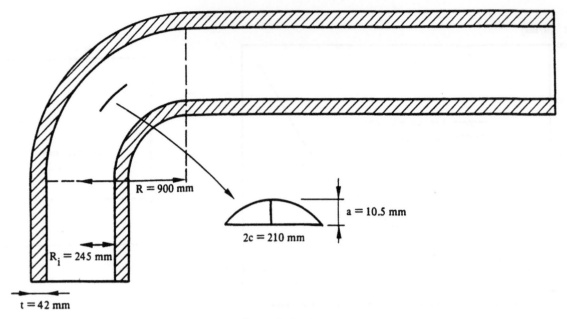

Figure 7 Cracked elbow

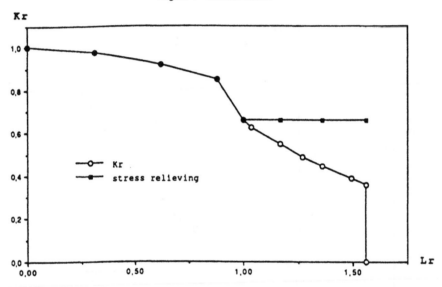

Figure 8   Failure Assessment Diagram Analysis

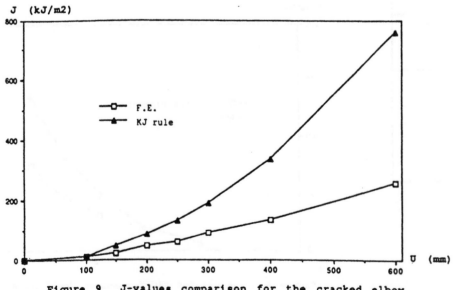

Figure 9   J-values comparison for the cracked elbow

Printed and bound by CPI Group (UK) Ltd, Croydon, CR0 4YY

03/10/2024

01040341-0020